FLUIDIZED BEDS
Combustion and Applications

PROFESSOR DOUGLAS ELLIOTT (1923–1976)

FLUIDIZED BEDS
Combustion and Applications

Edited by

J. R. HOWARD

Department of Mechanical Engineering, The University of Aston in Birmingham, UK

APPLIED SCIENCE PUBLISHERS
LONDON and NEW YORK

CHEMISTRY

7118-3926

5

APPLIED SCIENCE PUBLISHERS LTD
Ripple Road, Barking, Essex, England

Sole Distributor in the USA and Canada
ELSEVIER SCIENCE PUBLISHING CO., INC.
52 Vanderbilt Avenue, New York, NY 10017, USA

British Library Cataloguing in Publication Data

Fluidized beds.
  1. Fluidization
  I. Howard, J. R.
  660.2'84292   TP156.F5

  ISBN 0-85334-177-X

WITH 21 TABLES AND 125 ILLUSTRATIONS

© APPLIED SCIENCE PUBLISHERS LTD 1983

The selection and presentation of material and the opinions expressed in
this publication are the sole responsibility of the authors concerned.

Printed in Great Britain by Galliard (Printers) Ltd, Great Yarmouth

FOREWORD

It can be justifiably claimed that fluidized bed combustion is the only really new approach to combustion of fossil fuel that has emerged for at least six decades. The potential merits of burning coal in a fluidized bed of mineral matter, in which boiler tube surface is immersed, were recognized by the late Douglas Elliott as long ago as 1962 when he was a member of staff of one of the Central Electricity Generating Board's research laboratories. Although he was not the first to have thought of fluidized bed combustion (and it is by no means certain who did) the initiation of the current phase of activity on FBC in the Western world can most certainly be attributed to him. The climate in the early 1960s was not, however, favourable for application of the technology to power generation; there was a grow-ing momentum towards nuclear power, there were ample supplies of low cost oil, legislation overseas on sulphur dioxide and NO_x emissions had not begun to bite, and (before its problems were adequately recognized) the more glamorous magnetohydrodynamic system of power generation absorbed the major part of the effort available in the UK for research and development on new routes based on coal. His experimental programme at the CEGB became a casualty of this situation but he continued to have an influence on work elsewhere.

I first met Douglas in 1952 when we both were involved in the UK Ministry of Fuel and Power's programme on combustion of coal for gas turbines. Subsequently our activities in the fields of development in coal utilization frequently brought us together on topics such as gasification of coal for combined cycle power plant, high intensity combustion for MHD power generation and in particular during the later involvement at Leatherhead, of the British Coal Utilisation Research Association in fluidised bed combustion.

v

The one major exception to this was in the work on fluidized bed combustion for industrial steam generation. This began at Leatherhead in 1964 but we were unaware at that time of Douglas's already significant contributions to the technology.

Our main association was a consequence of a conversation in the car park at the CEGB Marchwood Laboratories between Douglas, Alan Roberts and myself towards the end of 1967. We there and then became convinced that, for power generation applications FBC would offer the maximum benefit if carried out under pressure in a combined gas and steam cycle. Douglas Elliott inspired the ensuing pilot-plant programme at Leatherhead and chaired the informal group comprising representatives of the CEGB, National Coal Board and BCURA that directed the work during the early stages and laid the foundation for the many subsequent research and development programmes in this field.

In the fertile environment of the University of Aston in Birmingham he worked on expanding the range of applications for FBC technology, laid the foundations for the commercial range of boilers and heat treatment equipment manufactured by Fluidfire, and provided education and training in FBC technology which will be of benefit to future generations.

Douglas Elliott was a remarkable individual who inspired all those privileged to claim his acquaintance. His advanced thinking and his enthusiasm to achieve an objective provided a spur for all those associated with him. I am not aware of any fluidized bed combustion system in vogue today that was not at one time or another commented upon by him.

His untimely death in 1976 prevented him from seeing the substantial application and commercialization of the technology. Although he would have been gratified at the progress so far made, he would probably be exasperated at the time that it has taken before some of the potentially more beneficial applications have been followed through; some indeed have yet to reach the demonstration stage.

It is to be hoped that this book, prepared as a tribute by some of his friends, will inspire more advances towards the achievement of his objectives.

H. R. Hoy, O.B.E.
Director,
National Coal Board Coal Utilisation Research Laboratory,
Randalls Road,
Leatherhead, Surrey KT22 7RZ

PREFACE

The untimely death of Professor Douglas Elliott on 16 June, 1976, was grievous to his wife and children, and his countless friends and colleagues; but it was grievous too for the engineering profession, just at a time when the challenges facing it were, and still are, formidable. Engineers like Douglas are a rare breed; he was one of the most perceptive and creative of his generation.

Production of this book was first thought about shortly after his death, when Douglas Probert, who succeeded him as Editor of the journal *Applied Energy*, suggested that five to six years hence might be a good time to launch a book to commemorate Douglas Elliott, because the technology to which he had contributed so much was on the threshold of industrial scale demonstration. Such an undertaking clearly required the help of many experts. This volume is the result of their labours. All of the authors knew him as a personal friend, colleague and collaborator. I am deeply grateful to all of them for giving their time and professional skill so generously at a time when, such is the interest in fluidized systems, their services are in great demand. Their eagerness is testament to the high esteem in which Douglas Elliott was held by his peers. Any views expressed in their chapters are those of the authors and not necessarily those of their employing organizations. Nonetheless, I am grateful to all these organizations for their cooperation in helping to bring this work to fruition.

The story of Douglas's own involvement in fluidized beds begins in the late 1950s and early 1960s when he was working at the Marchwood Engineering Laboratories of the Central Electricity Generating Board. He began some experiments which led him to re-discover fluidized bed combustion for himself and he saw its potential immediately. The objective of such work was to develop more economic systems for utilization of coal for electricity generation and for industrial boilers. All factors relating to the installation were meant to be covered by the adjective 'economic', including amortization of capital, maintenance and, very important, the environmental regulations.

Douglas's hopes for increased utilization of coal found renewed expression in 1969 in his inaugural lecture after he was appointed Professor of Mechanical Engineering at the University of Aston in Birmingham. The lecture, which examined some of the technical arguments which were at the root of electricity generation strategy at that time, was entitled 'Can Coal Compete?—The Struggle for Power'. Events and changes in economic and political conditions over a span of 13 years often confounds the predictions and policy recommendations of experts. Douglas's basic thesis remains pertinent today, for increased utilization of coal has since become an urgent priority, backed by governments and energy users. Fluidized bed technologies are an important element toward achievement of this objective particularly for the poorer qualities of coal, and low grade fuels. These are in abundance world-wide, are cheap, but are difficult to burn by other means particularly when clean, non-polluting combustion is essential.

However, the technology, as some chapters in this book show, has not been confined to combustion of coal; liquid fuels and gases can be burned and the technology exploited for incineration, drying, metallurgical furnaces and heat recovery.

The literature on application of fluidized beds to combustion, gasification and chemical engineering processes is vast. It is hoped however that this volume will be valuable to engineers, students and indeed anyone wanting to obtain a rapid grasp of the fundamental principles of fluidized beds, how they are applied to energy using processes and to conservation, while at the same time meeting the required environmental standards.

J. R. HOWARD

CONTENTS

LIST OF CONTRIBUTORS

J. S. M. BOTTERILL
Department of Chemical Engineering, The University of Birmingham, PO Box 363, Edgbaston, Birmingham B15 2TT, UK

J. BROUGHTON
GEC Gas Turbines Ltd, Cambridge Road, Whetstone, Leicester LE8 3LH, UK

J. HIGHLEY
National Coal Board, Coal Research Establishment, Stoke Orchard, Cheltenham, Glos. GL52 4RZ, UK

J. R. HOWARD
Department of Mechanical Engineering, The University of Aston in Birmingham, Gosta Green, Birmingham B4 7ET, UK

W. G. KAYE
National Coal Board, Coal Research Establishment, Stoke Orchard, Cheltenham, Glos. GL52 4RZ, UK

G. MOSS
Esso Research Centre, Esso Petroleum Co. Ltd, Abingdon, Oxfordshire OX13 6AE, UK

K. K. PILLAI
National Coal Board Coal Utilisation Research Laboratory, c/o BCURA Ltd, Randalls Road, Leatherhead, Surrey KT22 7RZ, UK

A. G. ROBERTS
*National Coal Board Coal Utilisation Research Laboratory, c/o
BCURA Ltd, Randalls Road, Leatherhead, Surrey KT22 7RZ, UK*

J. T. SHAW
*National Coal Board, Coal Research Establishment, Stoke Orchard,
Cheltenham, Glos. GL52 4RZ, UK*

A. M. SQUIRES
*Department of Chemical Engineering, Virginia Polytechnic Institute
and State University, Blacksburg, Virginia 24061, USA*

J. E. STANTAN
*National Coal Board Coal Utilisation Research Laboratory, c/o
BCURA Ltd, Randalls Road, Leatherhead, Surrey KT22 7RZ, UK*

M. J. VIRR
*Stone-Platt Fluidfire Ltd, 56 Second Avenue, Pensnett Trading Estate,
Brierley Hill, West Midlands DY6 7PP, UK*

Chapter 1

FLUIDIZED BED BEHAVIOUR

J. S. M. BOTTERILL

Department of Chemical Engineering, University of Birmingham, UK

NOMENCLATURE

A_b	Cross-sectional area of bed (m²)
C_g	Heat capacity of gas (J/kg K)
d_B	Diameter of bubble (m)
d_f	Arithmetic average of adjacent sieve apertures (m)
d_p	Particle diameter (m)
D_b	Diameter of bed (m)
F	Fraction of material in bed $<45\,\mu m$
g	Acceleration due to gravity (m/s²)
h_{gc}	Interphase gas convective component of bed-to-surface heat transfer coefficient (W/m² K)
h_{gp}	Gas-to-particle heat transfer coefficient (W/m² K)
h_{max}	Maximum bed-to-surface heat transfer coefficient (W/m² K)
h_{mf}	Bed-to-surface heat transfer coefficient at minimum fluidization (W/m² K)
$h_{overall}$	Overall bed-to-surface heat transfer coefficient (W/m² K)
h_{pc}	Particle convective component of bed-to-surface heat transfer coefficient (W/m² K)
h_{rad}	Radiative component of bed-to-surface heat transfer coefficient (W/m² K)
k	Constant
k_g	Gas thermal conductivity (W/m K)
m	Mass of particles in bed (kg)

1

ΔP_b Pressure drop across bed (N/m^2)
ΔP_D Pressure drop across distributor (N/m^2)
T_b Temperature of bed (K)
T_s Temperature of heat transfer surface (K)
U Superficial gas velocity (m/s)
U_B Rise velocity of single bubble (m/s)
U_m Gas velocity for maximum bed-to-surface heat transfer (m/s)
U_{max} Maximum operating velocity to avoid slugging (m/s)
U_{mB} Gas velocity at which bubbling first occurs (m/s)
U_{mf} Gas velocity at which bed becomes fluidized (m/s)
U_t Particle terminal velocity (m/s)
X Mass fraction
ε_b Emissivity of bed
ε_m Modified emissivity allowing for effect of immersed surface on adjacent bed temperature
ε_{mf} Voidage at minimum fluidization
ε_r Reduced emissivity
ε_s Emissivity of surface
ϕ Particle shape factor in Ergun equation (ref. 20)
ρ_f Density of fluid (kg/m^3)
ρ_g Density of gas (kg/m^3)
ρ_p Density of particle (kg/m^3)
μ_g Viscosity of gas (kg/m s)
Ar   Archimedes number, $d_p^3 \rho_g (\rho_p - \rho_g) g / \mu_g^2$
Nu   Nusselt number, $h d_p / k_g$
Nu_{gc} Nusselt number based on h_{gc}
Nu_{gp} Nusselt number based on h_{gp}
Nu_{max} Nusselt number based on h_{max}
Pr   Prandtl number, $C_g \mu_g / k_g$
Re   Reynolds number, $d_p U \rho_g / \mu_g$
Re_{mf} Reynolds number based on U_{mf}
Re_{opt} Reynolds number based on U_m

1 INTRODUCTION

1.1 The phenomenon of fluidization

The advantages of the technique of fluidization have often been listed (for example, see Botterill[1]). Principally they stem from the very large particle surface area exposed to the fluid, the ease with which the solids can be handled in the fluidized state and, in the heat transfer applications which are the concern of this book, the good heat transfer properties of a bubbling gas-fluidized bed. Inevitably there are also disadvantages. Operating rates are limited to within the range over which the bed can be fluidized. The cost of the pumping power to fluidize the bed may be excessive and particularly so with deep beds. There is a limit to the size and type of particle that can be handled by this technique. As will be stressed in this chapter, it is difficult to

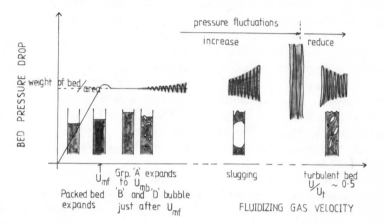

FIG. 1.   Schematic diagram of variation in bed pressure drop with gas flow rate.

characterize the particles themselves and there can be a very wide range of behaviour according to the conditions under which a fluidized bed is being operated. For a fuller consideration of bed behaviour and design see Geldart,[2] 'Gas Fluidization'.

A bed of particles offers resistance to fluid flow through it. As the velocity of flow increases, the drag force exerted on the particles increases. With upward flow through an unrestrained bed, the particles rearrange themselves to offer less resistance to the fluid flow and the bed will tend to expand unless it is composed of large particles (mean diameter > 1 mm). With further increase in the upward fluid velocity, the expansion continues until a stage is reached where the drag force exerted on the particles will be sufficient to support the weight of the particles in the bed (Fig. 1). The fluid/particle system then begins to exhibit fluid-like properties and it will flow under the influence of a hydrostatic head. This is the point of *incipient fluidization* and the gas velocity needed to achieve this is referred to as the *minimum fluidizing velocity*, U_{mf}. Beyond this velocity, the pressure drop across the bed will be approximately equal to the weight of the bed per unit area. Thus:

$$\Delta P_b A_b = \frac{m}{\rho_p}(\rho_p - \rho_f)g \tag{1}$$

where ρ_p is the effective density of the particle which may be porous, and ρ_f is that of the fluid. The effective ΔP excludes the hydrostatic pressure drop across the bed which can be neglected in gas-fluidized systems operating at atmospheric pressure. It is likely, however, that this pressure drop will be exceeded just prior to fluidization with gas-fluidized systems in order to overcome cohesive forces between the particles and break down the residual packing and interlocking of particles within the bed. It is also likely to be exceeded when operating under more extreme conditions at high gas flow rates when rising slugs of solids (see below) can be formed and energy is required in order to accelerate them. When the bed is composed of a powder with a wide size distribution, the pressure drop may never quite equal that of the weight of the bed per unit area thus indicating that not all of the bed is freely supported by the gas flow.

At the onset of fluidization the bed is more or less uniformly expanded and, up to this point, it makes little difference to the general bed behaviour whether the fluid is a gas or a liquid. However, whereas the liquid fluidized bed tends to continue to expand stably with further increase in upwards flow velocity until the particles are carried from the bed, the uniform expansion behaviour is soon lost as the gas velocity is increased except with fine powders. Thus, with the gas-fluidized system, instabilities develop and cavities containing few solids are formed. These look like bubbles of vapour in a boiling liquid and, as they rise through the bed, they are responsible for generating the solids mixing which is such an important feature of gas-fluidized bed behaviour.[3] Over the flow range between incipient fluidization and the onset of bubbling, the bed is said to be in a quiescent state.

At high velocities conditions are such that a lot of material is lost from the bed by elutriation even before the regime of pneumatic conveying is reached. Bubbles bursting at the surface generate pressure fluctuations and throw a spray of particles into the freeboard space above the bed. Many are carried away in the gas stream according to the particle size distribution and the gas velocity in the freeboard region above the bed. There is increasing interest in the possibility of operating beds at higher gas velocities in order to increase the throughput and to obtain more advantageous reaction conditions according to circumstances; the regimes of the *turbulent* and *fast* fluidized beds[4] (Fig. 1). Changes in bed behaviour on entering this regime are outlined in Section 1.4, below. Under these more extreme operating conditions it is obviously necessary to make provision for the constant return or replacement of the material carried from the bed.

1.2 Range of behaviour of gas-fluidized beds

Much of the range of behaviour encountered with gas-fluidized systems is a consequence of the different bubbling behaviour that can occur.[3] The rise rate of a single bubble of diameter d_B is given by:

$$U_B = k\sqrt{(gd_B/2)} \tag{2}$$

where the constant, k, depends on particle shape and size distribution, although its value is generally about 0·9. This velocity is usually greater than the interstitial gas velocity except for beds of Group 'D' materials (see below this section). A bubble within a chain or cloud of bubbles rises faster than one in isolation, except for those in beds of large mean particle diameter (~ 1 mm) when the rise rate tends to be slower. As bubbles rise through the bed they grow by collection of gas from the surrounding continuous phase and, most importantly, by coalescence. In this latter process, a smaller bubble comes within the influence of a faster rising one which overtakes it and draws it into its wake except, again, for the case of beds of large particles when cross-wise coalescence between bubbles predominates. On occasion, bubbles are also observed to shrink and this tendency is apparently stronger for bubbles below a critical size. Bubbles may divide spontaneously or on contact with an obstruction. General bubbling behaviour is affected by the design of the distributor by which the gas is introduced to the bed and by surfaces immersed within the bed. Strong, bubble-induced solids convection streams can become established which will tend to sustain themselves until something occurs to disturb the flow pattern and the system will switch to a new form.[5] As a bubble rises it displaces particles within its path and draws a streak after it giving rise to a drift profile. Also, apart from the case of bubbles in beds of large mean particle diameter, particles are carried within the bubble's wake and material is constantly collected by and shed from the wake as the bubble rises through the bed. To replace the material carried upwards through the bed by these processes there is the downwards return flow of solids through other regions of the bed.

With deep beds of high aspect ratio and at higher operational velocities, bubbles may grow until they occupy the whole cross-sectional area of the bed. These then carry a slug of particles ahead of them until instability occurs and the solids collapse back to the bed. This is the so-called *slugging* bed and, as mentioned above, energy has to be supplied to accelerate the slug of particles upwards so the overall pressure drop across the bed will then exceed the weight of the bed per unit area. Pressure fluctuations across the bed will increase[6] (Fig. 1). With a wide size distribution, there will be the tendency to lose a lot of the fines under the operating conditions required to keep the larger fractions within the fluidized state. It is also possible, of course, to fluidize larger particles within the dense fluid obtained by fluidizing the finer fractions. However, in either instance, on reduction of the gas throughput, there will be a tendency for segregation to occur with the deposition of the larger particles on to the distributor. This is generally to be avoided but it may be advantageous to encourage such deposition when it is desired to protect the distributor from the full rigour of exposure to a higher temperature bed.

It has often been suggested that some breadth of size distribution is desirable to promote stable fluidization. Geldart[7] has argued that it is the mean size on a surface to volume basis which is the important factor and, as he has pointed out, that the addition of a comparatively small amount of fines by weight can have a considerable influence on the average particle diameter expressed in this way and on bed behaviour. Later work shows that the actual proportion of fines (classed as material $< 45\,\mu m$) is of particular importance.[8] The significance of this effect can be better appreciated in relation to the classification[9] that Geldart suggested earlier which is primarily based on particle density and size from tests under ambient conditions (Fig. 2). Materials categorized within his *Group 'A'* generally have densities less than $1400\,kg/m^3$ and fall within the size range of 20 to 100 μm. These powders exhibit a pronounced degree of stable bed expansion when the minimum fluidization velocity is first exceeded, and it may be possible to sustain such uniform or *particulate* fluidization until the minimum fluidizing velocity has been exceeded by a factor of two to three times. However, with further increase in gas velocity a point will be reached when the bed will collapse back to a less expanded state approximating more closely to the degree of expansion under the minimum fluidizing conditions and most of the excess gas will flow through the bed as the bubble phase, i.e. the so-called *aggregative* fluidization. The gas velocity at which this bed collapse occurs is referred to as the minimum bubbling velocity, U_{mB}. Whilst the bed is in the quiescent state before it begins to bubble, only very limited particle mixing by a diffusive mechanism can take place.

Geldart's *Group 'B'* materials tend to have a mean size within the range

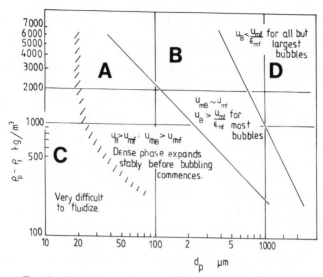

FIG. 2. Powder classification diagram. (After Geldart.[9])

40 to 500 μm and a density in the range 1400 to 4000 kg/m³. These exhibit much less stable bed expansion; free bubbling (aggregative fluidization) occurring at or a little above the minimum fluidization velocity. Abrahamsen and Geldart[8] gave the following relationship for prediction of the minimum bubbling velocity, U_{mB}:

$$U_{mB} = 2 \cdot 07 e^{0 \cdot 716 F} \left(\frac{d_p \rho_g^{0 \cdot 06}}{\mu_g^{0 \cdot 347}} \right) \qquad (3)$$

where F is the fraction of material less than 45 μm and the dimensional constant is for SI units. This was established from an extensive series of tests under ambient conditions. Earlier, Baeyens and Geldart[10] suggested that beds of particles would fall within Group 'B' rather than 'A' when their minimum fluidizing velocity exceeded the minimum bubbling velocity. From tests with air at ambient pressure and temperature the condition was approximately

$$(\rho_p - \rho_g)^{1 \cdot 17} d_p > 906\,000 \qquad (4)$$

Whereas bubbles tended to grow until their diameter was finally limited by the dimensions of the bed with materials of Group 'B', there seemed to be a definite maximum stable bubble size with beds of Group 'A' material and this size was influenced by the fines fraction, F. Thus, an equilibrium could apparently be reached between bubble growth by coalescence and spontaneous splitting to give a stable bubble size distribution within the upper levels of deep beds of Group 'A' materials.

Group 'C' materials are those of smaller mean size (< 30 μm) and/or of lower density so that interparticle forces have an effect greater than that of gravity. Such materials are very difficult to fluidize. It is unlikely here that the pressure drop across the bed will equal its weight per unit area. This shows that part of the weight, even if the bed is displaying quasi fluid-like properties, is supported by the interparticle forces and surface contacts. With these materials, *channelling* is very prone to occur, i.e. the flow of fluid opens up low resistance channels through the bed. Once such a path has been created, it tends to enlarge with further increase in gas velocity so that the gas is not properly distributed into the bed which never becomes truly fluidized. This can be overcome to some limited extent by using a mechanical stirring element moving close to the distributor which will break down the channels and so cause the gas flow to redistribute itself through the bed in the lower regions.

Group 'D' materials are usually of mean diameter greater than 600 μm and/or dense particles. Although a bubbling fluidized bed looks very turbulent and may be described as being turbulently fluidized when operated at higher fluidizing velocities, the gas flow condition within the interstices tends to be laminar or, at the most, transitional until one is dealing with materials of this Group D. With these materials, the interstitial flow rate of the gas through the continuous phase necessary to fluidize the bed is greater than the rise velocity of the bubbles. This gives rise to a through flow of gas without the gas circulation cloud round the bubble, as obtained with the other materials for which the bubble rise velocity is higher than the interstitial gas velocity.[3] Gas flow conditions tend now to be in the turbulent regime and the bubble coalescence pattern is different. Instead of a smaller bubble being caught up in the wake of a faster rising, larger bubble, cross-wise coalescence takes place between bubbles at a similar level in the bed. The ensuing solids mixing generated by this process is less effective than that by vertical coalescence.

Because of the difference in bubbling behaviour between different classes of material there can be very different changes when scaling up from small experimental beds to full-scale operational beds. This particularly comes about because bubbles are constrained in size by the scale of the equipment in beds of small diameter but have the opportunity to grow very much larger in larger scale equipment. With finer, less dense powders (Group A) the maximum stable bubble size is much smaller than with coarser, denser particles (Group B) so a stable bubble size distribution can be reached with smaller diameter beds with A materials than with B materials. Because larger bubbles rise faster than smaller ones, the gas associated with the bubbling process will escape from the bed more rapidly when the average bubble size is larger, so there can be much variation in the overall bed expansion brought about by change in the total bubble volume hold-up within the bed according to the bubble size distribution. This is

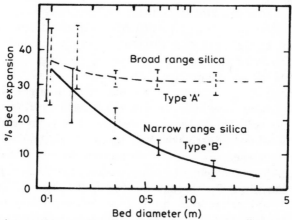

FIG. 3. Bed expansion as a function of bed diameter, only difference being size distribution of catalyst; expansion reflects bubble hold-up.[11]

well illustrated by some tests reported by de Groot[11] (Fig. 3). Here the only difference between the two series of tests on beds of a catalyst is the quantity of fines in the material and hence the mean particle diameter, a difference which is sufficient to alter the fluidization characteristics very markedly as Geldart noted;[7] the broad-range falling within the 'A' group and the narrow-range in the 'B' group of his classification. Associated with this change in behaviour will be changes in solids mixing within the bed and also in the efficiency of gas/solids contacting with important implications for reactor conversion efficiency.[2] The situation is further complicated by changes in behaviour when other operating variables, such as temperature and pressure, are changed (see Section 2 below). Much caution should therefore be used when applying the results of small-scale ambient temperature tests. Again, the immersion of heat transfer surfaces within the bed will affect bubbling behaviour but the disturbances caused by this are less likely to be scale dependent because the inserts will have the effect of breaking up the bed into smaller, repeated sections.

1.3 Components of fluidized beds

The bed must be contained within a suitable container (Fig. 4). The choice
of materials of construction is dependent upon the operating conditions,
ranging from refractory brick for large high-temperature kilns to steel
vessels, according to circumstances. A metal retaining surface may also be
used as a heat transfer surface if it is necessary to add or remove heat using
an external heat transfer medium. Often, however, the area of the
containing surface will be insufficient for this purpose and so it will be
necessary to immerse tubes within the bed to provide additional heat
transfer surface area. This is particularly so in the design of catalytic
reactors for highly exothermic reactions when the internals may be a
principal feature of the design.

In the defluidized condition the bed will be supported on the *distributor*.
This element of the equipment has the very important function of
presenting sufficient resistance to gas flow to stabilize the uniformity of gas
flow into the bed. This is necessary because there is no self-regulatory
change in pressure drop across the fluidized bed should gas flow increase in
one region at the expense of another; the pressure drop across the bed in the
fluidized condition remaining very closely equal to the bed weight per unit
area regardless of gas flow through it. However, the pumping power cost of
significant distributor pressure drop can be very expensive when dealing
with large gas flows and so excessive pressure drop has to be avoided. Rule
of thumb guides for deeper beds are that:

$$\frac{\Delta P_D}{\Delta P_b} \sim 0.15 \qquad \text{for} \quad \frac{U}{U_{mf}} \sim 1 \rightarrow 2 \tag{5}$$

and

$$\frac{\Delta P_D}{\Delta P_b} \sim 0.015 \qquad \text{for} \quad \frac{U}{U_{mf}} \gg 1 \tag{6}$$

For shallow beds (< 150 mm deep), however, the ratio is higher and often
close to 1. The simplest form of distributor is the pipe grid. Tuyeres are used
in some large installations (for example, Whitehead *et al.*[5]) where low
pressure drop is important. Perforated plates and bubble caps are also
commonly used for low pressure drop systems and porous ceramic tiles and
metal sinters for high pressure drop requirements. With perforated plates
and bubble caps, the orifice size and spacing is chosen to give the required
pressure drop. It is very expensive to make plates with orifices much less
than 1 mm in diameter yet the individual orifices should be no bigger than
$3d_p$ to reduce particle seepage into the plenum chamber beneath the

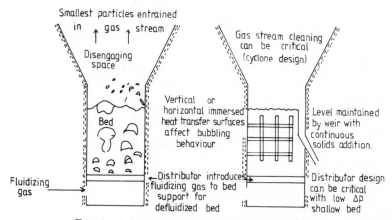

FIG. 4. Principal components of fluidized beds.

distributor. Should particles penetrate beneath the distributor they will mill around in the turbulent gas flow conditions there and suffer considerable attrition damage besides causing erosion damage to the plenum chamber and the underside of the distributor. Similarly, the bubble cap needs to be designed so that the natural angle of repose of the material will prevent significant seepage to below the distributor. The bubble cap type system has an additional advantage in high temperature systems in that stagnant material will be left between the caps and this will protect the distributor plate from the full rigours of bed temperature. Whereas orifice type systems promote gas jetting into the bed, porous plates produce a more even distribution of smaller bubbles. However, the underside of a porous tile or sinter will quickly become blocked should there be particulates in the fluidizing gas stream.

It is necessary to provide a space above the bed to promote the disengagement of particles thrown up by bubbles bursting at the bed surface. The smallest particles will still tend to be entrained into the gas stream even though the cross-sectional area in the space above the bed may be enlarged to reduce further the superficial gas velocity there. Accordingly, it will generally be necessary to provide cyclones or other disengaging devices for higher velocity systems.

Besides being included to provide additional heat transfer surface within the bed, the use of inserts has been advocated to promote more uniform bed behaviour by controlling bubble growth.[12] Their value for this latter purpose is questionable. Carelessly positioned internal surfaces (tubes, internal cyclone diplegs and the like) can engender particle damage and particularly so if gas jetting from the region of the distributor can cause particle streams to impinge upon them. The surfaces will then also be subject to erosion damage.

Taking advantage of the fluid-like character of a fluidized bed offers the means to operate on a continuous basis; continuously feeding and withdrawing material to and from the bed and also separating materials of different density. The bed level can readily be maintained by withdrawing material over a weir. Continuous solids addition to a bed of large area presents more serious problems and particularly so when a multiple feed injection system has to be employed. It is difficult then to monitor the uniformity of flow into the bed if the solid is fed from one source to a series of nozzles operating in parallel.

1.4 Choices in design

First, there are the velocity constraints within which the bed must be operated. Thus, the minimum operating velocity should be at least 1·3 to $2U_{mf}$ to ensure that the bed will be fluidized; it being safe to work closer towards the lower limit with larger particles when the actual superficial gas velocities involved are so much larger (of the order of metres per second). The upper limit should be less than the slugging velocity unless it is required to go beyond it into the regimes of turbulent or fast fluidized bed operation. Otherwise, the maximum velocity to avoid slugging, U_{max}, is given approximately[13] by:

$$U_{max} \simeq 0.07(gD_b)^{1/2} + U_{mf} \qquad (7)$$

There exists an obvious interconnection between the choice of particle size, d_p, (if this is a factor open to choice) which will determine the minimum fluidizing velocity, U_{mf}, under given operating conditions, the choice of operating velocity as a factor times U_{mf} and the necessary cross-sectional area of bed in order to be able to handle a given gas throughput. The prediction of U_{mf} is discussed in the following section.

Bed height is chosen to give sufficient gas/solid contacting time or in order that the bed shall contain a sufficient quantity of solids or a sufficient volume within which the necessary heat transfer surface may be immersed. As stressed above, pressure drop through the bed and the associated elements of the system can be very costly. This needs to be borne in mind when choosing the tube diameter and pitch in the design of immersed heat transfer surfaces. In large units, an optimum can be found between increasing containment costs for a larger area shallow bed and the extra cost of pumping power involved with a deeper bed of the same volume.[14]

There is also the increasing capital cost of the blower as one progresses beyond the capabilities of a simple fan ($5 \, \text{kN/m}^2$ head), to multistage blowers ($30 \, \text{kN/m}^2$), to Rootes blowers ($100 \, \text{kN/m}^2$), and finally to compressors, unless the process requirement necessitates high pressure operation when the pressure drop through the bed has less cost importance.

In small-scale beds of comparatively high depth/diameter ratio there is a predominant overall solids circulation pattern. Bubbles tend to draw solids upwards through the centre of the bed and there is the return flow of solids towards the wall of the bed. If the bed wall is transparent one can observe the return flow in characteristic stick-slip flow, although the degree of direct particle/wall interaction has never been determined. Only occasionally is the material close to the wall directly disturbed by rising bubbles although there is evidence that the particles may only be in direct contact intermittently with the wall or an immersed surface. In larger scale equipment there is greater uncertainty about the likely distribution of bubbles across the bed and of the direction of the solids convection.[5,15] One influences the other and semi-stable patterns become established; bubbles tend to move inward from a wall because of directional coalescence patterns and under the influence of the established return flow of solids.[16] The flow properties of 'A' type catalyst materials can be strongly influenced by the production or loss of fines from the bed. It is possible to generate strong convection patterns by suitably designing the system to encourage a particular circulation pattern (Fig. 5).

The pressure drop across a bed fluctuates more violently as the gas flow rate is increased and the surges occasioned by erupting bubbles become

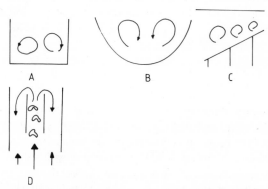

FIG. 5. Generation of circulation patterns: A, aspect ratio 1:2; B, variable depth bed; C, sectional distributor with separately controlled gas supplies; D, draft tube circulation generator.

more frequent until one passes into the regime of turbulent to fast fluidization. The character of bed behaviour then changes from one of slugging or bubbly flow to a much more uniform degree of turbulence with intense local mixing. The amplitude of pressure oscillations from being typically as much as 80 % of the average bed pressure drop falls to less than 40 % (Fig. 1) and material loss from the bed reduces with a bed free of immersed surfaces.[17] The peak pressure drop ratio at atmospheric pressure occurs typically at a velocity ratio $U/U_t \sim 0.5$ for 650 μm particles. At 10 atm the peak pressure drop ratio reduces and occurs over a relatively wide range of velocity ratios (~ 0.5 to 0.75). For larger particles the peak pressure drop ratio occurs at a lower velocity ratio. Immersion of a tube bank reduces the amplitude of pressure oscillation and shifts the turbulent regime to lower superficial velocities. The fast fluidized bed condition occurs at higher relative velocities and is marked by relatively high solid concentrations compared with the condition obtaining in dilute-phase flow; particles aggregate in clusters and strands and there is extensive back-mixing of solid.[4] It may only be possible to maintain a dense fast fluidized bed at atmospheric pressure and with velocities of around 3 to 4·5 m/s with fine powders (Group A).

2   THE MINIMUM FLUIDIZING VELOCITY AND TEMPERATURE AND PRESSURE EFFECTS

It must be appreciated that it is difficult to characterize particulate solids (see, for example, Allen,[18] Chapter 4). The mean particle size which is relevant in fluidized bed studies is that on a surface to volume basis. This is approximated to by the average obtained from sieve tests, viz:

$$d_p = \frac{1}{\sum x/d_f}$$

Where d_f is the arithmetic average of the adjacent sieve apertures and x is the mass fraction of the sieved particles. (Strictly speaking, d_p is only equal to the volume to surface mean when the particles are spherical.) The 50 % value on a plot of cumulative percentage undersize on a weight basis gives undue prominence to the larger particles within the distribution. Abrahamsen and Geldart[8] have further shown the importance of fines concentration when dealing with Group 'A' type materials. There is no satisfactory manner by which particle shape can be characterized numeri-cally for fluidization operations and it is even difficult to measure the voidage of a loosely packed bed and particularly so when the particles are porous. All are factors important in the estimation of minimum fluidizing velocity.

Although much is known about the behaviour of single particles (for example, see Clift *et al.*[19]), little is known about their behaviour with near neighbours. Also, whilst reliable equations have been developed for the prediction of pressure drop through packed beds when the fluid flow condition is laminar or turbulent, these are still subject to the limitations implicit in trying to allow for variation in particle shape and size distribution, and the empirical fit imposed to give continuity through the transitional regime is much more questionable. However, such correlations have often been used as the basis for the prediction of U_{mf} by taking the appropriate velocity to be that which, flowing through the packed bed of voidage equal to that of the incipiently fluidized bed, ε_{mf}, produces a drag force equal to the weight of the particles per unit area of the bed. (An alternative approach considers the opposite condition, namely the limiting case for particle settling velocities.) Thus, Ergun's correlation[20] takes the form:

$$Ar = 150\left(\frac{1-\varepsilon_{mf}}{\phi^2\varepsilon_{mf}^3}\right)Re_{mf} + \left(\frac{1\cdot75}{\phi\varepsilon_{mf}^3}\right)Re_{mf}^2 \qquad (8)$$

This reflects the very strong influence of voidage and the incipiently fluidized bed voidage is temperature dependent for 'B' group materials.[22] In the laminar regime an error of 10% in the voidage estimate would produce an error of 37% in the estimate of U_{mf}. In the turbulent regime the corresponding error would be 16%.

Various workers have attempted to avoid the problem of particle characterization by 'averaging out' the effects of shape and voidage and so deriving modified forms from the Ergun type correlation. Typically, Baeyens and Geldart[10] suggested the equation

$$Ar = 1823\,Re_{mf}^{1\cdot07} + 21\cdot7\,Re_{mf}^2 \qquad (9)$$

However, although this correlation can be recommended for use under ambient conditions and particularly for powders of Geldart's Group A when the flow will be consistently within the laminar regime so that only the first term, $1823\,Re_{mf}^{1\cdot07}$, is significant, it should only be used with caution beyond ambient conditions and for materials of Groups B and D. Indeed, because of the difficulties involved in characterizing the bed material and allowing for voidage it is better to measure U_{mf} directly when it is only required for ambient temperature operation. Non-cohesive, closely sized

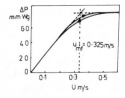
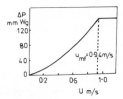
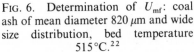

FIG. 6. Determination of U_{mf}: coal ash of mean diameter 820 μm and wide size distribution, bed temperature 515°C.[22]

FIG. 7. Determination of U_{mf}: bed of 2·32 mm mean diameter sand, turbulent/transitional gas flow conditions.[22]

particles display a sharp transition from the fixed to fluidized state and the fluidized bed pressure drop is close to the buoyant weight of the particles per unit area of bed. For laminar flow conditions the packed bed pressure drop increases linearly with increase in the gas flow rate, as would be expected. With a wide particle size distribution the transition from the packed to fluidized condition is less distinct because the fines tend to become fluidized before the larger particles. Measurements then tend to show a hysteresis effect between increasing and decreasing gas velocities (Fig. 6). Extrapolation of the pressure drop equivalent to the buoyant weight per unit area of bed to its intersection with the straight fixed bed pressure drop line, however, still gives a reliable measurement of the effective minimum fluidizing velocity but it is always inadvisable to attempt to operate a bed with a wide size distribution close to its minimum fluidizing velocity. Such beds should rather be brought to a bubbling condition as quickly as possible in order to reduce any tendency for segregation to occur to a minimum. Once particle segregation has occurred, it is very much more difficult to get the bed back into a well mixed state. When working with particles of larger mean diameter, the packed bed/gas flow rate relationship will show a degree of curvature indicative of transitional to turbulent flow conditions occurring within the bed (Fig. 7).

The Ergun type correlations suggest that U_{mf} should be independent of pressure whilst the flow through the bed is laminar but should be inversely proportional to the square root of the static pressure for turbulent gas flow conditions. This is approximately so. With temperature variation the situation becomes more complex. In the laminar flow regime, gas viscosity is the important factor. It increases with increasing temperature, so U_{mf} may be expected to fall. However, it does not fall in proportion to the viscosity increase because the bed voidage also increases; an effect noticed by Vreedenberg[21] but apparently ignored until recently.[22] For materials of Geldart's Groups B and D the Ergun equation predictions of U_{mf} were

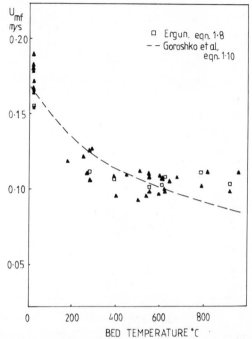

FIG. 8. Variation in U_{mf} with bed temperature, laminar gas flow conditions.

accurate over a range of experimental tests up to 900 °C when the appropriate gas properties, measured values of ε_{mf} and a value of ϕ chosen to give the best overall fit to the series of experiments were used. Unfortunately, no simple correlation is available for the prediction of the effect of temperature on bed voidage in the laminar flow regime nor can one reliably characterize the sphericity of the bed material without experimental test, so the Ergun equation is difficult to use. Goroshko *et al.*[23] proposed an approximate interpolation form of the equation in the form:

$$Re_{mf} = \frac{Ar}{150\left(\dfrac{1 - \varepsilon_{mf}}{\varepsilon_{mf}^3}\right) + \sqrt{\dfrac{1 \cdot 75 Ar}{\varepsilon_{mf}^3}}} \tag{10}$$

which fortuitously compensates for the change in ε_{mf} so that it allows moderately well for the effect of change in gas physical properties with temperature throughout the laminar flow regime if a value of ε_{mf} is chosen to fit the correlation to the readily measurable value of U_{mf} at ambient conditions (Fig. 8).

J. S. M. Botterill

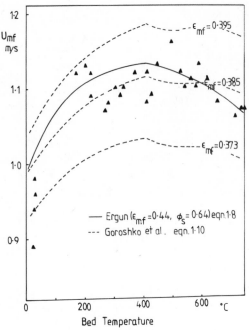

FIG. 9. Variation in U_{mf} with bed temperature, turbulent/transitional gas flow conditions.

In the turbulent flow regime where, with larger particles, voidage is independent of temperature, U_{mf} increases with increase in temperature because the gas density is then the important factor. However, the effect of increasing temperature on the gas physical properties leads to a reduction in operating Reynolds number which may be sufficient to reduce the flow condition from turbulent, through transitional to laminar according to mean bed particle size and range of operating temperature. Thus, this can readily be encountered with, for example, particles in the size range 3 mm $>$ $d_p > 800 \, \mu m$ and for beds operating between ambient and 900 °C. The transitional regime is also, of course, the range of the correlations for U_{mf} which is most in doubt. Typically for a 2·3 mm diameter particle fluidized in air, U_{mf} is first observed to increase with increasing operating temperature and then, beyond about 400 °C, to decrease. Application of the Goroshko equation presents difficulties because of the sensitivity to gas density change at lower temperatures and the effect of voidage variation at higher ones as the gas flow becomes increasingly laminar (Fig. 9). For a bed of sand of mean particle diameter $\sim 1300 \, \mu m$ there is a marked transition in

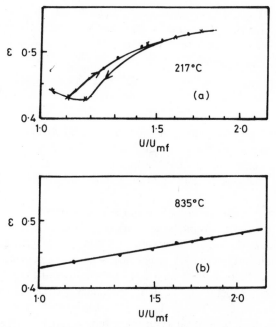

FIG. 10.   Voidage variation with U/U_{mf} for 1280 μm sand. (a) Transitional/laminar flow; (b) laminar flow.[22]

behaviour at about 450 °C or for alumina $d_p \sim 980\,\mu$m at 600 °C. Below these temperatures there is a distinct jump in the value of the bed voidage as the gas velocity is increased beyond that for U_{mf} (Fig. 10 for the sand) but this disappears at higher temperatures. This would seem to be a consequence of change in gas flow conditions at a value of $Re_{mf} \sim 12.5$ and $Ar \sim 26\,000$, so that the system crosses the boundary between the D and B class powders; a change from 'slow' to 'fast' bubbles most probably occurring with associated change in bed-to-surface heat transfer behaviour (see Section 3.2.2 below).

3   HEAT TRANSFER IN FLUIDIZED BEDS

3.1 Gas-to-particle heat transfer

Although gas-to-particle heat transfer coefficients based on the total particle surface area are often small,[24] a fluidized bed of particles is capable of exchanging heat very effectively with the fluidizing gas because of the

J. S. M. Botterill

very large surface area exposed by the particles (1 m³ of 100 μm diameter particles has a surface area greater than 30 000 m²). Typically, coefficients based on total area may be as low as 6 to 23 W/m² K and Kunii and Levenspiel[25] give a correlation in terms of a particle Reynolds number in the form:

$$Nu_{gp} = 0.03 \, Re^{1.3} \tag{11}$$

It is thus apparent that only a fraction of the total particle surface can be actually accessible to the fluidizing gas[26,27] but this rarely presents a limitation in achieving adequate heat transfer between the fluidizing gas and the bed particles because of the very large total particle surface area involved.[28]

It seems reasonable to suppose that the gas, with its relatively low heat capacity, will rapidly approach the temperature of the solids it is fluidizing and with which it is in contact across such a large surface area. However, work by Singh and Ferron[29] has detected marked temperature differences between them. Thus, in tests on a bed of spent cracking catalysts cooling from an initial temperature of 210 °C, they measured gas temperatures of the order of 100 °C when the bed had cooled to 170 °C. Nevertheless, because of its high heat capacity, it is the temperature of the solids which dominates the bed thermal behaviour and it is the gas temperature that follows that of the particles rather than the other way round.

On the basis of the gas-to-particle heat transfer correlation (eqn. (11)) the temperature difference between the entering gas and a well mixed bed of particles (so that it can be assumed that the particles are all at the same temperature) can be estimated to drop to half its initial value within a short distance for the gas entry. Thus, it was estimated[30] that the distance involved would be 1 mm for a bed of 450 μm particles in air at ambient conditions fluidized at $2U_{mf}$ and 2·4 mm at a temperature of 900 °C; the distance would be 1·5 mm for a bed of 1300 μm particles fluidized in air at ambient conditions.

3.2 Bed-to-surface heat transfer
To a first approximation, heat transfer between a bed and an immersed heat transfer surface under conditions of ordinary fluidization can be thought to consist of three additive components:

(i)   the particle convective component, h_{pc}, which is dependent upon heat transferred by particle circulation between the bulk of the bed and the region directly adjacent to the heat transfer surface;

(ii)   the interphase gas convective component, h_{gc}, by which particle-to-surface heat transfer is augmented by interphase gas convective heat transfer, and

(iii) the radiant component of heat transfer, h_{rad}.

Thus,

$$h_{overall} = h_{pc} + h_{gc} + h_{rad} \qquad (12)$$

| Approx. range of significance | $40\,\mu m \rightarrow 1\,mm$ | $> 800\,\mu m$ and at higher static pressure | Higher temperatures $(> 700\,°C)$ and differences |

The particle convective component, in particular, is affected by the volume flow of bubbles close to the heat transfer surface. Bubbles can also have a marked effect if they should shroud the surface. Some workers[31] would allow for direct heat transfer between the bubbles and surface as well as their effect in reducing the particle convective component but this bubble gas heat transfer component is rarely of significance.

When the bed is operating in the 'turbulent' or 'fast fluidization' regimes, different mechanisms can be expected to be operative and it would seem to be more profitable, as Staub suggested,[32] to apply the correlation methods earlier developed for flowing dense solid/gas suspensions (see Section 3.3 below).

3.2.1 Particle Convective Component, h_{pc}

The particle convective component is responsible for the characteristic marked increase in bed-to-surface heat transfer coefficient which is observed when a gas-fluidized bed passes from the quiescent into the bubbly state (Fig. 11). It was Mickley and his co-workers who first appreciated the part played by the high volumetric heat capacity of the particles in this process (~ 1000 times that of the gas at atmospheric pressure). It is the particles, then, which have the capacity to transport heat from the bulk of the bed to the region adjacent to the transfer surface. Particles in the bulk of the bed exchange heat with the fluidizing gas and, by conduction through it, with each other. They usually stay there long enough to come to the same temperature as their neighbours. A 'packet' of particles at the bulk bed temperature is swept into the vicinity of the heat transfer surface under the action of the bubble-induced mixing processes. When it first arrives close to the exchange surface there is a high local temperature gradient and heat transfers rapidly from bed to surface or vice versa. The longer the packet of particles resides close to the transfer surface the more closely local particle and surface temperatures approach. Highest mean temperature differences are therefore to be expected under those bed operating conditions which

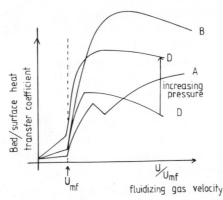

FIG. 11.   Variation in bed-to-surface heat transfer coefficient with gas flow rate for Groups 'A', 'B' and 'D' materials, and illustrating the effect of operating pressure on Group 'D' materials.

give a rapid circulation of material between the bulk of the bed and the transfer surface, i.e. with low particle residence times adjacent to the transfer surface as suggested by the basic Mickley and Fairbanks model.[33]

Particle-to-particle and particle-to-surface contact areas are too small for significant heat transfer to occur through points of solid contact.[1] Heat has to transfer by conduction through the gas phase which constitutes the chief resistance to heat transfer in this mechanism because of its low thermal conductivity. The increase in bed-to-surface heat transfer coefficients with operating temperature is a consequence of the increase in gas thermal conductivity with increasing temperature. The strong inverse dependence of the heat transfer coefficient upon particle size results from the increase in the effective proportion of the surface area through which heat can flow by *short transfer paths* between particles and surface as particle size reduces until eventually interparticle forces restrict particle circulation (Group 'C' materials). The coefficient then falls with further reduction in mean diameter. A simple model suggests that particle packing close to the transfer surface will also be important. Bed-to-surface coefficients were found to fall as predicted for a reduction in particle packing density at the surface when a heat transfer element was stirred through a stably expanded bed to give low effective particle/surface residence times.[34]

With finer powders of Geldart's Group 'A' there will be a degree of stable bed expansion with increasing fluidizing gas velocity[9] before the minimum bubbling velocity is exceeded. With such materials there is an additional minor peak in the rising portion of the curve[35] (Fig. 11). This is a consequence of some diffusive particle mixing before the bubbling condition is reached, the continuous phase then expands further with increase in gas velocity and the heat transfer coefficient decreases until the minimum bubbling velocity is exceeded. The expanded phase then collapses back and bubble-generated mixing develops and the coefficient increases to reach the principal maximum value.

For both Group 'A' and 'B' materials, a maximum is reached in bed-to-surface heat transfer as the fluidizing gas velocity is progressively increased. Beyond this the blanketing effect of increasing bubble flow across the transfer surface counters any effect of increasing bubble-generated particle circulation. Todes[36] has given a correlation for the optimum flow conditions as:

$$\text{Re}_{opt} = \frac{\text{Ar}}{18 + 5 \cdot 22 \sqrt{\text{Ar}}} \tag{13}$$

This was developed from the Goroshko correlation for U_{mf} (eqn. (10)) with assumptions about bed voidage. It gives reasonable predictions for beds of larger particles when the optimum gas velocity for maximum bed-to-surface heat transfer is closer to that for minimum fluidization, but it is not to be recommended for use with powders of mean diameters $< 400 \, \mu m$ when there is an increasing difference between these two velocities.

Whereas an increase in static operating pressure could be expected to lead to some reduction in the overall heat transfer coefficient with Group 'A' type materials because of a possible increase in continuous phase voidage and hence reduction in particle packing density at the heat transfer surface,[1] little effect is to be expected with Group 'B' class materials. It is only for conditions when the interphase gas convective component of heat transfer, h_{gc}, begins to be significant (see Section 3.2.2 below), i.e. when the gas flow conditions are in the transitional or turbulent flow regime and the effect of gas density on heat transfer becomes important, that there is an increase in heat transfer coefficient with increasing static operating pressure (Fig. 11).

The work reported by de Groot[11] referred to above (Section 1.2) illustrates the pronounced change in bed behaviour consequent upon different bubble development patterns resulting from change in equipment scale. Most of the published bed-to-surface heat transfer correlations were obtained in small-scale tests in which there was a strong definite solids circulation pattern. They reflect the particular behaviour of the beds used and the wide spread of their predictions is not therefore surprising. Bubbles tend to rise through the centre of the bed and there is the return flow of solids in a 'stick–slip' motion down the bed wall. Thus, such key factors as particle residence time will depend on the length of the exposed heat transfer element and particularly so if it is a section of the vertical containing bed wall. The tests have also usually only been carried out at ambient temperature yet the results are presented in the form of generalized correlations. Van Heerden[37] tested the then available correlations and found a range of 400-fold in the predicted range with one hypothetical system. Make a guess at the magnitude of the coefficient and you would not be so wildly wrong as by misapplying such correlations! Nevertheless, the approximate correlation by Zabrodsky *et al.*[38] for maximum bed-to-surface heat transfer coefficients, h_{max}, has both the merit of giving reasonable predictions for Group 'B' type materials and being simple. The form of the correlation is:

$$h_{max} = 35 \cdot 8 \rho_p^{0 \cdot 2} k_g^{0 \cdot 6} d_p^{-0 \cdot 36} \qquad (14)$$

This dimensional correlation contains the particle density, ρ_p, which is related to its heat capacity and the other two key variables: particle diameter, d_p, and gas thermal conductivity, k_g. One could conservatively expect to obtain values of 70 % of this maximum under reasonable operating conditions. It can be seen (Fig. 12) that the correlation allows very well for the effect of change of gas thermal conductivity with temperature on maximum bed-to-surface heat transfer until, above about 600 °C, radiant heat transfer begins to become significant. (In tests with a low emissivity and gold transfer surface (open circles), it can be seen that the Zabrodsky correlation continues to give good prediction to higher temperatures.)

For the finer, less dense Group 'A' powders, Khan *et al.*[35] give the correlation:

$$Nu_{max} = 0 \cdot 157 \, Ar^{0 \cdot 475} \qquad (15)$$

This has not been tested over a range of operating temperatures and it could be expected that changes in gas physical properties resulting from changing temperature would affect the bed's fluidization behaviour and hence the bed-to-surface heat transfer coefficients. It is also common knowledge that changes in size distribution through attrition damage or loss of fines can strongly influence the flow behaviour of such Group 'A' type powders close to surfaces, so causing marked change in bed-to-surface heat transfer.

Many fundamental models of the basic particle convective heat transfer process have been proposed[1] but these can not generally be used because

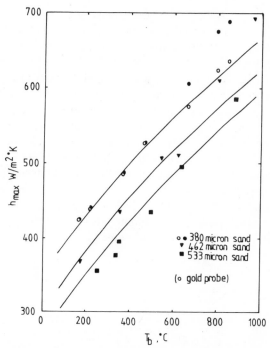

FIG. 12. Temperature effect on maximum bed-to-surface heat transfer coefficient Group B material.[39] Comparison between experimental measurements and predictions of Zabrodsky equation; radiative component significant above 600 °C.

the conditions adjacent to the transfer surface and the values of the basic parameters are not known. The most realistic simple model involves a packet replacement model modified to incorporate an additional wall transfer resistance to allow for the greater resistance to heat transfer in the region of increased voidage adjacent to the heat transfer surface. Good correspondence has been obtained on this basis between observed and predicted coefficients in fluidized bed[40] and flowing packed bed[41] tests when the particle residence time adjacent to the transfer surface could be determined. The value of the 'wall resistance' corresponds closely to that of an effective gas gap of 1/10 of a particle diameter. Examination of many detailed mechanistic models reveals that it is necessary to fall back on limited experimental tests to evaluate a 'constant' of the model and this generally hides questionable assumptions about bed behaviour which become more apparent as the implications of the model are followed through.

3.2.2 *Interphase Gas Convective Component*, h_{gc}

It is not until the interstitial gas flow conditions become transitional to turbulent that there is appreciable transfer of heat by convection through the gas. The threshold of significance for this factor is therefore dependent on mean particle diameter ($> 800 \, \mu m$) and static operating pressure (see Fig. 11). The finer tail fraction within the size distribution may be able to damp out turbulence within the gas stream because the results reported by Golan et al.[42] show negligible effect of mean particle diameter over tests with four beds of mean particle diameters between $800 \, \mu m$ and $2700 \, \mu m$ and with top fractions ranging from $1410 \, \mu m$ to $5600 \, \mu m$ but with a similar size distribution below $1410 \, \mu m$. However, over this mean size range the bed-to-surface coefficients are not very sensitive to particle diameter in any case (Fig. 13).

Baskakov and Suprun[43] first estimated this component by analogy from mass transfer measurements and recommended the following correlations:

$$\mathrm{Nu}_{gc} = 0 \cdot 0175 \mathrm{Ar}^{0 \cdot 46} \mathrm{Pr}^{0 \cdot 33} \qquad \text{for } U > U_m \qquad (16)$$

and

$$\mathrm{Nu}_{gc} = 0 \cdot 0175 \mathrm{Ar}^{0 \cdot 46} \mathrm{Pr}^{0 \cdot 33} \left(\frac{U}{U_m} \right)^{0 \cdot 3} \qquad \text{for } U_{mf} < U < U_m \qquad (17)$$

where U is the fluidizing gas velocity and U_m is the corresponding value for maximum bed-to-surface heat transfer. An alternative empirical correlation has been derived by Denloye and Botterill[44] over a range of operating conditions and operating pressures up to 10 atm, on the assumption that the bed-to-surface heat transfer coefficient at the onset of fluidization, h_{mf}, is a measure of h_{gc}. (Although the bed is not then mobile, there will be some conductive heat transfer through the gas between the particle and surface to augment the interphase gas convective component, so the assumption is not strictly true.) This correlation took the form:

$$\frac{h_{gc} d_p^{1/2}}{k_g} = 0 \cdot 86 \mathrm{Ar}^{0 \cdot 39} \qquad \text{for } 10^3 < \mathrm{Ar} < 2 \times 10^6 \qquad (18)$$

having dimensions of $\mathrm{m}^{-1/2}$. The corresponding maximum particle convective component, h_{pc}, was correlated against the Archimedes number by:

$$\frac{h_{pc_{max}} d_p}{k_g} = 0 \cdot 843 \mathrm{Ar}^{0 \cdot 15} \qquad (19)$$

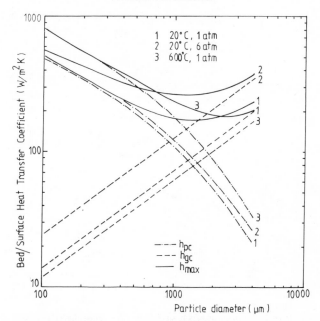

FIG. 13. Effect of particle size on maximum bed-to-surface heat transfer coefficient, h_{max}, the quiescent bed-to-surface coefficient taken to be the interphase gas convective component, h_{gc} and the difference taken to be the maximum particle convective component, $h_{pc_{max}}$.[39]

on the assumption that the two components are additive. Their relative magnitude as a function of particle diameter is illustrated in Fig. 13. Correspondence was good between the atmospheric pressure measurements of h_{mf} and the predictions of the Baskakov and Suprun correlation but there was increasing deviation as the operating pressure was increased; conditions under which h_{mf} more closely approximates to h_{gc}. The form of the correlations assumes that the gas convective component will be relatively insensitive to increases in gas velocity. This is reasonable because there will be little change in the actual gas flow conditions through the continuous phase or adjacent to the transfer surface with such Group 'D' bubbling beds.[9]

With particles approximately in the size range 500 μm to 3 mm, bed-to-surface heat transfer can be expected to be comparatively little dependent on particle size and decreasingly dependent on superficial operating velocity.[39] This is because of the complementary contributions of the particle convective and interphase gas convective components of heat

J. S. M. Botterill

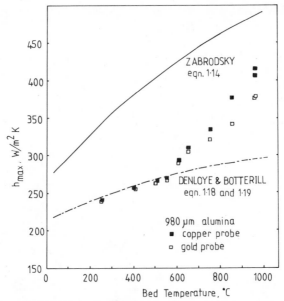

FIG. 14. Results reflecting change in dominant bed-to-surface heat transfer mechanism at an operating temperature of 560 °C with a bed of alumina mean diameter 980 μm; above 600 °C it can be seen that the radiant heat transfer contribution is increasingly important.[39]

transfer (Fig. 13). Figure 13 also illustrates that the comparative independence of particle size moves to somewhat higher mean particle sizes as the operating temperature increases.

As noted above (Section 2), change in bed voidage variation was observed with a bed of ~ 1300 μm mean diameter sand as the operating temperature increased above 450 °C (Fig. 10). Corresponding variation in heat transfer behaviour was also observed[22,39] (see similar test conditions with alumina in Fig. 14). This suggests a change in behaviour consistent with the bed having had the characteristics of a Group 'D' material at lower temperature so that the interphase gas convective component was then important, to behaviour characteristic of a 'B' type material at higher operating temperatures when the variation in coefficient reflected the increased importance of change in gas conductivity (eqn. (14)). This is likely to have been associated with a change in bed mixing consequent upon a change in bubbling behaviour from the condition where the bubble rises slower than the interstitial gas to faster than it (see comment on Group 'D' materials, Section 1.2 above). It may be expected that there will be conditions where bed behaviour will change markedly over a limited range of operating temperatures but further work is required to delineate this critical region more precisely.

3.2.3 Radiative Component, h_{rad}

The estimation of the radiative component presents the most difficulty. For conditions where this component is significant (above $\sim 600\,°C$), particle packing and gas conductivity tend to become of somewhat lesser importance and there is evidence[1] that the radiant mode tends to transfer some heat at the expense of the particle convective mechanism so that these two components (eqn. (12)) are not strictly additive. The effect is greater with bigger particles because the heat transfer surface receives and radiates energy from and to the whole of the particle's surface visible to it. Thus, relatively less heat is left to flow by conduction through the particle to pass to the exchange surface by conduction through the shortest gas transfer paths close to the 'point of contact' between particle and surface as particle size increases.

In high temperature systems, the bed materials are likely to be of a refractory character and these generally have low emissivities. However, as Zabrodsky pointed out,[45] the overall bed will have a different effective emissivity, ε_b, to that of its constituent materials which will be much closer to unity. Thus, Makhorin *et al.*[46] suggest values of ~ 0.9 for a bed of 1–1.5 mm sand of particle emissivity 0.6 at 1000 °C and something like 0.65 for corundum of 1.5–2 mm diameter and particle emissivity of 0.27 from their measurements.

Even allowing for the variation in effective bed emissivity and the effect of temperature on this, there also remains an additional factor—the influence of the heat transfer surface on the bed temperature directly adjacent to it. The presence of a cooler surface, for example, modifies the temperature of the adjacent bed consequently reducing the local radiant heat flux. Baskakov and his co-workers[47] have attempted to deal with this using a modified emissivity, ε_m. Thus, in Fig. 15, for heat removal through the transfer surface, the modified emissivity will be lower than the true reduced emissivity between bed and surface. It reduces further as the temperature difference between the surface and the bulk of the bed increases in order to compensate for the increased cooling influence of the surface on the bed directly adjacent to it and thereby to allow for the local reduction in radiant heat flux. (Makhorin *et al.*[46] reported that the local bed temperature initially fell at a rate of 350 to 860 K/s, close to a surface maintained at a temperature 370 °C lower than the bulk bed temperature of 1100 °C and

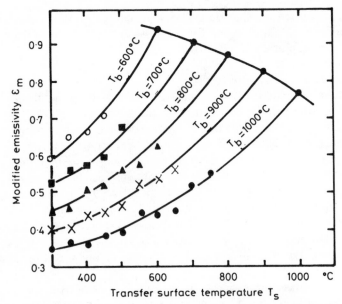

FIG. 15. Variation in effective emissivity, ε_m, of a non-isothermal alumina bed of small particles with both the temperature of transfer surface, T_s, and bed, T_b. (After Baskakov *et al.*;[47] values of ε_m would be somewhat lower for sand.)

stabilized with local particle temperatures 50 to 100 °C lower than those in the bulk of the bed.) It may be expected that the total bed-to-surface heat transfer coefficient, h, will be of the order of 600 W/m^2 K for an immersed tube in a bed operating at a temperature of 800 °C, which is of similar magnitude to the likely steam side coefficient for a unit raising steam at 300 °C and 20 atm pressure. Under such circumstances the tube surface temperature may approximate to 550 to 600 °C, leading to an effective emissivity of ~0·6 which seems a reasonable value.

The heat transfer surface will obtain a sight deeper into the bed when the bed is bubbling vigorously. This will further favour the radiant component because the bed interior so exposed will be less influenced by the temperature of the transfer surface and will also behave more like a black body radiator. Because of this, the maximum in the variation in coefficient with fluidization conditions should be less prominent.

For rule of thumb estimates, the radiative component can be estimated using absolute temperatures and an adaptation of the Stefan–Boltzmann equation in the form:

$$h_{rad} = \frac{5 \cdot 673 \times 10^{-8} \varepsilon_r (T_b^4 - T_s^4)}{T_b - T_s} \qquad (20)$$

where ε_r is a reduced emissivity to take into account the different emissive properties of the surface ε_s, and the bed, ε_b, given by:

$$\varepsilon_r = \frac{1}{\left(\dfrac{1}{\varepsilon_s} + \dfrac{1}{\varepsilon_b}\right) - 1} \tag{21}$$

and T_b is the bed temperature adjacent to the surface. Alternatively, because this is not generally known, a modified emissivity, ε_m, may be used in place of ε_r together with the bulk bed temperature (Fig. 15) following Baskakov *et al.*[47]

3.3 Immersed tubes and other surfaces

Immersion of components for heat treatment or provision of tubes to provide heat transfer surface if the containment surface affords inadequate transfer area, will obviously affect the local fluidization behaviour. This, in turn, affects the particle convective component of heat transfer, h_{pc} (Section 3.2.1). Thus, bed-to-surface heat transfer coefficients are higher to a small object (e.g. a piece of coal) which is free to circulate within the bed than to a stationary surface. When horizontal tubes are immersed within a bed, although much heat transfer surface is exposed to the cross-flow of solids which will lead to low local particle residence times and higher particle convective components of heat transfer, a stagnant layer of defluidized particles will form on top of the tube and bubbles may tend to shroud the downward facing surface reducing transfer in those regions. The situation is more acute still if a flat plate is immersed at an angle. Whilst, according to circumstances, the underside can be quite effectively washed by flowing solids carried upwards by ascending bubbles, the average residence time may still become quite extended and the upperside, itself, is likely to be blanketed by semi-stagnant solids.

When horizontal tubes are mounted on a staggered array there is a greater tendency for the rising bubbles to periodically displace the stagnant cap of solids which inevitably forms. Very close spacing seriously restricts the particle circulation and, in the limit, the bed may rise above the tubes which then would form an unintentional distributor for the bed. However, McLaren and Williams[48] found that heat transfer coefficients do not decrease dramatically until the tubes are very close together. They found that their heat transfer results for both in-line and staggered arrays fell on a single curve when plotted against the narrowest gap between the tubes. Coefficients only fell by about 25 % as the gap reduced from 282 to 15 mm in tests with a fluidized ash of wide size distribution and mean $\sim 450\,\mu\text{m}$ but containing a small fraction greater than 1·7 mm. They also commented on obtaining coefficients some 20 % lower when the tube bundle was located close to the distributor, a region where behaviour is strongly influenced by distributor design and where the bubbles generating solids circulation are still generally small.

Vertical tubular inserts have been advocated as a means of preventing slugging in deeper beds[12] but their utility in this is doubtful. With vertical heat transfer tubes, the prevailing vertical solids mixing pattern tends to give longer particle residence times at the transfer surface than with the cross-flow situation of horizontal tubes. The final choice between horizontal or vertical tubes, however, would seem to depend largely upon constructional convenience. Because of the influence of tubes and other inserts on bed behaviour, it is to be expected that those with such internals will be less sensitive to the overall effects of changing scale (a problem stressed in Sections 1.2 and 3.2.1 above).

The very complete review of bed-to-tube heat transfer by Saxena *et al.*[49] draws attention to the wide range of results that have been reported. As stressed above (Section 3.2.1), any of the correlations should only be used with caution and only after having confirmed that they relate to relevant operating conditions. It is unlikely that greater error would be incurred from an estimate based on the simpler empirical equations. Thus, a conservative estimate for the bed-to-tube coefficient within beds of Group 'B' type materials would be about $0.7 \times h_{\mathrm{pc_{max}}}$ as predicted by the Zabrodsky equation (eqn. (14)) and, because the maximum is much flatter for Group 'B' type powders, using between 70 and 100 % of the sum of the predictions from correlations eqns. (18) and (19) for 'D' type materials.[39] At temperatures above 600 °C some estimate must also be made for the radiative component, h_{rad}. This can be done using eqn. (20) and an effective emissivity for the bed adjacent to the transfer surface between 0·6 and 0·7 following Baskakov *et al.*[47] would seem reasonable.

It would be prudent to carry out full-scale 'unit cell' tests in any large design based on repeated modules and particularly so because the effects of lateral mixing may become very important when there is a large number of tubes immersed within the bed (see also Chapter 3). Because of the variation in bed behaviour with operating temperature[22] (Section 2 above) there is a need also to use considerable care in the interpretation and extrapolation of the results of cold tests when the operational system incorporates 'large' local heat sources. However, it is interesting to note that 'single tube' tests are reported to give results[31,40] comparable to those obtained to tubes within arrays.

A different situation is encountered with so-called 'turbulently' fluidized beds.[17] Reported results showed a maximum heat transfer coefficient at gas velocities about one-quarter of the particle terminal velocity. Typical values in tests with beds of mean particle diameter between 600 and 2600 μm, operating at static pressures up to 10 atm, were ~ 230 W/m^2 K and the coefficient varied as the 0·2 to 0·3 power of the gas density depending on particle size and tube geometry.[17] Staub[32] has subsequently developed a credible heat transfer model deriving from the correlations for flowing dense solids/gas suspensions to which this system more closely approximates. This uses gas and solid superficial velocities pertinent to the upflow and downflow conditions prevailing under the turbulently fluidized conditions and as predicted within his flow model.[32]

By and large, the controlling mechanisms for fluid-bed heat transfer can be envisaged (Section 3.2) but the challenge is to design the system so that advantage can be taken of bed behaviour to obtain high or regulatable bed/surface coefficients through control of the fluidization conditions. In principle, it is possible to design a system in order to force a rapid particle exchange between the vicinity of the heat transfer surface and the bulk of the bed in order to obtain the maximum particle convective component if a comparatively low mean particle size should make this advantageous. It can be done either by using stirring elements to reduce particle residence time at the transfer surface or by designing the system to obtain a fast bed circulation rate past immersed transfer surfaces. However, the utility of this is doubtful[1] and fluidized solids lose their fluid-like properties on impaction at a surface.

The use of finned tube transfer elements has been advocated for conditions where the bed side coefficient may be limiting. In deep bed applications, bed-to-surface coefficients based on the total exposed surface area are expectedly reduced over those obtainable per unit area of bare tube. Thus, Staub and Canada[17] give values showing reductions of the order of 30% after allowing a correction for the fin efficiency (see also Saxena *et al.*[49]). Overall effectiveness will depend on fin configuration, the relative size of particle and fin spacing and on the tube arrangement itself. Nevertheless, rough calculations suggest that the effective heat transfer capacity within a given volume can be increased by a factor of three by using finned tubes. It was the late Professor Elliott who particularly appreciated the cost of deep atmospheric pressure beds through their pumping power requirement and advocated the application of shallow bed heat exchange elements using finned tubes, a topic considered in Chapter 11. Based on the outside area of the tube, coefficients as high as 1 to 4 kW/m^2 K were obtainable. Although such shallow bed units are operated with bed expansions as high as 400%, bubbles remain restricted in size by the closeness of the fin spacing (which may be as small as 15 to 20 particle diameters) and very short particle residence times are obtained at the heat transfer surface; particle velocities being possibly an order of magnitude higher than those obtaining in deeper beds. Very similar heat transfer coefficients per actual unit area of total surface are then obtainable between the two systems. This was but one of the many innovative ideas that Professor Elliott developed and it is not inappropriate to conclude this introductory review chapter with a personal acknowledgement of what our friendship meant to me. All who knew him found his enthusiasm so infectious and I greatly miss the stimulation and pleasure of our work together.

REFERENCES

1. BOTTERILL, J. S. M., *Fluid-Bed Heat Transfer*, Academic Press, London, 1975.
2. GELDART, D. (ed.), *Gas Fluidization*, John Wiley and Sons, New York, 1983.
3. ROWE, P. N., in *Fluidization*, Davidson J. F. and Harrison, D. (eds.), Academic Press, London, 1971, Ch. 4.
4. YERUSHALMI, J. and CANKURT, N. T., *Powder Technol.*, **24**, 187 (1979).
5. WHITEHEAD, B., DENT, D. C. and MCADAM, J. C. H., *Powder Technol.*, **18**, 231 (1977).
6. GELDART, D., HURT, J. M. and WADIA, P. H., *A.I.Ch.E. Symp. Series*, **74**(176), 60 (1978).
7. GELDART, D., *Powder Technol.*, **6**, 201 (1972).
8. ABRAHAMSEN, A. R. and GELDART, D., *Powder Technol.*, **26**, 36 (1980).
9. GELDART, D., *Powder Technol.*, **7**, 285 (1973).
10. BAEYENS, J. and GELDART, D. in *Fluidisation et ses Applications*, Societé Chimie Industrielle, Toulouse, 1973, p. 263.
11. DE GROOT, J. H., *International Symposium on Fluidization*, Drinkenberg, A. A. H. (ed.), Netherlands University Press, Amsterdam, 1967, p. 348.
12. VOLK, W., JOHNSON, C. A. and STOTLER, H. H., *Chem. Eng, Prog.*, **58**, 44 (1962).
13. STEWART, P. S. B. and DAVIDSON, J. F., *Powder Technol.*, **1**, 61 (1967).
14. ELLIOTT, D. E., HEALEY, E. M. and ROBERTS, A. G., *Conference of the Institute of Fuel and l'Institut Francais des Combustibles et de l'Energie*, Paris, 1971.
15. WERTHER, J., *Powder Technol.*, **15**, 155 (1976).
16. GRACE, J. R. and HARRISON, D., *Fluidization*, Pirie, J. M. (ed.), *Tripartite Chemical Engineering Conference*, Montreal, Inst. Chem. Eng., London, 1968, p. 105.
17. STAUB, F. W. and CANADA, G. S., *Fluidization—Proceedings of the Second Engineering Foundation Conference*, Davidson, J. F. and Keairns, D. L. (eds.), Cambridge University Press, 1978, p. 340.
18. ALLEN, T., *Particle Size Measurement*, Chapman and Hall, London, 1975.

19. CLIFT, R., GRACE, J. R. and WEBER, M. E., *Bubbles, Drops and Particles*, Academic Press, London, 1978.
20. ERGUN, S., *Chem. Eng. Prog.*, **48**, 89 (1952).
21. VREEDENBERG, H. A., *Chem. Eng. Sci.*, **9**, 52 (1958).
22. BOTTERILL, J. S. M., TEOMAN, Y. and YÜREGIR, K. R., *Powder Technol.*, **31**, 101 (1982).
23. GOROSHKO, V. D., ROSENBAUM, R. B. and TODES, O. H., *Izvestiya Vuzov, Neft'i Gaz*, **1**, 125 (1958).
24. JUVELAND, A. C., DOUGHTY, J. E. and DEINKIN, H. P., *Ind. Eng. Chem.*, (*Fund.*), **5**, 439 (1966).
25. KUNII, D. and LEVENSPIEL, O., *Fluidization Engineering*, Wiley, New York, 1969.
26. LITTMAN, H. and SLIVA, D. E., *Int. Heat Transfer Conf.*, *Heat Transfer*, Versailles, Elsevier Publishing Co., Amsterdam, 1970, Paper C.T.1.4.4.
27. ZABRODSKY, S. S., *Int. J. Heat and Mass Transfer*, **6**, 23 (1963).
28. BARILE, R. G., SETH, H. K. and WILLIAMS, K. A., *Chem. Eng. J.*, **1**, 263 (1970).
29. SINGH, A. N. and FERRON, J. R., *Chem. Eng. J.*, **15**, 169 (1978).
30. BOTTERILL, J. S. M., *Gas Fluidization*, Geldart, D. (ed.), John Wiley and Sons, 1983, Ch. 9.
31. CATIPOVIC, N. M., JOVANOVIC, G. N., FITZGERALD, T. J. and LEVENSPIEL, O., *Proceedings of 1980 Fluidization Conference*, Grace, J. R. and Matson, J. M. (eds.), Plenum Publishing Corporation, NY, 1980, p. 225.
32. STAUB, F. W., *Trans. ASME. J. Heat Transfer*, **101**, 391 (1979).
33. MICKLEY, H. S. and FAIRBANKS, D. F., *A.I.Ch.E. Journal*, **1**, 374 (1955).
34. BOTTERILL, J. S. M., BRUNDRETT, G. W., CAIN, G. L. and ELLIOTT, D. E., *Chem. Eng. Prog. Symp. Series*, **62**(62), 1 (1966).
35. KHAN, A. R., RICHARDSON, J. F. and SHAKIRI, K. J., *Fluidization—Proceedings of the Second Engineering Foundation Conference*, Davidson, J. F. and Keairns, D. L. (eds.), Cambridge University Press, 1978.
36. TODES, O. M., *Applications of Fluidized Beds in the Chemical Industry*, Part II, Izd Znanie, Leningrad, 1965, pp. 4–27.
37. VAN HEERDEN, C., *J. Appl. Chem.*, **2**, Supplement Issue No. 1, S7 (1952).
38. ZABRODSKY, S. S., ANTONISHIN, N. V. and PARNAS, A. L., *Canad. J. Chem. Eng.*, **54**, 52 (1976).
39. BOTTERILL, J. S. M., TEOMAN, Y. and YÜREGIR, K. R., *A.I.Ch.E. Symp. Series*, **77**(208), 330 (1981).
40. BOCK, H. J. and MOLERUS, O., *Proceedings of 1980 Fluidization Conference*, Grace, J. R. and Matson, J. M. (eds.), Plenum Publishing Corporation, NY, 1980, p. 217.
41. DENLOYE, A. O. O. and BOTTERILL, J. S. M., *Chem. Eng. Sci.*, **32**, 461 (1977).
42. GOLAN, L. P., CHERRINGTON, D. C., DIENER, R., SCARBOROUGH, C. E. and WIENER, S. C., *Chem. Eng. Prog.*, **75**, 63 (July 1979).
43. BASKAKOV, A. P. and SUPRUN, V. M., *Int. Chem. Eng.*, **12**, 119 (1972).
44. DENLOYE, A. O. O. and BOTTERILL, J. S. M., *Powder Technol.*, **19**, 197 (1978).
45. ZABRODSKY, S. S., *Int. J. Heat and Mass Transfer*, **16**, 241 (1973).
46. MAKHORIN, K. E., PIKASHOV, V. S. and KUCHIN, G. P., *Fluidization—Proceedings of the Second Engineering Foundation Conference*, Davidson, J. F. and Keairns, D. L. (eds.), Cambridge University Press, 1978, p. 93.

47. BASKAKOV, A. P., BERG, B. V., VITT, O. K., FILIPPOVSKY, N. F., KIRAKOSYAN, V. A., GOLDOBIN, J. M. and MASKAEV, V. V., *Powder Technol.*, **8**, 273 (1973).
48. MCLAREN, J. and WILLIAMS, D. F., *J. Inst. Fuel*, **42**, 303 (1969).
49. SAXENA, S. C., GREWAL, N. S., GABOR, J. D., ZABRODSKY, S. S. and GALERSHTEIN, D. M., in *Advances in Heat Transfer*, Vol. 14, Academic Press Inc., New York, 1978, p. 149.

Chapter 2

COMBUSTION OF COAL IN FLUIDIZED BEDS

J. BROUGHTON

GEC Gas Turbines Ltd, Whetstone, Leicester, UK

and

J. R. HOWARD

Department of Mechanical Engineering, The University of Aston in Birmingham, UK

NOMENCLATURE

A	Bed (plan) area
A_B	Frontal area of bubble
A_C	Frequency factor in Arrhenius equation (eqn. (41))
A_p	Area of particle surface
Ar	Archimedes number
C	Concentration of gas ($kmol/m^3$)
C_e	Concentration of component in freeboard
C_{CO_2}	Equilibrium concentration of carbon dioxide
C_0	Concentration of component in inlet air
C_p, C_s	Oxygen concentration in dense phase, at surface
$\bar{C}_{pa}, \bar{C}_{pc}$	Mean specific heat of ash, coal
$\bar{C}_{pg}, \bar{C}_{ps}$	Mean specific heat of gas, mean particle specific heat
D_B	Bubble diameter
D_H	Bubble diameter at height H above distributor
D_0	Initial bubble diameter
D_R	Bed containment diameter
$\mathscr{D}$	Mass diffusion coefficient

$\mathscr{D}_i$	Molar diffusion coefficient (species i)
d_c	Char particle size
d_p	Particle size
$\bar{d}_p$	Mean particle size
E_A	Reaction activation energy
F_A	Air flow rate
F_c	Coal feed rate
F_i	Molar flow rate of gas to or from surface
f	Fraction of heat removed
$f_1(\varepsilon)$	Voidage function (eqn. (2))
$f_2(\varepsilon)$	Voidage function (eqn. (3))
g	Acceleration due to gravity
H	Height above distributor
H^c	Heat of combustion
H_{mf}	Bed height at minimum fluidizing velocity
h_c	Bed-to-particle heat transfer coefficient
h_0	Mass transfer coefficient
K	Overall mass transfer coefficient
K_{OR}	Orifice coefficient
k	Gas thermal conductivity
k^1	Dimensionless rate constant
k_{BC}	Mass transfer coefficient between bubble and cloud
k_C	First-order reaction rate constant
k_{CD}	Mass transfer coefficient between cloud and dense-phase
k_R	Reaction rate
M_B	Mass of char at equilibrium
M_c	kmol of char in batch
M_0	Initial batch mass (kmol)
m_p	Mass of a single particle
N_{OR}	Number of orifices per unit cross-sectional area of distributor
N_p	Number of particles
P_o	Partial pressure of gas
Q_B	Bubble volumetric throughput
Q_w	Heat removed by cooling surfaces
Q_w^*	Heat removed by cooling surfaces at design temperature and flow
R	Gas constant for air
R_A	Air/Fuel ratio
Re_{mf}	Reynolds number at minimum fluidization
Re_T	Reynolds number at particle terminal velocity
r	Radial distance from centre of carbon particle

r_c	Char particle radius
S	Air/Fuel ratio at stoichiometric composition
Sh	Sherwood number $= (h_0 d_c / \mathscr{D})$
T_B	Bed temperature
T_0	Initial air temperature
T_S	Particle surface temperature
T_w	Wall temperature
t	Time
U	Gas velocity (superficial)
U_B	Bubble velocity
U_I	Isolated bubble velocity
U_{mf}	Minimum fluidizing velocity
U_{SI}	Isolated slug velocity
U_T	Particle terminal velocity
V_B	Bubble volume
X	Bubble exchange factor
X_A	Excess air
x_i	Fraction of particle sample of size d_{p_i}
z	Function of gas velocity $= (U - U_{mf})/U$

Asterisk * against any symbol (thus Q^*) means 'the design value of' as distinct from the actual value achieved.

Greek symbols

α	Constant of integration
α_A	Actual air flow divided by design air flow
α_C	Actual coal flow divided by design value
β	Constant of integration
ε	Bed voidage
ε_{mf}	Bed voidage at minimum fluidization
ζ	Reaction mechanism factor
η_c	Combustion efficiency
λ	Function defined in eqn. (71)
μ	Gas viscosity
ρ_c	Char density
ρ_p	Particle density
σ^1	Function defined in eqn. (78)
σ	Function defined in eqn. (72)
τ	Burn-out time
ψ	Function defined in eqn. (74)

1 INTRODUCTION

The significantly larger magnitude of the world's reserves of coals compared with those of oils and natural gas renders them of primary importance both as fuel and a feedstock for chemicals and steel-making. After a steady decline in its use during the past 25 years due to an era of cheap oil and gas, coal has become attractive as a fuel again because of successive price increases in crude oil and political uncertainties about the security of supply.

The very wide range of substances commonly referred to as 'coal' has led to the need to classify coals according to their characteristics and use to which they are put. This matter is enlarged upon in Section 5.2. However, this chapter is primarily concerned with outlining the way in which coal burns within fluidized beds of inert particles, the factors which control combustion and some of the considerations which have to be made by the coal combustion system designer and development engineer in order to produce a reliable, efficient and economic fluidized bed combustor.

2 HOW COAL BURNS

2.1 Combustion in a hot air stream

If a piece of coal is placed in a hot air stream whose temperature exceeds that required to ignite the coal and whose speed is fast enough to ensure that a copious supply of oxygen is available for combustion then three events occur, namely:

(i)   the coal temperature will rise to its ignition temperature and beyond;

(ii) for a short period of time, volatile matter will be evolved and can burn at or beyond the particle;

(iii) the remaining solid combustible matter (mostly carbon), will be oxidized relatively slowly with the evolution of heat until only the incombustible ash remains.

Figure 1 shows a history of the process and it will be seen that most of the burn-out time of the coal particle is occupied by the oxidation of the carbon. Some important points concerning evolution of volatiles should be understood.

1. The volatilization process is *not* simple physical vaporization such as occurs when a liquid is heated, but the consequence of chemical reactions.

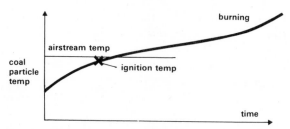

FIG. 1. Temperature history of burning coal particle.

2. The volatiles given off are hydrocarbon gases arising from irreversible chemical reactions, i.e. thermal decomposition, which occur when the coal is heated. Condensation and further cooling of these gases will not reconstitute their original chemical and physical form.

3. Volatiles are emitted from the interior of the coal particle from which they seep out to the exposed surface.

4. The chemical reactions may be exothermic.

5. There is no clearly defined temperature at which the volatiles are released, but the higher the surface heat flux into the coal particle, the shorter the time taken for all the volatiles to be given off.

6. The volatiles burn in much the same way as liquid fuel vapours burn, with a flame located a short distance away from the solid surface. The flame provides heat to the solid surface to assist further volatilization until all the volatiles have been evolved.

7. The evolved hydrocarbon gases may not burn completely in the region around the coal particle often appearing as a smoky yellow flame and deposit soot on the surfaces downstream.

8. Incompletely burnt volatiles can only be burnt to completion if they are passed through a high temperature zone containing excess oxygen.

2.2 Combustion of a single coal particle in a fluidized bed

All of the above applies to some extent to combustion of a coal particle when dropped into a fluidized bed of hot (say 850 °C) inert particles fluidized with air. However, in the case of a coal particle burning in a hot free air stream, the coal surface temperature can greatly exceed the fusion temperature of the coal ash so that the ash melts and can stick to surfaces downstream. In the fluidized bed, however, the burning carbon surface can be controlled at a temperature below the ash fusion temperature, due to a greater rate of heat transfer away from the burning surface by the bed inert particles than the surface-to-gas convection and radiation that occurs in combustion in a free stream of air. (Note that the heat transfer coefficient between a fluidized bed and a surface immersed in it is commonly more than five times as large as between a moving gas stream and a surface.)

2.3 Implications of fluidized bed combustion

There is no softened ash to condense on to heat transfer surfaces and foul them. Much of the ash will be in the form of small entrained particles (*c.* 100 μm) in the exhaust gases, where it can be removed by normal particulate removal equipment, e.g. cyclones, while some large ash particles may remain in the bed; indeed, ash may be used as the bed inert material. Methods for dealing with the ash from combustion of coal in a fluidized bed are described in Chapter 3. Because the temperature at the exposed surface of a burning carbon particle is relatively low in a fluidized bed (commonly < 1300 K), emissions of oxides of nitrogen formed by oxidation of atmospheric nitrogen are greatly reduced relative to other methods. However, the presence of nitrogen in the coal molecule leads to nitrogen oxides emission but at much lower levels than when fixation occurs. Furthermore, the temperature of the carbon surface is sufficiently low to ensure that the vapour pressures of alkali metal impurities in the coal or ash, a serious cause of fire-side corrosion with coal-fired plant, are not significant in the products of combustion.

3   GENERAL CONSIDERATIONS OF FLUIDIZED BED COMBUSTORS

3.1 Design approaches

There are many differences between the approaches taken to fluidized bed combustor design. Thus, one type of combustor operates under pressures of the order of 10 bar with bed depths of over 1 m, while another type operates at atmospheric pressure with bed depths of less than 0·2 m. Some designs feature over-bed feed of uncrushed coal while in others crushed coal is fed near to the distributor plate. This wide range of designs, dealt with in other chapters, is worth qualitative consideration before coal burning mechanisms are considered quantitatively, since many factors can influence how the coal reacts in a fluidized bed.

3.2 Factors influencing combustion

The most important characteristics defining a fluidized bed combustor are described below.

(i) *Bed Depth*. This can be set by many considerations such as

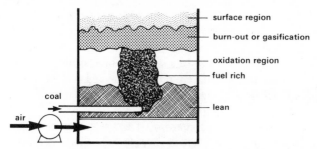

FIG. 2.   Typical zones of fluidized bed reactors.

containment of tubing or adequate depth for fines burn-out in the bed. A wide range of reaction environments can occur (as shown schematically in Fig. 2), ranging from gasification to oxygen rich combustion.

(ii) *Fluidizing Velocity.* A bed operating near the incipient fluidizing velocity is characterized by little mixing and high temperature gradients. A prediction of the peak particle temperature is essential in this region. At very high fluidizing velocities (above the terminal velocity of the particles) a lean phase region exists above the fluidized bed and the bed can become a recirculating (or fast) fluidized bed. The range between these extremes is described further in Chapter 1.

(iii) *Particle Size.* The particle size dominates the fluidized bed characteristics since many of the gas-transfer properties are determined by this. Size distribution can influence segregation and particle packing characteristics and in some instances can also influence combustion.

(iv) *Pressure.* The operating pressure in the combustor and centripetal acceleration of bed particles, if the bed is of the rotating type, influence the fluidization properties and affect the reaction mechanisms.

(v) *Fuel and Ash Properties.* These properties and those of additives such as limestone can influence the way a fluidized bed combustor is operated; thus, should the ash have a low softening temperature, the bed temperature must be low enough to avoid sintering. Coal of high fines content may lead to burning over the bed in some designs; volatile matter can burn-out in or above the bed; a coal of high sulphur content may have lime added to the coal feed so as to reduce emissions of sulphur dioxide; a coal of high ash content can lead to build-up of solids in the bed, changes of bed material properties and fluidization behaviour.

(vi) *The Bed Temperature.* This is controlled to avoid sintering and sometimes to optimize additive performance. At low temperatures, burning rates can be substantially reduced since reactions become much slower;

this aspect of combustion is important for such processes as burning carbon off cryolite (a material with a melting point of less than 800 °C) from aluminium manufacturing cells.

(vii) *Fuel Feeding Method.* In some fluid bed combustors dry, crushed coal is fed at a number of different points in the bed and the spacing of these feed-points must be optimized; in other designs coal is fed over the bed surface and contains large pieces.

(viii) *Geometry of the Combustor.* The geometry of a fluidized bed influences the behaviour of the bed very greatly; thus, a bed can operate in the slugging regime if it has a depth-to-diameter ratio greater than unity whilst in a shallow bed a wide variety of bed material circulation patterns can exist.

(ix) *Air Feeding Method.* Various methods are used for introducing air into fluidized beds. The simplest method is the use of a horizontal flat plate which may be pierced or be of a porous material; however, the plate can be inclined or shaped if it is desired to influence the properties of the bed. A few other methods in common use are conical base sections, sparge-pipes, bubble-cap plates and packed beds of dense, large particles. The large number of patents filed reflects the multitude of distributor designs which exist and their importance in operation.

3.3 Combustion models—What and why?
Combustion is modelled for several reasons:

(i)   Scale-up of burners can be made with greater certainty.

(ii)   Design can be based on sound principles rather than purely on experience—which can be misleading and inefficient.

(iii)   Problems can be solved with a minimum of costly experimental work.

(iv)   Control systems can be designed most accurately if the systems involved can be described mathematically.

(v)   Human curiosity; man does not readily accept pure empiricism.

The models used vary from relatively simple, partial models which describe one aspect of the process only, e.g. single char particle combustion models, to lumped parameter models of sets of processes, such as coal burning in a bed with several coal feed-points and ash and coal elutriation occurring. The models are usually sets of equations which are solved to predict aspects of behaviour; the equations may be algebraic or differential and several will be developed in this chapter.

The effects of varying any of the factors discussed in Section 3.2 can best be considered in terms of mathematical models since the interactions which occur can make empirical scale-up very misleading. However, the complexity of the chemistry and physics often severely limits the models that can be formulated; in fact, it is common experience that practice leads theory by many years and much fluidized bed engineering is based purely upon experience and empiricism. Nevertheless, considerable progress has been made in formulating models and their use is growing rapidly.

The alternative of pure empiricism can be misleading, is time-consuming and expensive. Models aid the designer and developer to predict changes and exploit the results of test work more fully.

The relatively simple models developed here are useful in many practical situations. For example, the carbon hold-up in a fluid bed is greatly affected when the air flow alters and several trends can be predicted, as can excess air level, the maximum particle temperature which relates to sintering and bed temperature. The algebraic simplicity of the simple models can, in many cases, more than make up for their mechanistic limitations.

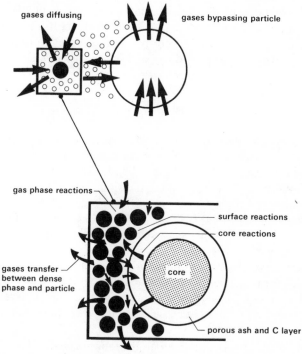

FIG. 3. Mechanisms of reaction in fluidized beds.

More complicated models to explain NO_x emissions, sulphur retention/limestone behaviour, devolatilization, etc., are important to the research worker and require a more specialized study than is offered here.

3.4 General considerations of fluidized bed reactions

The process under consideration is basically the transfer of gases to and from the particle surface and the reactions which occur at the surface, within the particle and in the surrounding gas phase. The wide range of reactions which can occur in a relatively simple bed design is shown in Figs. 2 and 3. The sizes of the zones and their effects are very dependent on fluidizing velocity since increase in this causes more bubbling of the bed and consequently improves the solids mixing. At low fluidizing velocities bed temperature gradients can become very significant. Reactions in all of the zones have several features in common and these can be considered in conjunction with Fig. 3, where the general reaction can be seen to consist of: transfer of gas into and out of bubbles; between bubbles and the dense phase; between dense phase and particle and reaction at the particle.

The types of model used to describe coal char combustion are developed in Section 6; however, first we shall review these aspects of fluidized bed behaviour and coal science that are required in order to show how the models have evolved and provide a basis for further models.

4   FLUIDIZATION

Chapter 1 describes the wide range of behaviour encountered in fluidized beds. What follows here is an indication of how reactions in fluidized beds may be modelled so that the results may be used successfully for the design and development of fluidized bed combustion systems.

4.1 Range of considerations

The modelling of reactions in fluidized beds requires an appreciation of the fluid mechanics of fluidized beds, since part of the problem is to describe how gases move within the bed. The most powerful tool in this modelling is undoubtedly the use of bubble motion and transfer theories which are now fairly well advanced. These theories are dealt with in detail in several major texts and progress is still being made in refining these theories further (see refs. 1–4).

It is not possible to go into detail on the range of bubble-based theories here; however, a brief summary of some main aspects of the two-phase theory is given to enable some of the fluid mechanics used in modelling reactions to be appreciated.

4.2 Fluidization range

The concepts of minimum fluidization and terminal velocities have been discussed in Chapter 1. For most purposes these are conveniently expressed in the form:

$$\text{Re}_{mf} = \frac{\text{Ar}}{f_1(\varepsilon) + f_2(\varepsilon)\sqrt{\text{Ar}}} \tag{1}$$

where

$$f_1(\varepsilon) = \frac{150(1 - \varepsilon)}{\varepsilon^3} \tag{2}$$

$$f_2(\varepsilon) = \sqrt{\frac{1\cdot 75}{\varepsilon^3}} \tag{3}$$

and,

$$\text{Re}_T = \frac{\text{Ar}}{18 + 0\ 61\sqrt{\text{Ar}}} \tag{4}$$

For low Archimedes number systems, which are fairly common in high temperature reactors, these expressions can be used in simplified dimensional approximations ($\text{Ar} < 1000$)

$$U_{mf} = \frac{\rho_p d_p^2 g}{1400\mu} \quad \text{(m/s)} \tag{5}$$

For $\text{Ar} < 200$

$$U_T = \frac{\rho_p d_p^2 g}{18\mu} \quad \text{(m/s)} \tag{6}$$

When a fluidized bed reactor operates with velocities in excess of the terminal velocity of some of its constituent particles these will be entrained in the freeboard and, unless they are returned, the bed will lose material. The regime of operation above U_T is used in many practical situations and is referred to as 'Fast Fluidization', 'High Velocity Fluidization' and 'Circulating Fluidization'. At high velocities the behaviour of the lean phase must be taken into account, usually as a co-current flow system. However, lean-phase systems are not considered further here; some information is given in refs. 1 to 5.

4.3 The two-phase theory of fluidization

A fluidized bed (operating significantly above incipient fluidization) can be observed to be similar in appearance to a boiling liquid, and since it exhibits many of the properties of a liquid the analogy between motion of bubbles in fluidized beds and in liquids became apparent many years ago. The two-phase theory was proposed by Toomey and Johnson in 1952 and has subsequently been improved by many workers; see Davidson and Harrison[1] and Kunii and Levenspiel.[3]

Basic Statement of Theory

The basic theory states that all gas in excess of that required for incipient fluidization of the particles flows as bubbles. This can be stated mathematically as:

$$\sum A_B U_B = (U - U_{mf})A \tag{7}$$

4.3.1 Bubble Velocity

The motion of a single bubble in a fluidized bed can be described according to the traditional inviscid fluid equation

$$U_1 = 0{\cdot}71\sqrt{gD_B} \quad (m/s) \tag{8}$$

or, should the bubble diameter be larger than 30 % of the bed containment it behaves like a slug in a liquid, and,

$$U_{Sl} = 0{\cdot}35\sqrt{gD_R} \quad (m/s) \tag{9}$$

When many bubbles are present simultaneously, the isolated bubble velocity is no longer applicable, since interaction and coalescence increase the velocity substantially above that predicted by eqn. (8). A semi-empirical approach is usually taken in which the isolated bubble velocity is related to the bubble size through eqn. (8) and

$$U_B = k(U - U_{mf}) + U_1 \tag{10}$$

where $k \simeq 1$ in the region well above a distributor plate and is a function of the bed voidage.

The importance of slug flow models is that in many small-scale experiments the bed operates in a slugging regime, in which transfer is different from that in the bubbling regime. The equations are similar in form to those used in bubbling bed models; the differences are discussed in detail in ref. 1.

4.3.2 Bubble Hold-up

The presence of bubbles causes the bed to expand and the prediction of bubble hold-up is possible using the two-phase theory, assuming uniform bubble sizes:

$$\frac{H - H_{mf}}{H} = \frac{U - U_{mf}}{U_B} \tag{11}$$

Thus, using eqn. (10):

$$\frac{H}{H_{mf}} \simeq \frac{(U - U_{mf}) + U_I}{U_I} \tag{12}$$

4.3.3 Prediction of Bubble Size

The prediction of bubble size in fluidized beds is important to the full understanding of reactions. In general terms this can be expressed as:

$$D_B = f\{(U - U_{mf}), \text{Distributor design}\}$$

But, it is also affected by gross bed flow patterns such as gulf-streaming or deliberate circulation.

Davidson and Schüler[5] proposed a theory for bubble formation that is widely applicable to fluidized beds with orifice plate type distributors:

$$D_0 = (U - U_{mf})^{0.4} K_{OR}$$
$$K_{OR} = f\{g, N_{OR}\} \tag{13}$$

Bubble formation with porous tile plates is not so well understood but a value for K_{OR} of 0·915 has been suggested by Geldart.[6]

The growth of bubbles by coalescence is described empirically by equations such as:

$$D_H = D_0 + K_{OR}(U - U_{mf})^m H^n \tag{14}$$

and this form is suitable for many practical uses.

4.3.4 Mass Transfer Between Bubbles and Dense-Phase Solids

The gas in the bubbles is continually exchanging with that in the dense phase by diffusion and mass flow into the bottom of bubbles and out of their tops. The transfer mechanics are still not completely resolved, but it is possible to show that the equations of transfer are:

(1) Bubble to cloud,

$$k_{BC} = \frac{Q}{V_B} = \frac{4 \cdot 5 U_{mf}}{D_B} + \frac{5 \cdot 85 \mathcal{D}^{1/2} g^{1/4}}{D_B^{5/4}} \tag{15}$$

(2) Cloud to dense phase,

$$k_{CD} = 6 \cdot 78 \, \frac{\mathcal{D}\varepsilon_{mf} U_B^{1/2}}{D_B^3} \tag{16}$$

(3) Total transfer to dense phase,

$$\frac{1}{K} = \frac{1}{k_{BC}} + \frac{1}{k_{CD}} \tag{17}$$

An approximation developed for this transfer coefficient that has proved useful in some models is:

$$\frac{Q}{V_B} = \frac{0 \cdot 11}{D_B} \tag{18}$$

4.3.5 Bubble Purging

The number of times a bubble is purged in passing through the bed is a very useful parameter in fluidized bed models. This is defined as:

$$X = \frac{Q H_{mf}}{U_B V_B} \tag{19}$$

Since bubble size and hence velocity vary greatly in a deep bed, X varies continuously and an incremental approach is often used

$$X = \int_0^H \left(\frac{Q}{U_B V_B} \right) dH$$

However, for many cases a mean bubble can be defined as an adequate approximation.

4.3.6 Simple Reactions

The basic two-phase theory as developed by Davidson and Harrison[1] produces an equation for the exit gas composition when *first-order* chemical reactions occur, namely

$$\frac{C_e}{C_0} = z e^{-X} + \frac{(1 - z e^{-X})^2}{k^1 + (1 - z e^{-X})} \tag{20}$$

assuming perfect mixing in the dense phase. Here, from eqns. (15) and (18) we find:

$$X = \frac{H_{mf} 6 \cdot 34 U_{mf}}{D_B \sqrt{g D_B}} \left\{ 1 + \frac{1 \cdot 3 \mathcal{D}^{1/2}}{U_{mf}} \left(\frac{g}{D_B} \right)^{1/4} \right\} \tag{21}$$

$$z = \frac{U - U_{mf}}{U} \tag{22}$$

$$k^1 = \frac{k_C H_{mf} \varepsilon_{mf}}{U} \tag{23}$$

This equation, despite being an approximation, is very useful and its limits are worthy of further consideration.

(i) Infinite reaction rate $(k_C^1 \to \infty)$

$$\frac{C_e}{C_0} = ze^{-X} \tag{24}$$

Here the extent of reaction is purely fluid mechanically determined.

(ii) Extremely rapid exchange $(X \to \infty)$

$$\frac{C_e}{C_0} = \frac{1}{1 + k^1} \tag{25}$$

A great number of models to predict exit compositions exist, depending upon the assumptions made about dense-phase composition and other factors such as bubble property variation in the bed. The interested reader is referred to the references. It is especially significant that no bubble models can yet predict precisely the behaviour of simple reactions from first principles, and the selection of the best model is a matter of experience.

4.4 Some further considerations

It is impossible to more than touch upon the main features of the fluid mechanics of fluidized beds here; the objective being to provide a consistent basis for the consideration of combustion in fluidized beds. Although the equations given here are useful tools, the limitations are worth some further considerations:

(a) Bubble behaviour can still only be confidently predicted in fluidized beds at relatively low values of $(U - U_{mf})$. Coalescence of bubbles and interactions between them complicates the understanding of transfer processes at higher velocities or in shallow beds.

(b) The equations for bubble-to-dense phase exchange are approximate.

(c) The use of the minimum fluidizing velocity in the equations may be incorrect, since in some systems dense-phase expansion can occur and the superficial velocity can exceed U_{mf}.

(d) Distributor effects can dominate bed behaviour in some circumstances, such as when very shallow beds are used or circulation patterns are deliberately induced.

(e) Bubble cloud and wake behaviour can dominate fine particle systems.

(f) At high fluidizing velocities the bed surface becomes a significant lean-phase region and influences contacting.

(g)   A topic that has not been considered here is that of solids mixing
       and separation in fluidized beds, an effect that is highly sensitive to
       density differences. The subject is not well understood quanti-
       tatively and a few references are given for further study (see refs. 1,
       2, 4 and 22).

In conclusion, although the two-phase theory provides a powerful tool for
the understanding of fluidized bed phenomena, much further refinement is
needed before it will be applicable to practical problems without
experience-based corrections being required. Purely predictive design is
rarely possible.

5   COAL SCIENCE CONSIDERATIONS

5.1 Range
The emphasis will be upon coal combustion, although the treatment used
can be applied to the many other solid fuel types, such as bagasse, waste,
wood and plastics. The coal science literature is vast and only those aspects
of relevance to fluidized bed combustion are considered; refs. 7 to 11
provide more detail.

5.2 Coal structure and types
The wide range of coal types was formed by biochemical and geothermal
processes called coalification; the extent of the coalification determining
the approach to pure carbon. The extent of coalification is partly described
by the Rank of the coal; some of the main groupings of coal being shown in
Table 1. All coal contains ash and moisture which do not contribute to the
heat release when the coal is burned. For ease of classification, coal types
are commonly compared on a dry, mineral-matter free basis (DMMF). In
this the main elements are carbon, hydrogen, oxygen, nitrogen and
sulphur.
      The ash content of a coal as-mined is a mixture of inherent ash which is in
the form of very small particles (typically 2 μm size), bound within the coal
pieces, and adventitious ash. In coal cleaning plant most of the adventitious
inerts are removed and the coal can be sized if necessary. The treated coals
have lower sulphur and ash contents and higher calorific values than the
untreated coals.
      The moisture content depends upon the treatment processes used; only
the surface moisture being affected by climatic conditions, e.g. air drying.
The surface moisture can have tremendous implications on the coal

TABLE 1
Properties of typical British coals

| Coal type | Rank | DMMF properties | | | | | | | Typical washed smalls | | | | |
| | | Element percentages | | | | | VM (%) | GCV (MJ/kg) | Percentage | | As fired | | |
		C	H	N	S	O			Moisture	Ash	GCV (MJ/kg)	NCV (MJ/kg)	F_A/F_C
Anthracite	101	95·0	3·0	1·0	0·7	0·3	5	35·82	8	8	29·66	28·94	9·84
Dry steam coal	201	92·4	4·0	1·4	0·9	1·3	11·5	36·52	7	8	30·59	29·66	10·1
Medium volume coking coal	305	88·8	5·0	2·0	0·8	3·4	25·0	36·29	8	8	30·35	29·26	10·03
Very strongly caking	401	87·5	5·3	1·9	1·0	4·3	35·0	36·05	9	8	29·54	28·38	9·74
Strongly caking	501	86·8	5·3	1·7	1·0	5·2	35·0	35·70	9	8	29·19	28·03	9·61
Medium caking	601	85·0	5·3	1·8	0·8	7·1	35·0	34·89	11	8	27·80	26·61	9·16
Weakly caking	701	84·5	5·2	1·8	0·8	7·7	35·0	34·42	13	8	26·75	25·54	8·84
Very weakly caking	802	82·0	5·3	1·7	1·0	10·0	39·0	33·73	16	8	25·24	23·96	8·26
Non-caking	902	81·0	5·1	1·6	0·7	11·6	40·0	33·03	18	8	23·34	22·59	7·85
Lignite	—	70·0	5·0	1·0	1·0	23·0	53·0	26·8	15	5	21·44	20·19	7·09
Wood	—	50·0	6·0	1·0	0	43·0	80·0	18·63	15	0	15·82	14·35	5·1

DMMF = Dry, mineral-matter free basis.   VM = volatile matter.

handling plant needed; for example, wet coal can freeze solid in winter conditions or wet fines can stick to handling plant causing clogging. Moisture bound into the coal particle by molecular forces cannot be removed by simple air drying. Its influence on combustion of coal is not properly understood quantitatively. Qualitatively, bound moisture is released when the coal is heated; the expansion during evaporation creates pores in the coal particle increasing the surface area available for combustion.

An industrial coal can be fully specified by reference to the cleaning plant used, the coal rank and the size of the coal. Coal size is described by the topsize and bottomsize. A wide range of graded coals exist such as doubles, singles, cobbles, nuts, beans and peas. Singles (usually + 12·5 mm, − 25 mm) and washed smalls (below 12·5 mm) are common industrial grades in the UK.

5.3 Coal costs

The cost of coal depends upon many factors: distance from the mine; annual tonnage purchased; coal specification required; sulphur, ash and water contents; area of the country; and the country concerned. In general, an untreated open-cast coal is the lowest in cost; however, the ash content and size range can make this impractical for industrial use without further treatment. Washed smalls are a commonly used type since the specification is reproducible, fines content, ash and sulphur are low. Graded coals such as singles have the great advantages of close uniformity and ease of handling that can more than compensate for their extra cost.

5.4 Calorific value and coal analysis

The calorific value of coal types varies widely and the most useful values are the 'as-fired' Higher Heating Value (or Gross Calorific Value—GVC) and Lower Heating Value (Net Calorific Value—NCV). The usual range of GCV is 21–31 MJ/kg and quoted figures for coal-fired equipment are usually based on 26 MJ/kg coal. Coal analyses are usually quoted in two forms:

(i) Ultimate Analysis—in this the percentage by mass of the main elements is given (dry mineral-matter free basis).

(ii) Proximate Analysis—in this, moisture, ash, volatile matter and fixed carbon (calculated by difference) are quoted.

Ultimate analysis allows accurate flue gas compositions to be predicted but proximate analysis is extremely useful.

The stoichiometric air/fuel ratio for a coal can be calculated accurately given the ultimate analysis; while where proximate analysis only is available, an approximation such as:

$$\text{Mass air/Mass fuel} = 0.32\,\text{GCV} \qquad (26)$$

when GCV is in MJ/kg, can be used.

5.5 Fluidized bed combustion implications

The most important coal properties from the point of view of fluidized bed combustion are: sulphur content; volatile matter, nitrogen and ash content; ash fusion temperatures; and size distribution of the coal. Properties such as swelling index are of little relevance. In common with other types of coal burning, when coal is burnt in fluidized beds the volatile matter is emitted rapidly while the char which is left after emission of volatiles burns slowly.

There remains considerable debate about the mechanisms of volatile emission. The reactions which occur to form the volatiles from the cyclic macromolecules present in coal are numerous and complex; most of these reactions are thermal cracking processes. The liberated volatiles then escape from the interior of the coal particle via the internal pores, some of which are created by the formation of volatiles. The mechanism can be treated empirically in the manner common to drying technology; that is a constant mass transfer rate process (limited by the heating rate and coal structure) followed by a falling mass transfer rate period, where the decomposition reaction rate (usually first-order with respect to the mass of volatile matter) dominates the emission rate. The constant rate volatile emission takes relatively little time (seconds) but the falling rate period occupies a much longer time, although less than the carbon burnout time.

6 COMBUSTION MODELS AND THEIR SIGNIFICANCE

6.1 Introduction

A wide variety of aspects of combustion in fluidized beds have been studied and modelled, see refs. 2, 12, 13, 14 and 19. Before proceeding to more detailed treatment of combustion models, some general comments (see below) are appropriate on the following:

(i)   Volatile emission and combustion.
(ii) Single particle combustion rates.

(iii)   Single particle temperature rise.
(iv)   Batch combustion of char.
(v)   Continuous combustion.

6.1.1 General Comments

(*i*) *Volatile Emission and Combustion.* If the volatiles burn in the freeboard space above the fluidized bed they cause the exhaust gas temperature to rise, which can cause softening of elutriated ash as it passes through the freeboard region. Since the volatile matter can represent as much as 40 % of the calorific value of coal its role is crucial. The theory of volatile emissions and combustion is dealt with in some detail in refs. 14 and 15.

(*ii*) *and* (*iii*) *Single Particle Reactions.* An appreciation of how a single coal particle reacts is essential to the understanding of how large numbers of particles burn. The maximum temperature rise of the coal particle is important because of ash softening as well as to the reactions occurring at the particle surface or within the particle itself.

(*iv*) *Batch Combustion.* This is a useful experimental tool which can be used to predict how more complicated systems will behave or to compare the behaviour of different coals.

(*v*) *Continuous Combustion.* This leads towards modelling of real systems with discrete coal feed points, ash removal, elutriation, recycle of elutriated solids, heat removal, and gross circulation patterns within the bed.

6.2 Combustion of volatile matter

The volatile matter content of coal can account for up to 40 % of the heat released in combustion; whether or not this burns out in the fluidized bed depends upon such features as bed depth, the method of coal feed, etc. As was discussed in Section 5.5, volatile matter emission is not readily described mathematically, so the approach adopted here is that of 'mapping' the regimes that can occur. By using simple heat balances it is possible to consider several features of a furnace. For example, in shallow bed fired boilers it is desired to minimize the excess air needed to maintain the bed temperature without having to use immersed cooling tubes, while in deep beds non-combustion of the volatiles can produce over-cooling. The extremes are shown in Fig. 4.

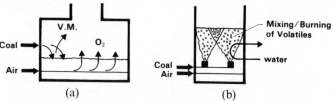

FIG. 4. Emission and combustion of volatile matter in fluidized beds. (a) Combustion in freeboard (shallow bed). (b) Combustion within bed (deep bed).

Heat Balance

Figure 5 shows the control volume containing the fluidized bed, together with the flows of heat and mass across the boundary.

A simplified overall heat balance which applies to each type of system in the steady-state can be written:

Heat Released in the Bed = Air Enthalpy Increase + Solids Heat Removal
+ Heat Removed by Cooling Surfaces

That is,

$$F_c H^c \eta_c = F_A \bar{C}_{pg}(T_B - T_0) + F_c \bar{C}_{ps}(T_B - T_0) + Q_w \tag{27}$$

where

$$Q_w = Q_w^* \frac{[T_B - T_w]}{[T_B^* - T_w]} \tag{28}$$

Note: Asterisked values represent basic design values.

For convenience we can introduce,

$$Q_w^* = f F_c^* H^c \eta_c \tag{29}$$

$$\alpha_C = \frac{F_c}{F_c^*} \tag{30}$$

$$\alpha_A = \frac{F_A}{F_A^*} \tag{31}$$

$$R_A^* = S(1 + X_A^*) \tag{32}$$

where S is the stoichiometric fuel/air ratio. So that,

$$\alpha_C H^c \eta_c = (\alpha_A R_A^* \bar{C}_{pg} + \alpha_C \bar{C}_{ps})(T_B - T_0) + f H^c \eta_c \frac{(T_B - T_w)}{(T_B^* - T_w)} \tag{33}$$

In typical uses the design values (asterisked) and $\alpha_A = \alpha_C = 1$ will be set by

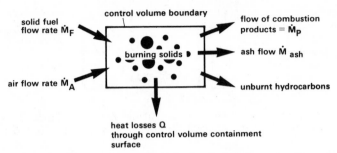

FIG. 5.　Heat and mass flows in fluidized bed combustor.

the fluidization characteristics; the air flow rates allowable being set by poor fluidization and excessive elutriation. In deep beds the volatile matter usually burns in the bed so η_c tends to unity; in shallow beds it would be as low as 0·6. The heat removal fraction, f, is set to obtain the desired bed temperature at design conditions.

Example 1—Boiler with In-bed Evaporator Tubes

Wall temperature	$T_w = 200\,°C$
NCV	$H^c = 26\,000\,kJ/kg$
Bed temperature (design)	$T_B = 950\,°C$
Excess air	$X_A = 20\,\%$
Combustion efficiency	$\eta_c = 1$
Stoichiometric air	$S = 9$
In-bed heat removal	$f = 0·5$
Flue-gas specific heat	$\bar{C}_{pg} = 1·2\,kJ/kg$
Ash specific heat	$\bar{C}_{pa} = 0·7\,kJ/kg$
Air inlet temperature	$T_0 = 15\,°C$

Equation (33) becomes:

$$26\,000 \left(\alpha_C - \frac{0·5(T_B - 200)}{750} \right) = (13·2\alpha_A + 0·7\alpha_C)(T_B - 15) \qquad (34)$$

This equation is mapped in Figs. 6(a) and (b). For constant excess-air operation the bed temperature changes substantially or the air flow rate changes; for example, taking $\alpha_C = \alpha_A = 0·6$, we obtain a bed temperature of 747 °C. The problem of turn-down of a combustor with tubes immersed are illustrated by the figures and have led to designs in which sections of the bed are slumped to achieve turn-down.

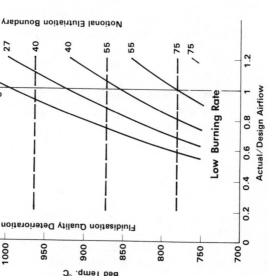

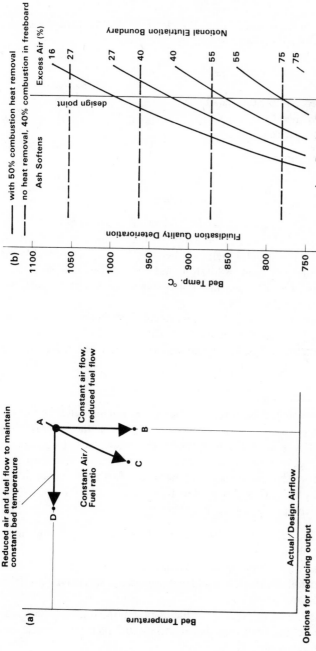

FIG. 6. (a) Change of operating point diagram. (b) Operating regime map.

Example 2—In a Furnace Design Which Does not Include Immersed Tubing and Has Over-bed Feeding of Coal, Causing Over-bed Combustion

Bed temperature (design)	$T_B^* = 960\,°C$
NCV	$H^c = 25\,000\,kJ/kg$
In-bed combustion efficiency	$\eta_c = 0·6$
In-bed heat removal	$f = 0$
Excess air (design)	$X_A = 40\%$

Equation (33) becomes:

$$T_B = 15 + \frac{990(\alpha_C/\alpha_A)}{1 + 0·046(\alpha_C/\alpha_A)} \tag{35}$$

The operational excess-air level thus determines the bed temperature; this is determined by the coal-feed technique and coal size and type used. Thus, bituminous coals are needed if high over-bed combustion is to be achieved along with high fines content. Radiation and wall cooling usually make f greater than zero. The after burning behaviour needs a separate analysis analogous to the spreader stoker.

This model does not lend itself to fundamental predictions since the in-bed combustion is design dependent, as is the excess-air level. The main uses are in predicting off-design behaviour trends and the need for bed slumping.

6.3 Char combustion models

The char particles remaining after volatile matter release can take several minutes to burn out. The theoretical models to describe the processes are developed here. The models are developed through several steps in each of which more complications are allowed for. Furthermore, the uses and limitations of each are discussed to show where further work could be justified and why more complex models have been produced.

The stages developed in logical progression are:

(i)   Combustion of a single isolated ash-free particle in a high temperature air stream.

(ii) Combustion of single particle in a low velocity fluidized bed. This is an extension of case (i) in which the presence of inert bed material is allowed for.

(iii) Combustion of a batch of particles in a bubbling bed. Here the transfer of gases between dense and bubble phase is allowed for.

(iv) Continuous combustion models.

(v)   Particle temperatures. The prediction of the relationship between particle size, bed temperature and fluidizing conditions is considered.

(vi) Applications of models to practical systems which have feedpoints, elutriation, volatiles, etc.

6.4 Single isolated particle combustion

This topic has been extensively studied over the years, excellent reviews being given in refs. 8 and 9.

The set of reactions that can occur at and near to a char particle surface is still not fully understood; for present purposes no significant ionization is considered and the reactions are:

$$C + O_2 \rightarrow CO_2 \tag{36}$$

$$C + \tfrac{1}{2}O_2 \rightarrow CO \tag{37}$$

$$CO_2 + C \rightarrow 2CO \tag{38}$$

$$CO + \tfrac{1}{2}O_2 \rightarrow CO_2 \tag{39}$$

The effects of the presence of other molecules, such as water, are discussed in ref. 16. Only reaction (39) occurs in the gas phase. During combustion oxygen has to diffuse to the particle and the carbon dioxide must escape from the particle; both oxygen and carbon dioxide react with the carbon. This leads to the consideration of two extreme cases:

(i)   Infinite mass transfer rate of oxygen direct to the carbon surface where the burning rate is determined by the degree of chemical reactivity of the carbon and its temperature. This situation occurs at very high air velocities.

(ii) Infinite rate of reaction where oxygen and carbon dioxide movement dominate—this occurs at high particle temperatures or combustion in free-space. It has been shown by experiments that this mechanism dominates the combustion of particles larger than about $100\,\mu m$ at low Reynolds carbon particle number and temperature of about $1000\,°C$.

Any carbon monoxide formed by eqns. (37) and (38) tends to be consumed very close to the surface except for very small particles. The gas phase reaction (39) can be suppressed by additives and the ratio of carbon monoxide to carbon dioxide concentration at temperatures between 400 and $900\,°C$ for graphite and char has been shown to fit the following:

$$\frac{C\langle CO\rangle}{C\langle CO_2\rangle} = 2400\exp-\frac{(51\,830)}{(RT_s)} \tag{40}$$

Thus, with $R = 8.314\,\text{kJ/kmol}$, at surface temperatures above $1273\,\text{K}$, the primary product is expected to be carbon monoxide.

6.5 Kinetic rates

Experiments where diffusion rates are extremely high have been fitted by the Arrhenius form, assuming a first-order reaction.

$$k_R = A_C \exp\left(\frac{-E_A}{RT_S}\right) \tag{41}$$

with constants $E_A = 150\,\text{MJ/kg\,mol}$ and $A_C = 7260\,\text{kmol/m}^2\,\text{s\,atm}$ ($R = 8.314\,\text{kJ/kmol}$).

The rate of reaction is then given by:

$$F_i = k_R A_p P_o \quad (\text{kmol/s}) \tag{42}$$

Thus, the burning rate is proportional to d_c^2 (the particle area). The time for burn-out can be found by writing:

$$F_i = \frac{\text{d}}{\text{d}t}\left(\frac{\rho_p}{12}\frac{\pi}{6}\,d_c^2\right)$$

$$= \frac{\rho_c}{12}\frac{\pi}{6}\,3d_c^2\,\frac{\text{d}(d_c)}{\text{d}t} \tag{43}$$

so that,

$$\frac{\text{d}(d_c)}{\text{d}t} = \frac{24}{\rho_c}k_R P_o \quad (\text{m/s}) \tag{44}$$

Integrating for burn-out of a particle

$$\tau = \frac{\rho_c d_c}{24 k_R P_o} \tag{45}$$

That is, the burn-out time is proportional to particle diameter.

Example 3—Combustion at 900 °C with No Particle Temperature Rise

$$\rho_c = 1400\,\text{kg/m}^3$$
$$P_o = 0.21\,\text{atm}$$
$$k_R = 1.52 \times 10^{-3}\,\text{kmol/m}^2\,\text{s\,atm}$$

From eqn. (45) we obtain:

$$\tau = 1.82 \times 10^5 d_c$$

so that a 1 mm particle would take about 3 min to burn-out under these hypothetical conditions. The specific burning rate [mass/(area × time)]

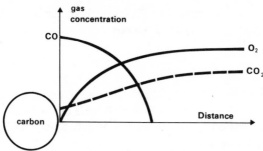

FIG. 7. Concentration distribution of gases within the region near a burning carbon surface.

would be $1{\cdot}52 \times 10^{-3} \times 12 \times 0{\cdot}21 \text{ kg/m}^2 \text{ s}$, that is $3{\cdot}83 \text{ g/m}^2 \text{ s}$. Note that for a surface temperature of $1100\,^{\circ}\text{C}$, k_R would be greater so that the burn-out time would be reduced to 19 s; that is, the kinetically controlled reaction is highly temperature sensitive.

6.6 Mass transfer rates

Combustion controlled by mass transfer rates is often what determines how droplets or particles burn. There are two extreme cases of diffusion limited reaction.

(i) Where oxygen reacts directly at the carbon surface to produce carbon dioxide, according to eqn. (36).

(ii) Where carbon dioxide reacts with carbon to form carbon monoxide which diffuses away and burns with oxygen according to eqns. (37), (38) and (39).

In both cases it can be assumed that the concentration profiles, as shown in Fig. 7, are steady as the rate of particle reaction is low, thus the concentration field is Laplacian. For $r > d/2$

$$\nabla^2 C = 0 \qquad (46)$$

or, in spherical coordinates,

$$\frac{d}{dr}\left(r^2 \frac{dC}{dr}\right) = 0 \qquad (47)$$

The mass flow of gas is given by Fick's Law as:

$$F_i = \mathscr{D}_i \left(\frac{dC}{dr}\right) A_p \qquad (48)$$

at $r = d_c/2$.

Assuming that the diffusion coefficients of oxygen and carbon dioxide are equal at the particle surface, we can write at $r = d_c/2$:

$$\left(\frac{dC}{dr}\right)_{O_2} = -\left(\frac{dC}{dr}\right)_{CO_2} \qquad (49)$$

Model (a) Oxygen Producing Carbon/Dioxide at the Surface
Concentration profiles are easily found by solving eqn. (47), so with the boundary conditions at $d/2$

$$C\langle O_2 \rangle = 0$$

$$C\langle CO_2 \rangle = C_{CO_2}$$

At $r \to \infty$,

$$C\langle O_2 \rangle = C_0$$

$$C\langle CO_2 \rangle = 0$$

The solution of eqn. (47) is:

$$C = \alpha - \frac{\beta}{r} \qquad (50)$$

Thus, the oxygen profile is:

$$C\langle O_2 \rangle = C_0(1 - d/2r)$$

and the carbon dioxide profile is:

$$C\langle CO_2 \rangle = C_{CO_2}(d/2r)$$

The rate of oxygen transfer to the surface is given by:

$$F_i = h_0 A_p C_0 \qquad (51)$$

From eqn. (48):

$$F_i = \mathscr{D}(\pi d_c^2)\left(\frac{2C_0}{d_c}\right) \qquad (52)$$

$$= 2\pi \mathscr{D} d_c C_0 \qquad (53)$$

Equating this to eqn. (51), we obtain:

$$\frac{h_0 d_c}{\mathscr{D}} = 2 \qquad (54)$$

This group is the Sherwood number (Sh). Equation (54) is a classical solution for mass transfer through a stagnant medium.[10]

Model (b) Carbon Monoxide Formation

When carbon monoxide is the main product at the carbon surface, the boundary conditions at $d_c/2$ become

$$C\langle O_2 \rangle = 0$$

$$C\langle CO \rangle = 2C_0$$

(since every mole of oxygen produces two moles of carbon monoxide from eqn. (37)).

At $r \to \infty$,

$$C\langle O_2 \rangle = C_0$$

$$C\langle CO \rangle \to 0$$

Thus, proceeding as before,

$$C\langle O_2 \rangle = C_0 \left(1 - \frac{d_c}{2r} \right) \tag{55}$$

$$C\langle CO \rangle = C_0 \left(\frac{d_c}{r} \right) \tag{56}$$

As before, the overall reaction rate is found by applying Fick's Law (eqn. (48)) at the surface,

$$F_i = 4\pi \mathcal{D} d_c C_0 \tag{57}$$

Comparison of eqns. (57) and (53) shows that carbon reacts twice as fast as when carbon dioxide is the primary product at the carbon surface.

In general, it is common to use the form:

$$F_i = 2\pi \mathcal{D} d_c C_0 \zeta \tag{58}$$

as the burning rate, where ζ is a mechanism factor between 1 and 2 depending upon the carbon monoxide/carbon dioxide formation ratio. There is an alternative approach to the carbon monoxide formation process in which the carbon monoxide is considered to burn in a diffusion flame around the particle. This mechanism is dealt with in refs. 1 and 10; it produces the same result for the burning rate, although the radial distribution of oxygen and carbon dioxide concentration around the carbon particle is different from the above.

The simple theoretical approach to the reactions between carbon and oxygen given here are little more than an introduction to this fascinating, apparently simple process.

The mass of a carbon particle is given by $\pi \rho_p d^3/6$. The rate of decrease of diameter can be solved to enable the relationship between particle size and time to be found.

For each kmol of oxygen consumed a kg atom of carbon reacts, so that:

$$\frac{dm_p}{dt} = \frac{\pi}{6} \rho_c \frac{d(d_c^3)}{dt} \quad \text{(kg/s)} \tag{59}$$

Therefore,

$$\frac{d(d_c)}{dt} = 12 F_i \frac{2}{\pi \rho_c d_c^2} \quad \text{(m/s)}$$

$$= 48 \zeta \frac{\mathscr{D} C_0 t}{\rho_c d_c} \quad \text{(m/s)} \tag{60}$$

Solving this between d_c and d_2 we obtain:

$$(d_c^2 - d_2^2) = 96 \zeta \frac{\mathscr{D} C_0 t}{\rho_c}$$

or for $d_2 = 0$ at complete burn-out,

$$\tau = \left(\frac{\rho_c}{96 \zeta \mathscr{D} C_0} \right) d_c^2 \tag{61}$$

This, again, is a classical expression for burn-out of oil droplets or coal particles under diffusion control.

Example 4—Combustion at 900°C

$$\rho_c = 1400 \, \text{kg/m}^3$$
$$\mathscr{D} = 320 \times 10^{-6} \, \text{m}^2/\text{s}$$
$$C_0 = 2 \cdot 18 \times 10^{-3} \, \text{kmol/m}^3$$

Inserting the typical values given above into eqn. (61), we obtain:

$$\tau = \frac{10 \cdot 45 \times 10^6 d_c^2}{\zeta} \quad \text{(s)}$$

so that a 1 mm particle would take between 5 and 11 s to burn-out under these conditions. This would mean that unless the surface temperature were much higher than 900°C, the reaction rate would be limited by the reactivity of the carbon. In practice, the carbon surface temperature is commonly sufficiently high to give a very rapid burning rate, but the rate at which oxygen reaches carbon is insufficient to permit such rapid burning. The burning rate is thus said to be 'mass transfer limited'.

Many further extensions of the basic theories given here are possible; for example, it is easy to show that under conditions where carbon monoxide is the primary product at the surface the presence of carbon dioxide in the gas stream increases the reaction rate (as carbon dioxide is effectively an oxidant).

Excellent reviews of the vast literature on the reactions occurring in the element set carbon, hydrogen, oxygen, nitrogen and sulphur are to be found in refs. 16 and 17.

6.7 Combustion of a single particle in a high temperature fluidized bed

This is an extension of the previous set of models where an allowance is made for the resistance to diffusion that occurs in the inert particles surrounding the burning particle. (A further resistance to diffusion can arise due to an ash layer attached to the coal particle which has to be penetrated by the air if the coal is to burn completely. For simplicity however this resistance will be discounted.) In this case the Laplace equation for spherically symmetrical diffusion becomes,

$$\frac{d}{dr}\left(\varepsilon r^2 \frac{dC}{dr}\right) = 0 \tag{62}$$

The solution of this is:

$$C_p - C_s = \int_R^\infty \alpha \, \frac{dr}{\varepsilon r^2} \tag{63}$$

where the bed voidage ε near the coal particle surface is a function of r depending upon the coal particle size and the bed material size. The simplest case is that when the bed material size is much smaller than the boundary layer thickness; when it is reasonable to assume that the voidage is constant at ε_{mf} at the surface. In this case, following the earlier analysis in Section 6.6 we can obtain:

$$Sh = 2\varepsilon_{mf} \tag{64}$$

The more general case is that when the function $\varepsilon\langle r \rangle$ is derived. A good curve-fit for the case when the coal particle is larger than the inert material over the range between r_c and $(r_c + d_p)$ that can be used in eqn. (62) is:

$$\varepsilon = 1 - 1\cdot8 \frac{(r - r_c)}{d_p} + 1\cdot2 \frac{(r - r_c)^2}{d_p^2} \tag{65}$$

For $r > (d_p + r_c)$ the voidage can be approximated by the dense-phase value:

$$\varepsilon = \varepsilon_{mf} = 0.4$$

Equation (65) can be solved for a given value of r_c and produces a higher Sherwood number than does the constant voidage assumption. The combustion of a char particle that is much smaller than the bed material is equivalent to free-space combustion.

6.8 Combustion of a batch of char in a fluidized bed

When a batch of char burns the oxygen is progressively consumed from the distributor; a fraction always bypassing through the bed in bubbles. The rate of combustion can be controlled by the oxygen transfer from bubble to dense phase, the mechanism of reaction at the active char surface or both. The main features of the process are as shown in Fig. 3.

The burning rate can be equated to the oxygen feed rate minus the rate of oxygen leaving the dense phase minus the rate of oxygen leaving the bubble phase. Thus, assuming full backmixing of the dense phase, we can apply eqn. (24) as an approximation (strictly only true when the rate of reaction is infinite).

$$\frac{dM_c}{dt} = UAC_0 - UAC_p - UA(C_0 - C_p)ze^{-X} \tag{66}$$

$$= (C_0 - C_p)A\{U - (U - U_{mf})e^{-X}\} \quad \text{(kmol/s)} \tag{67}$$

Some oxygen breakthrough is inevitable unless $X \to \infty$ and $C_p = 0$, when the burning rate equals the oxygen supply rate. In order to calculate the burning rate the values of X and C_p are needed. In order to calculate the burning rate a second equation is needed; this can be found from the particle reaction rate

$$\frac{dM_c}{dt} = N_p h_c A_p C_p \zeta \tag{68}$$

where ζ is the mechanism factor and has values between 1, for carbon dioxide formation at the surface, and 2 for carbon monoxide formation. Therefore,

$$\frac{dM_c}{dt} = N_p \, \text{Sh} \, \mathscr{D} \pi d_c \zeta C_p \tag{69}$$

Eliminating C_p we obtain,

$$\frac{dM_c}{dt} = \frac{C_0}{\lambda + \sigma/d_c} \tag{70}$$

where

$$\lambda = \frac{1}{A(U - (U - U_{mf}))e^{-X}} \quad (71)$$

and

$$\sigma = \frac{1}{N_p \mathscr{D} \, \text{Sh} \, \pi \zeta} \quad (72)$$

In order to find the burn-out time of a batch of char of uniform size, as before we note that

$$\frac{dM_c}{dt} = \frac{N_p \pi \rho_c d_c^2}{24} \frac{dd_c}{dt} \quad \text{(kmol/s)} \quad (73)$$

Putting,

$$\psi = \frac{N_p \pi \rho_c}{24} \quad (74)$$

$$\frac{dd_c}{dt} = \frac{C_0}{\psi d_c^2 \lambda + \sigma \psi d_c} \quad (75)$$

Therefore,

$$\tau = \frac{\frac{1}{3}\lambda \psi d_c^3 + \frac{1}{2}\sigma \psi d_c^2}{C_0} \quad (76)$$

$$= \frac{\lambda M_0}{C_0} + \frac{\rho_c d_c^2}{48 \mathscr{D} \, \text{Sh} \, \zeta C_0} \quad (77)$$

where M_0 is the initial mass of C (kmol).

Again, it must be stated that this is a gross oversimplification; assuming that the reaction is diffusion limited, elutriation is negligible, the two-phase theory of fluidization applies reasonably, etc.

Example 5—Low Velocity Fluidized Bed

$$X = 1 \qquad U = 0 \cdot 5 \, \text{m/s} \qquad U_{mf} = 0 \cdot 1 \, \text{m/s}$$
$$A = 1 \, \text{m}^2 \qquad \text{Sh} = 1 \cdot 0 \qquad \mathscr{D} = 320 \times 10^{-6} \, \text{m}^2/\text{s}$$
$$\zeta = 2 \qquad M_0 = 0 \cdot 2 \, \text{kmol} \qquad C_0 = 2 \cdot 2 \times 10^{-3} \, \text{kmol/m}^3$$

Thus, inserting the above typical values in eqn. (71) we find $\lambda = 2 \cdot 83 \, \text{s/m}^3$. Substitution into eqn. (77) gives $\tau = 258 + 2 \cdot 07 \times 10^7 d_c^2$, so that a batch of $2 \cdot 4 \, \text{kg}$ of 1 mm char particles will take 279 s to burn out. With infinite bubble phase to dense phase transfer ($X = \infty$), λ is only reduced to 2 and the burn-out time is reduced to 203 s; that is, the burn-out time is increased by about 50 % above that obtained from the oxygen supply rate, and up to 30 % of the oxygen does not react in the bed.

6.9 Continuous combustion with uniform fuel distribution

Consider a relatively simple case of continuous combustion; coal char is fed into a small area bed and the ash is completely elutriated. In this case the rate of change of carbon concentration is zero, so

$$\frac{\mathrm{d}M_\mathrm{c}}{\mathrm{d}t} = F_\mathrm{c}$$

i.e. the burning rate is equal to the coal feed rate. The following assumptions are made in order to avoid overcomplication:

   (i)   Particle size is characterized by the harmonic mean coal feed size (in the bed this is variable).

$$\bar{d}_\mathrm{p} = \frac{1}{\sum\limits_i \left(\dfrac{x_i}{d_{\mathrm{p}i}}\right)}$$

  (ii)   The reaction rate is considered to be limited by mass transfer.
 (iii)   The equations used are those developed for single particle and batch combustion (see Sections 6, 7 and 8).

Thus, as for eqn. (70),

$$F_\mathrm{c} = \frac{C_0}{\lambda + \sigma/d_\mathrm{c}} \tag{78}$$

In order to apply this for practical use we note that σ, from eqn. (72), can be written as:

$$\frac{\sigma}{d_\mathrm{c}} = \sigma^1 = \frac{\rho_\mathrm{c} d_\mathrm{c}^2}{M_\mathrm{B} 6 \mathscr{D} \operatorname{Sh} \zeta} \tag{79}$$

where M_B is the mass of char contained in the bed under equilibrium conditions. Using this equation it is possible to relate the char feed rate, the equilibrium char hold-up of the bed and the bubble exchange factors.

    This approach, despite the numerous simplifications, can be used in many engineering situations when linked to tests.

Example 6
Using the data from Example 5 in Section 6.8 (apart from M_0), for coal

feed rates of 7·2, 18, 25·2 and 33·5 kg/h, taking a mean feed size of 5 mm and coal particle density of 1400 kg/m³, we can calculate the equilibrium char hold-up values. Note that these coal feed rates are equivalent to 167, 417, 583 and 775 mmol/s, respectively.

Thus, from eqns. (78) and (79), with $\lambda = 2·83 \, \text{s/m}^3$,

$$\sigma^1 = \frac{2·2 \times 10^{-3}}{F_c} - 2·83 \quad (\text{s/m}^3)$$

(when F_c is in kmol/s) and from eqn. (79),

$$M_B = \frac{9·11}{\sigma^1}$$

so that the respective char hold-up values are 0·88, 3·72, 9·68 and 1000 kg. Also, we can see that the limiting value of F_c is 33·6 kg/h and an excess air value of 39 % is the lowest possible if all the char is burnt in the bed under these conditions, without gasification or emission of volatiles.

6.10 Coal particle temperature

When the combustion mechanism is strongly influenced by reaction kinetics the temperature of the burning surface is an important consideration. In addition, the agglomeration tendency of the ash is highly temperature dependent so that the peak temperature reached by the coal particle during burn-out needs to be estimated in order to ascertain whether or not agglomeration will occur. The temperature attained by a particle is determined by a heat balance. In the steady-state,

$$
\begin{aligned}
\text{Heat released by reaction} = \; & \text{Heat removed by conduction} \\
& + \text{Heat removed by convection} \\
& + \text{Heat removed by radiation} \quad (80)
\end{aligned}
$$

$$Q = h_c A_p (T_S - T_B) \tag{81}$$

and

$$Q = 2\pi \mathscr{D} d_c C_0 \zeta H^c \tag{82}$$

The heat of combustion of the char is given by[10]

$$H^c = (557\,000/\zeta - 161\,000) \quad (\text{kJ/kmol}) \tag{83}$$

Thus, with $\zeta = 1$ for the maximum rise,

$$T_S = T_B + \frac{2 \mathscr{D} C_0 H^c \zeta}{h_c d_c} \tag{84}$$

For large coal particles burning in a bed of fine particles an approximation is:

$$h_c = \frac{0 \cdot 016}{\sqrt{d_p}} \quad (\text{kW/m}^2 \, \text{K}) \qquad (85)$$

For fine particles the approximation Nu = 2 can be used, with the thermal conductivity modified to take account of radiation, thus:

$$h_c = \frac{2k}{d_c}$$

The equation for the full range of d_c and d_p values cannot be written in an exact form and for present purposes the two contributions can be summed, so that:

$$h_c = \frac{2k}{d_c} + \frac{0 \cdot 016}{\sqrt{d_p}} \quad (\text{kW/m}^2 \, \text{K}) \qquad (86)$$

Example 7
The highest temperature reached by char particles of 0·1, 1·0 and 10 mm in a bed of inert particles of 0·5 mm diameter at 950 °C can be calculated using the data given below.

$$d_p = 0 \cdot 5 \, \text{mm} \qquad\qquad T_B = 950 \, °\text{C}$$
$$\mathscr{D} = 320 \times 10^{-6} \, \text{m}^2/\text{s} \qquad C_0 = 2 \cdot 18 \times 10^{-3} \, \text{kmol/m}^3$$
$$\zeta = 1 \qquad\qquad\qquad H^c = 396\,000 \, \text{kJ/kg}$$

The conductivity for use in the Nusselt number is the packed bed value with an allowance for radiation, so that $k = 0 \cdot 9 \, \text{W/m K}$ (cf. 0·45 without radiation).

Thus, substituting in eqn. (84),

$$T_S = 950 + \frac{0 \cdot 552}{h_c d_c} \qquad (87)$$

The results are:

	d_c (mm)		
	0·1	1·0	10
h_c (kW/m^2 K)	18·7	2·57	0·89
T_{max} (°C)	1 245	1 170	1 012

These results are indicative only, being based on simplifications of the real situation. It is easy to see how more realistic factors can be included in the heat balance; for example, the effect of surface temperature on the reactions can be included as can more accurate heat transfer relationships. When the higher-order effects are allowed for it is usually impossible to derive an explicit relationship between T_s and the combustion parameters, so that graphical or numerical approaches are necessary; more details of these are given in refs. 10 and 18.

There are few experimental results on the temperature of burning particles in fluidized beds (this is due to the great difficulty in producing unambiguous results). However, some work has been carried out on systems with high excess air[18,19] which gives an indication of the maximum carbon particle temperature. A temperature of 100–200 K above that of the bed appears to be the highest obtainable, although there is some evidence that fine particles (below 0·1 mm) can exceed this in bubbles or the freeboard region. When coal is burning continuously, the coal particle temperature depends upon the excess air level used. The above expressions do not contain the excess air as a variable; it may be helpful to consider what happens when the excess air is increased by reducing the coal feed rate, leaving the mass air flow through the fluidized bed unchanged. Assuming an initial operating condition with small excess air, if the heat removal from the bed is then adjusted so as to maintain the bed temperature unchanged then since the oxygen concentration in the supply air is unchanged the oxygen concentration in the exhaust combustion products is greater than before the coal feed rate was reduced. Thus, the space average concentration of oxygen within the bed is higher; for a first-order reaction, the reaction rate would then increase. A new heat balance for the burning particle would then be reached dependent upon the coal particle-to-bed heat transfer coefficient, h_c. For an unchanged heat transfer coefficient the coal particle temperature must rise.

This sequence of events would be repeated for further reduction in coal feed rate until the bed temperature could no longer be maintained.

7 MORE ADVANCED MODELS

The models considered in the previous sections are very useful in understanding fluidized bed combustion and in the assessment of results from fluidized beds; however, some models have gone much further than these and a brief discussion of them is worthwhile. The more advanced chemistry and physics aspects of coal combustion are discussed in refs. 9, 10 and 11. In calculating combustion rates, particle size distributions, etc., several further aspects have been considered.

(i) *Particle size distribution.* The size distribution of bed material and effluent solids are important in the design of the reactor and gas cleaning system. The methods of treating size distributions are thoroughly discussed in ref. 2 and are applied to fluidized bed combustion in ref. 20. The algebra becomes more complicated but the principles used are not affected.

(ii) *Convective mass transfer.* The Sherwood number, Sh, is increased at high fluidizing velocity due to forced convection. Little work has been done on this, but a useful approach is to use the form that has been established for simple forced convection:[21]

$$Sh = 2\varepsilon + 0.6Re_{mf}^{0.5}Sc^{0.4} \tag{88}$$

Reaction rates and particle temperatures at high fluidizing velocities have recently been studied and the increase in rates of combustion is considered further in refs. 12, 18 and 19.

(iii) *Particle heating-up and devolatilization.* The rate of heating-up of a particle in a fluidized bed can be estimated from simple theory although detailed analysis is not simple and is not considered here. The heat balance used is:

$$M_p\bar{C}_{pc}\frac{dT_S}{dt} = h_cA_p(T_B - T_S) \tag{89}$$

so

$$T_S = T_B\left(1 - \exp\left(-\frac{6h_ct}{d_p\bar{C}_{pc}}\right)\right) \tag{90}$$

This oversimplification is useful in the absence of better models but the interested reader is referred to refs. 10 and 14.

(iv) *Influence of ash and coal properties on combustion rate.* The effects of ash on combustion rate are considered in detail in ref. 10. However, one coal property which is not yet fully appreciated is the tendency of some coals to shatter into small pieces shortly after being introduced into the bed. Shattering alters the size distribution of the coal in the bed, affecting both volatiles' emission and the burning rate of residual carbon.

(v) *Feedpoint disposition and elutriation.* These are considered in ref. 22. The most important conclusions from these are that, solids mixing is much more rapid in the vertical plane than in the horizontal plane and that the processes are most readily approached by diffusion models.

(vi) *Freeboard reactions.* These are very important for three main reasons: the gas temperature can exceed the ash softening temperature; carbon monoxide can be formed; and the fines burn-out improves combustion efficiency. The reactions in this phase can be very important in overall system modelling, especially when high fire loadings are involved.

8 SUMMARY

Some simple methods of estimating the burning rates, burn-out times and temperatures of char particles have been given. It is important for the combustor designer to have such knowledge. For example, if the char particle temperature is greater than that at which the coal ash or bed inerts melt or becomes soft, then bed sintering will occur. If the char particle temperature is low then the rate of oxidation of the char and hence the rate of heat release by each particle is low and in order to sustain output a very large amount of char must be present in the bed. This raises problems of control of bed temperature.

The contribution made by volatiles in coal to the performance of fluidized bed combustion systems is not well understood. It is clear that volatiles are important because they account for a large fraction of the calorific value of some coals and much of their heat content is released above the bed. Work is currently in progress to try to shed further light on such matters and their consequences for designers.

Successful design of fluidized bed combustion systems combines several disciplines, ranging through fluid mechanics, heat transfer, behaviour of fluidized solids, chemical thermodynamics, fuel technology and mathematical modelling. These alone, however, are not enough; in the end the combustion system must have high reliability while their capital, running and maintenance costs must show at least the same return on investment as competing systems. One of Douglas Elliott's many virtues was that he never lost sight of such practical constraints, but he had the imagination and facility for finding the way around them.

REFERENCES

1. DAVIDSON, J. F. and HARRISON, D., *Fluidisation*, Academic Press, London, 1971.
2. Fluidized Combustion Inst. F. *Symp. Ser. No. 1*, London, 1975.

3. KUNII, D. and LEVENSPIEL, O., *Fluidization Engineering*, John Wiley & Sons Inc., New York, 1969.
4. BOTTERILL, J. S. M., *Fluidised Bed Heat Transfer*, Academic Press, London, 1975.
5. DAVIDSON, J. F. and SCHÜLER, B. O. G., The formation and initial motion of gas bubbles in an inviscid liquid, *J.I.Ch.E.*, **38**, 335 (1960).
6. GELDART, D., The effect of particle size distribution on the behaviour of gas-fluidized beds, *Powder Technol.*, **6**, 207 (1972).
7. PERRY, R. H. and CHILTON, C. H., *Chemical Engineer Handbook*, 5th Edn, McGraw-Hill, Tokyo, 1973, Ch. 9.
8. LAURENDEAN, N. M., Heterogeneous kinetics of coal char gasification and combustion, *Prog. in Energy and Combustion Science*, **4**, 220 (1978).
9. COOPER, J. H. and ROSE, J. W., *Technical Data on Fuel*, 7th Edn, British National Committee, World Energy Conference, London, 1977.
10. FIELD, M. P., GILL, D. W., MORGAN, B. B. and HAWKSLEY, P. G. W., *Combustion of Pulverised Coal*, BCURA, Leatherhead, 1967.
11. SKINNER, D. G., *The Fluidised Combustion of Coal*, Mills & Boon, London, 1971.
12. CHAKRABORTY, R. K. and HOWARD, J. R., Combustion of char in shallow fluidised bed combustors, *J. Inst. Energy*, 48 (1981).
13. DAVIDSON, J. F. and CAMPBELL, E. K., The combustion of coal in fluidised beds, ref. 2, p. A2-1.
14. PILLAI, K. K., The influence of coal type on revolatilisation and combustion in fluidized beds, *J. Inst. Energy*, 142 (1981).
15. STUBINGTON, J. F., The role of coal volatiles in fluidized bed combustion, *J. Inst. Energy*, 191 (1980).
16. GAYDON, A. G. and WOLFHARD, H. G., *Flames—Their Structure, Radiation and Temperature*, 3rd Edn., Chapman and Hall, London, 1970.
17. HARKER, J. H., *Physical Properties*, H.T.F.S. Report No. HFTS. DR-40, Part 9, AERE Harwell, 1980.
18. ROSS, I. D., PATEL, M. S. and DAVIDSON, J. F., The temperature of burning carbon particles in fluidized beds, *J.I.Ch.E.*, **59**(2), 83 (1981).
19. CHAKRABORTY, R. K. and HOWARD, J. R., Burning rates and temperatures of carbon particles in a shallow fluidized bed combustor, *J. Inst. Fuel*, **51**, 220 (1978).
20. WELLS, J. W., KRISHNAN, R. P. and BALL, C. E., *A Mathematical Model for Simulation of AFBC Systems. Proc. 6th Int. Conf. on Fluidized Bed Combustion*, Atlanta, Georgia (USA), Vol. III, 1980, p. 773.
21. BASU, P., BROUGHTON, J. and ELLIOTT, D. E., Combustion of single coal particles in fluidised beds, ref. 2, p. A3-1.
22. HIGHLEY, J. and MERRICK, D., The effect of the spacing between solids feedpoints on the performance of a large fluidized bed reactor, *A.I.Ch.E. Symp. Series*, **67**(116), 219 (1971).

Chapter 3

FLUIDIZED BED INDUSTRIAL BOILERS AND FURNACES

J. HIGHLEY and W. G. KAYE

National Coal Board, Coal Research Establishment, Cheltenham, UK

NOMENCLATURE

A	Surface area in contact with bed
A_t	Area of immersed tubing
C	Fuel calorific value
F_a	Air feed rate
F_f	Fuel feed rate
F_h	Feed rate of heat carrier
f	Fraction of heat input available in the bed
H_a	Enthalpy of air
H_f	Enthalpy of fuel
H_g	Enthalpy of dust-laden gas at bed surface
H_{hb}	Enthalpy of heat carrier at bed temperature
H_{hi}	Enthalpy of heat carrier at feed temperature
h	Convective heat transfer coefficient
Q_b	Heat supply to bed
Q_e	Heat available for transfer directly to in-bed tube surfaces
Q_g	Heat required to bring air and fuel to bed temperature
Q_i	Heat input to boiler
Q_o	Heat output of boiler
Q_r	Heat transfer by radiation from bed surface
Q_s	Net heat requirement for sulphur retention additive
Q_t	Heat transfer to in-bed surfaces
Q_w	Heat transfer to wall

S_a Stoichiometric air kg/kg of fuel
T_b Bed temperature
T_t Temperature of heat transfer surface
X Excess air level
ε Emissivity
η Boiler efficiency
σ Stefan–Boltzmann constant

1 INTRODUCTION

Rising prices of oil and gas reflect the limited supplies of these fuels in the context of increasing world demand for energy, and a return to the use of coal by industry is seen as inevitable. However, there is concern that increased use of coal should not have an adverse effect on the environment in the factory or its surrounds. Also, there is concern that the capital cost of a fully-automatic coal-burning installation is considerably more (two to three times in some applications) than the oil-fired equivalent, offsetting the price advantage of coal so that oil continues to be used. Fluidized combustion is a technique for using coal which offers significant advantages over conventional industrial coal-burning systems,[1,2] as follows:

(1)   The performance is less sensitive to coal type and quality, giving improved flexibility of fuel supply.
(2)   Disruption of production for boiler maintenance is less frequent because ash deposition and fouling are virtually eliminated and there are no moving parts in the combustion chamber.
(3)   Start-up and load following are amenable to automation.
(4)   It is possible to control emissions to the atmosphere.
(5)   The boiler is more compact and less expensive, although this is partly offset by more expensive ancillary equipment.

To appreciate these advantages, it is first necessary to describe the various types of industrial boiler and to discuss the constraints imposed by the boiler design with conventional firing methods for coal, oil and gas. Accordingly, this chapter starts with a description of boiler types and conventional firing methods (Sections 2 and 3). This is followed by a discussion of the factors involved in selecting the optimum boiler performance for a given application (Section 4). Then the various features of a FBC boiler are discussed, including start-up, load-following and mechanical design (Section 5). These design principles are illustrated by describing several FBC boiler installations and discussing their operating performance (Sections 6 and 7).

Industrial furnaces and kilns are major energy users but, in general, the substitution of coal is more difficult than in boilers. However, there are some applications where significant progress is being made in applying FBC, particularly in the production of hot gas for drying processes. FBC furnaces are discussed in Section 8.

2 CONVENTIONAL BOILER TYPES

2.1 Water-tube boilers

Industrial boilers cover the output range from 500 kW, for heating factories and commercial buildings, to 50 MW or more, for combined heat and power (CHP) schemes in the process industries.[3,4] The larger units within this range are of the 'water-tube' type and consist mainly of tubing containing the evaporating water and the steam. By containing the pressurized water and steam inside tubes, the mechanical stresses in the metal are substantially less than they would be in the large diameter pressure vessels of other boiler types. Water-tube construction is therefore particularly appropriate for operation at the high pressures associated with the operation of steam turbines for generating electricity in CHP schemes.

As shown in Fig. 1, a few of the tubes are used to form a combustion chamber by welding fins between them to form walls to contain the combustion gases. Heat transfer to the combustion chamber walls is mainly by radiation. The other tubes are arranged in an array through which the combustion gases pass, to be cooled by convective heat transfer to stack temperature. One of the simplest designs is the two-drum (often called bi-drum) D-type of Fig. 1. Evaporation of water to steam in the vertical pipes causes a natural water circulation from the lower 'mud drum' to the upper 'steam drum'. The boiler must be designed to ensure that the circulation is sufficient to leave at least 10% water, by volume, in the steam leaving the tube bank in order to provide adequate cooling of the tube metal. The steam is then separated in the steam drum and the water, which fills the lower half of the drum, is returned to the mud drum. Feed water is pumped into the steam drum at the rate necessary to maintain a constant water level and at the pressure required for the steam. The steam leaving the drum is at saturation temperature for the operating pressure, in the range from 195 °C at 14 bar to 278 °C at 62 bar. For most turbines it will then be superheated by passing through a tube

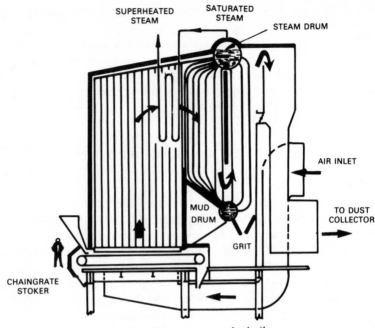

FIG. 1.   D-type water-tube boiler.

bank positioned at the exit from the combustion chamber. Typical superheat temperatures are from 315°C at 14 bar to 480°C at 60 bar. The
evaporator tubes are manufactured using carbon steel, but higher grade
steel is required for the superheater tubes since they operate at a high
temperature because of the lower cooling effect of steam in comparison
with water.

An alternative boiler configuration for natural water circulation is the
A-type shown in Fig. 2. This incorporates two mud drums with one steam
drum to provide a central combustion chamber from which the gas separates
into two equal flows through the side convection sections. Most other
configurations require a pump to circulate the boiling water through the
tubes, as in the designs for larger boilers with outputs up to 2000 MW
(700 MWe) for power generation. One example is the once-through coil
boiler in which the water is almost completely evaporated in a single
long coiled tube and then, after separation of the remaining droplets
containing concentrated dissolved solids, passes through a few long coiled
superheater tubes (see Fig. 13, Section 6.5).

The combustion chamber must be of a size which is sufficient for near-

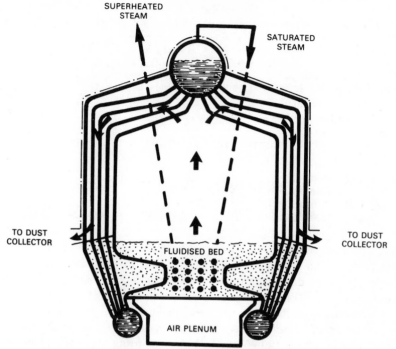

Fig. 2. Combustion Engineering's A-type boiler.

complete combustion and, in the case of coal, to cool the combustion gases sufficiently to re-solidify molten ash droplets. With oil or gas firing the size will be based on a heat release rate in the range 450 to 700 kW/m^3 (Table 1). Heat transfer to the walls is predominantly by radiation and with a typical heat flux of 200 kW/m^2, the combustion gases leave at about 1200 °C. With coal, the combustion chamber size is usually dictated by the need to cool the gases to about 1000 °C so that the fine ash will have cooled from liquid droplets to solid, in order to minimize corrosion and fouling of the convective tubes. This criteria is usually sufficient to provide the space and residence time to complete coal combustion at the lower rate of 300 to 400 kW/m^3, for both stoker and pulverized fuel firing (see Section 3.3).

The convective heat transfer section is usually designed to cool the gases to within 50 K of steam saturation temperature. Further heat is then recovered in an economizer operating at low water temperature or an air heater. The design of the convective section is a compromise between capital

TABLE 1
Typical heat-release rates (coal, oil and gas)

	Fixed bed (Coal)	Oil	Gas
Shell boilers			
per unit volume of furnace tube (kW/m^3)	300–750	600–2 000	500–2 250
per unit area of total furnace tube surface (kW/m^2)	90–100	180–210	150–170
Water-tube boilers			
per unit volume of furnace chamber (kW/m^3)	300–400	450–700	450–700
per unit area of projected water-tube surface (kW/m^2)	200–220	200–220	200–220

and operating costs. Increasing the gas velocity improves the heat transfer coefficient so that fewer tubes are needed, but the associated penalty of higher pressure drop requires a more expensive fan and gives rise to additional power consumption in operation. In the case of oil and gas firing the tubes are typically arranged on a spacing which gives an inter-tube gas velocity of about 30 m/s, resulting in a convective heat transfer coefficient of about 120 W/m^2 K. With coal firing there would be un-acceptable erosion by the hard fused ash particles at this velocity. How-ever, the rate of erosion is approximately proportional to the fourth power of velocity, and so a relatively small reduction in velocity results in substantially less erosion. The usual maximum velocity with conventional coal-firing is about 15 m/s, giving a heat transfer coefficient of about 75 W/m^2 K.

Thus, with coal firing, the need to prevent liquid ash fouling the con-vective tubes, the lower combustion intensity and the lower gas velocity to avoid erosion, result in a boiler about double the volume of oil or gas firing with almost double the number of convective tubes. Consequently, it is more expensive to manufacture and transport to site, and a larger boiler house is required to accommodate it.

2.2 Fire-tube boilers
In a fire-tube boiler (often called 'shell boiler') the water and steam are contained in a cylindrical pressure vessel, with the fire-tube and convection tubes, through which the hot gases pass, immersed in the water[3,4] (Fig. 3). With automated production, fire-tube boilers are less expensive to manu-

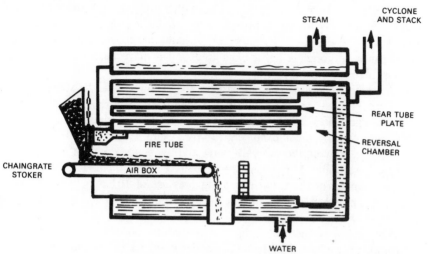

STEAM

CYCLONE
AND STACK

REAR TUBE
PLATE

REVERSAL
CHAMBER

FIRE TUBE

CHAINGRATE
STOKER

AIR BOX

WATER

FIG. 3.   Shell boiler with chain-grate stoker.

facture than small water-tube units and they are the type usually selected
for outputs from 1 to 10 MW. However, they require very careful design
since they are large complex pressure vessels, and there are limitations on
output and steam pressure. During the past 20 years, developments in oil
firing have increased the maximum output from 10 MW (with coal) to
20 MW and, in the UK, fire-tubes have virtually superseded water-tube
boilers in new installations for saturated steam up to this output. However,
in many other countries, including the US, water-tube boilers are still
usually preferred for outputs above about 6 MW, although the use of large
shells is increasing. Most fire-tube boilers are supplied as factory assembled
packages, complete with firing equipment, water pumps and controls.

The output limitations with shell boilers are caused by the need to
increase the metal plate thickness of the fire-tube and containment shell as
their diameters are increased. There is a size and corresponding output at
which the plate thickness required makes the boiler uncompetitive with
water-tube construction or multiple smaller units, and this size is dependent
on the operating pressure. However, an important limitation is that there
is a maximum permissible plate thickness of 22 mm for the fire-tube under
British Standard Specification BS 2790. At this plate thickness, the
maximum tube diameter (under compression) is about 1·8 m at a steam
pressure of 10 bar, and somewhat less at higher pressure. A fire-tube of
1·8 m diameter is sufficient to achieve an output of 10 MW with oil firing,

but only 6 MW with a stoker. For higher outputs it is necessary to incorporate two fire-tubes into one shell, thus allowing a total output of 20 MW with oil firing. It is usually uneconomic to use more than two fire-tubes, although a few stoker-fired boilers have been manufactured with three.

As the diameter of the containment shell is increased, the stress in the end-plates is greater than that in the shell itself. The required thickness of the end-plates is proportional to the product of pressure and shell diameter, with some allowance given for strengthening given by welding 'stay bars' between the two end-plates and by incorporating a proportion of thick wall tubes as part of the tube bank between them. The maximum shell diameter at the usual operating pressure of 10 bar is about 4·5 m, and the practical ceiling on pressure is about 24 bar, at a somewhat smaller size.

The length of the boiler represents a compromise between saving in capital cost, achieved by using a smaller diameter of shell and fire-tube with fewer convective tubes, and the associated increase in operating cost for a more powerful fan, required because of the increased pressure drop at a higher gas velocity through longer tubes. In practice, both length and diameter are progressively increased as output is raised, and the length at the maximum output is typically 6 m.

The steam generated around the fire-tube and convection tubes rises in bubbles to the water surface and into the steam space, creating considerable turbulence and splashing. In order to avoid excessive carryover of water droplets with the steam, the mean velocity of steam leaving the water surface is usually kept below 0·06 m/s. Most conventional fire-tubes are of the horizontal type, as in Fig. 3, and sufficient water surface is available to prevent wet steam. However, there are attractions in adopting a vertical configuration for fluidized bed firing (see Section 6.2). Although there is less water surface area than in a horizontal design, there is sufficient for low output units. However, steam production in a vertical design is limited to a maximum of 5 kg/s (10 MW) at the largest shell diameter of about 4·5 m.

In horizontal shell boilers the rear tube plate and entries to the tubes are subject to a high heat flux, caused by the gas turbulence in the reversal chamber at the end of the fire-tube. Any scale formation on the water side of these surfaces would lead to a high metal temperature, with a loss of mechanical strength, and it is therefore essential that the feed water is chemically treated to a high standard. To minimize this risk of mechanical failure of the rear tube plate, boilers are usually designed so that the gas temperature at the plate does not exceed 1000 °C. In early designs of 'dry back' fire-tube boilers the reversal chamber was a refractory box, but as combustion intensities have increased, particularly with oil and gas, the gas temperature leaving the fire-tube has exceeded 1000 °C. Additional heat transfer surface has been incorporated to cool the gas before the tube plate by incorporating the reversal chamber into the boiler shell, to provide 'wet-back' conditions as in Fig. 3. Even with this design, it is heat transfer rather than combustion intensity which limits the output of the larger oil and gas-fired boilers. The average heat flux in the fire-tube with oil is typically 180 to 210 kW/m^2; and this allows a heat release in the range 600 to 1000 kW/m^3 in large boilers and a maximum output of 9 MW from a single fire-tube. With gas the less radiant flame reduces the heat flux to between 150 and 170 kW/m^2 and, consequently, outputs with gas are somewhat less than with oil. With smaller boilers, the ratio of surface area to volume in the fire-tube increases, permitting an increased combustion intensity. Thus, the fire-tube size in the smallest boilers is determined by the maximum combustion intensity which can be achieved, typically up to 2000 kW/m^3 with oil and somewhat higher with gas.

Conventional coal firing in a fire-tube boiler is by a mechanical stoker, and this occupies the lower half of the tube for about two-thirds of its length. Because of this, the available heat transfer surface area is only 66 % of that with oil or gas firing and the maximum output with the largest fire-tube is reduced from 10 MW to 6 MW. The highest combustion intensity which can be achieved with a stoker is much less than with an oil or gas flame, typically about 750 kW/m^3, and this, together with the need to compensate for the volume occupied by the stoker, necessitates substantially larger fire-tubes in lower output boilers where combustion is the limiting factor.

With oil and gas firing the velocity through the convection tubes is up to about 45 m/s but with coal it must be limited to about 30 m/s to avoid erosion. Thus, 50 % more convection tubes are required with coal. Furthermore, the tube surface area must be increased somewhat to compensate for the reduced heat flux due to the lower heat transfer coefficient and inevitable ash deposits inside the tubes. Because of the need to accommodate the larger fire-tube and additional convective tubes, the shell diameter of a coal-fired boiler is always larger than the oil-fired equivalent, but the boiler length is shorter. However, as previously discussed, the main limitation on the boiler output is fire-tube diameter and, therefore, the output of the coal-fired boiler of the maximum size is only 10 MW, with twin fire-tubes, in comparison with 20 MW with oil.

From this discussion, it will be clear that shell boilers for coal firing by mechanical stokers are at a capital cost disadvantage in comparison with boilers for oil or gas throughout the output range. For outputs up to 6 MW the coal-fired boiler is larger due to the lower combustion intensity and reduced gas velocities. From 6 to 10 MW there is an additional penalty due to requiring a second fire-tube and stoker, with associated coal feed and controls. Above 10 MW, multiple boilers are necessary with coal whereas outputs up to 20 MW can be achieved with two oil-fired fire-tubes. Thus, coal-fired shell boilers are at the greatest capital cost disadvantage for the larger installations, which are usually continuous process applications where the fuel price advantage of coal would otherwise make it particularly attractive.

2.3 Small heating boilers
Most boilers smaller than 1 MW are for low pressure steam or hot water. Traditionally these are constructed by on-site assembly of cast iron sections, which can be handled without needing lifting equipment. However, they are being supplied increasingly as factory assembled packages and many designs are now steel fabrications. The convenience of oil and gas makes them attractive fuels, but fully automated coal stokers are also available. Fluidized bed combustion, with its expensive start-up and control systems, is unlikely to be attractive for those boilers unless oil and gas are not available and the available coal is of too low a grade for stokers.

3   CONVENTIONAL COMBUSTION EQUIPMENT

3.1 Gas burners
The combustion of gas in industrial boilers and furnaces is governed by the rate of mixing of the gas and air. Burner design is therefore primarily concerned with ensuring rapid gas mixing.[3] Nozzle mixing burners are used in most boilers above 1 MW and these utilize air supplied by a fan. The fan is normally incorporated into the burner unit for outputs up to 15 MW, but is separate for higher outputs. The air is discharged through an annulus at the burner face and the gas is admitted into the air stream through ports in a central nozzle. Most of the combustion occurs either immediately at the burner face or in the 'tunnel' and efficient combustion is obtained with less than 15% excess air above theoretical (stoichiometric) requirements. Often a burner will be required to fire oil or gas, and in this

case the oil injector will be at the centre and the gas supplied to a ring around it.

3.2 Oil firing

Combustion of oil is also primarily a matter of ensuring rapid mixing with air and this requires atomization to fine droplets, typically sized between 0·03 and 0·15 mm. Distillate oils (kerosine and gas oil) can be atomized at ambient temperature, except in the coldest winters, but residual oils (light, medium and heavy fuel oil) need to be heated to reduce their viscosity. Residual oils must also be heated during storage (10 to 35 °C according to grade) and for pumping (10 to 45 °C). Atomizing burners are prone to blockage and it is essential that oil is filtered efficiently before firing.

The oil burner consists of a central atomizing device with an annular air inlet surrounding it. There are two main types of atomizer, pressure-jet and two-fluid. In the pressure-jet type, oil is supplied under pressure (500 kPa gauge and above) through tangential inlets to a conical chamber, where it acquires a high rotational velocity. The swirling oil is then forced out through an orifice at the end, and it forms a thin film which breaks into droplets. The momentum of the oil spray is low relative to that of the air and careful design of the air inlet is necessary to ensure mixing and flame stabilization.

In the two-fluid designs the oil is atomized by part of the combustion air or by steam. With low pressure air from a fan, about 25% of the total is required for atomization; under 5% is needed if compressed air is available. A popular type is the rotary-cup in which the oil is fed to the inner surface of a conical cup rotating at 4000 to 6000 rpm. As a thin film of oil leaves the outer edge of the cup under centrifugal force it is atomized by an air jet from an annular nozzle around it. Combustion is then completed by secondary air.

3.3 Coal firing

3.3.1 Coal Types

No stoker will burn all types of coal and it is necessary to know the type of coal, size grading and ash content of the coal that it is intended to burn. In the UK, the National Coal Board classifies coal into numbered types (Table 2) according to the volatile matter, expressed on a dry ash-free basis, and the caking properties of the clean coal, i.e. the extent to which it swells and fuses when heated. The coals with low NCB numbers are referred to as high rank coals, and vice versa.

Almost all the coal sold in the UK is prepared by one of two systems which are aimed at regulating the 'ash' content of the fuel. The ash in coal is present in two forms: the inherent mineral matter which is part of the coal structure (typically about 5 % of the coal by weight); and the adventitious mineral matter from bands of rock above, within and below the coal seam. The ash content of coal as-mined usually ranges from between about 20 and 50 %; only the relatively few coals with below 20 % ash are marketed as-mined. With mechanical mining, most of the coal is sized below 50 mm.

TABLE 2
National Coal Board (UK) coal classification

NCB rank	Type of coal	Caking properties	Gray–King coke type	Volatile matter (daf, %)	Typical calorific value (MJ/kg, daf)
Wood	—	—	—	85–90	19·8
Peat	—	—	—	65–70	22·0
Lignite (brown coal)	—	—	A	55–60	26·7
900	High volatile	Non-caking	A–B	>30	32·5
800	High volatile	Very weakly caking	C–D	>30	—
700	High volatile	Weakly caking	E–G	>30	—
600	High volatile	Medium caking	G_1–G_4	>30	—
500	High volatile	Strongly caking	G_5–G_8	>30	—
400	High volatile	Very strongly caking	>G_8	>30	35·0
301	Medium volatile	Strongly caking	>G_6	20–30	36·4
300	Medium volatile	Non-to-weakly caking	A–G	20–30	—
206	Low volatile	Non-caking	A–D	10–20	—
204	Low volatile	Strongly caking	G_5–G_8	$17\frac{1}{2}$–20	36·9
203	Low volatile	Medium caking	G_1–G_4	$15\frac{1}{2}$–$17\frac{1}{2}$	—
202	Low volatile	Weakly-caking	C–G	14–$15\frac{1}{2}$	—
201	Low volatile	Non-caking	A–B	$9\frac{1}{2}$–14	36·4
100	Anthracite	Non-caking	A	<$9\frac{1}{2}$	36·4

In the first preparation system, all of the coal is washed to remove most of the stone and rock, giving an ash content of 6 to 10 %. The coal is then separated into size grades according to demand. The premium quality coals for the smaller industrial users are washed doubles (25 to 50 mm) and washed singles (13 to 25 mm) which are convenient to store and handle. The larger users burn washed 'smalls' coal containing fines and for which only the top size is specified (that is the coal left after removal of everything larger than 13 mm).

The second type of coal preparation is used at collieries supplying coal for power generation and the very large industrial users. The aim is to produce a coal with an ash content of about 17 %, controlled at a specified level if it is to be burned as pulverized fuel. This is achieved by screening the coal, typically at 13 mm, removing the stones from the oversize, and then reblending. The actual screen size is selected according to the size distribution of the coal and its ash content. This technique avoids the more difficult separation of fine stone from fine coal, and is a relatively inexpensive method of reducing transport costs. The coal is marketed as 'part-treated smalls'.

The ease of handling coals, and also combustion efficiency with stokers, is affected mainly by the amount of fine coal (less than 3 mm) and moisture present. Irrespective of the size distribution, coal will flow readily providing that it contains less than 6 % free moisture, but it becomes more difficult to handle at somewhat higher moisture contents. With a coal having between 10 and 14 % free moisture and a high fines content (more than 10 % below 0·5 mm) there are severe problems with bridging and 'ratholing' in bunkers and sticking in feeders. The flow properties improve again with more than 14 % moisture.

In some countries, the coal has a much higher inherent ash content than in the UK, sometimes up to 20 % and, exceptionally up to 60 % (e.g. Arigna coal in Ireland). Large reserves of high inherent ash coals exist in India, China, Brazil and other developing countries. It is not possible to reduce the ash content below the inherent level and this has often precluded the use of these coals in conventional stokers. Fluidized combustion offers a method of utilizing these coals efficiently. It also provides a means of utilizing the small proportion of coal in the rejects from coal preparation, but this will only be economic in the immediate vicinity of the mines.

3.3.2 Pulverized Fuel Firing

Nearly all the coal consumed in modern, large power stations is burnt in pulverized-coal fired boilers. In this system the coal is dried, pulverized so that about 90 % is below 75 μm and then carried in an air suspension to an array of large burners firing into the combustion chamber.[3,4] The combustion rate is determined initially by the mixing of fuel and air, as with oil, but the flame is longer than with oil because of the time required to burn the char (coke) particles. The high temperature within these flames is such that the ash softens and may even melt to form small droplets and some sodium and other alkali metal salts vaporize. The combustion chamber must be sufficiently large to allow the combustion gases to cool to a temperature at which the ash droplets solidify, and consequently the ash fusion temperature is an important design parameter. Typical values are in the range 1000 to 1200 °C. Subsequently, as the gases cool in passing through the boiler, sodium salts (especially sodium sulphate formed by gas-phase reactions) condense on to the boiler tubes in the form of a viscous liquid to which ash sticks, giving rise to tube 'fouling' which inhibits heat transfer. With most coals this occurs slowly and much of the ash can be dislodged by using 'soot blowers'. However, if the coal has a high sodium content, typically associated with a chlorine content of 0·4% and above (chlorine is more readily measured), tube fouling can require regular shut-downs for cleaning and also give rise to high rates of corrosion at the high metal temperature of superheater tubes.

Most of the ash remains suspended in the combustion gases as they pass through the boiler, although provision must be made to remove some ash from the combustion chamber and bottom of the convection passes. A high-efficiency (99%) electrostatic precipitator is required to separate the ash dust before the gas is discharged to the atmosphere.

In addition to use at power generating stations, pulverized firing is the standard coal combustion system for large industrial boilers with outputs above about 50 MW. Mechanical stokers are usually preferred for smaller outputs than this, the moving grate becoming less expensive than the mills and electrostatic precipitator.

3.3.3 Over-bed Fired Stokers

All other conventional coal-burning systems are of the 'fixed bed' type and are derived from the burning of a pile of coal on a grate.[3,4] The commercial singles (13–25 mm) and smalls (13 mm–zero) are fired without any on-site preparation (other than spraying dry coal with water). In the case of an overfired stoker, the coal is burning at the bottom of the bed and raw coal is fed on to the top (Fig. 4). This gives the advantage that the flame front moves through the bed in the same direction as the air, ensuring ignition of the raw coal. However, a disadvantage is that volatiles released above the flame front are entrained by the combustion gases without passing through a hot combustion zone. Secondary air is supplied above the bed under turbulent conditions in order to ensure combustion of these volatiles.

The traditional over-bed fired unit is the spreader stoker in which the coal is sprinkled over the surface of the bed by means of a rotating impeller. The finer coal particles burn in suspension, while larger pieces fall on to a grate to form the bed. The grate moves across the floor of the combustion chamber to discharge the ash, usually below the coal feeder. Spreader stokers are used in water-tube boilers up to 80 MW, burning smalls of all ranks except anthracite, which is relatively unreactive. In shell boilers,

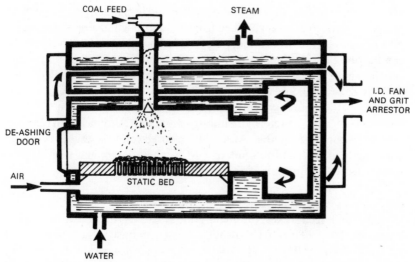

COAL FEED   STEAM

I.D. FAN AND GRIT ARRESTOR

DE-ASHING DOOR

AIR

STATIC BED

WATER

FIG. 4. Shell boiler with over-bed firing (Vekos Powermaster).

there is insufficient volume in the combustion chamber to complete the combustion of fines in suspension and, therefore, spreaders in shells are limited to the use of singles coal (13 to 25 mm). A particularly successful unit in the UK has been the Vekos Powermaster boiler (Fig. 4), manufactured by G. W. B. Parkinson Cowan Ltd, which is a packaged horizontal shell boiler incorporating a simple spreader from a central coal feed and a static grate for outputs up to 5 MW (or 10 MW with twin fire-tubes). Cyclone grit arresters are sufficient to meet the UK clean air requirements. Depending on the fines content, the combustion efficiency is in the range 94 to 97%.

3.3.4 Underfeed Stokers

This type of stoker has the combustion zone at the top of the fire bed and raw coal is supplied from below. The ignition plane has to travel down against the direction of the air flow and ignition is thus more difficult to establish. However, in this system the volatiles have to pass through the burning coke zone above the ignition plane and are more likely to be burned. As in all fixed bed stokers the gases leaving the bed contain carbon monoxide and often gaseous hydrocarbons which must be burned by providing secondary air above the bed surface.

In the traditional underfeed stoker, the raw coal is fed by a horizontal screw to the bottom of a retort, displacing the burning coal upwards. The ash forms a clinker at the top of the fire bed and is manually removed. The primary combustion air is supplied from a fan through tuyeres in the upper part of the retort, and secondary air is fed above the fire bed to complete combustion. This is the simplest and cheapest type of stoker and is the standard method of firing commercial and small industrial boilers with outputs up to about 2 MW thermal. A disadvantage is that ash removal is a manual operation for conventional designs, usually once per day with the low-ash content coal preferred, although units with automatic ash removal are now becoming available. Although underfeed stokers were at one time manufactured with outputs up to 6 MW, these larger units have been superseded by chain-grate stokers.

3.3.5 Chain-grate Stokers

Like the underfeed stoker, the chain-grate stoker has the coal ignited from above so that volatiles pass through a hot combustion zone. A typical design is shown in Fig. 3. The coal is fed from a hopper on to a moving grate, forming a layer 75 to 150 mm thick. Above the grate, at the entry to the combustion chamber, there is a refractory arch which radiates heat down to the surface of the incoming coal to establish ignition. As the grate carries the coal across the combustion chamber, the ignition front moves downward, reaching the bottom of the coal layer about two-thirds along. The bed above the ignition plane consists of burning coke and this is almost completely burnt by the end of the grate, where the ash falls off, usually on to an extraction conveyor belt.

Chain-grate stokers have a relatively high capital cost and sometimes have high maintenance requirements, but they offer the important advantage of burning a smalls coal, either 13 mm to zero or 25 mm to zero. Traditionally, chain-grates were limited to coal with more than 7 % ash to protect the grate from overheating, but lower ash contents can be burned with white iron grate bars. The coal must be sufficiently wet to hold the fines together, otherwise there is excess unburnt carbon loss through elutriation and 'riddlings' through the grate bars. About 1–1·5 % free moisture is required for every 10 % of fines below 3 mm. The maximum burning rate (about 2 MW/m^2) is limited by bed instability as the finer particles begin to fluidize. Chain-grates have been manufactured up to 70 MW for large water-tube boilers, but they are mostly supplied as factory assembled package units for shell boilers, with a maximum output of 6 MW per fire-tube. A non-caking coal is preferred for a chain-grate, but it is possible to fire coal of any rank, although a reduced burning rate is necessary with some types.

3.3.6 Coking Stokers

In the coking stoker, the coal is fed from a hopper by a reciprocating ram on to a 'coking plate' at the entry to the combustion chamber. The volatiles are released as it heats there, and these pass over the fire bed and burn with secondary air supplied above the grate. Further additions of coal push the coke on to a reciprocating grate, where it burns as it moves along. A slightly caking coal is preferred, although the reciprocating grate helps to break up sheets of coke and allow caking coals to be burned if necessary. As with chain-grates, ash is required to protect the grate from overheating and moisture is necessary with smalls. In common with all stokers, the combustion efficiency is in the range 94 to 98 %, depending on the fines content and excess air level.

4   INTEGRATING COMBUSTION REQUIREMENTS WITH BOILER DESIGN

In designing a boiler installation for a particular application the objective is to produce a design which will satisfy the site requirements with minimum capital and operating costs. In general, it is possible to reduce the capital cost by accepting a lower efficiency, a restricted turn-down range (ratio of maximum to minimum output) and a slower dynamic response to load changes. Large boilers used for power generation, operated at a high load factor (i.e. throughout the year at near their maximum output), require a high efficiency at their maximum output but turn-down range efficiency at reduced load and dynamic response are usually of secondary importance. Industrial steam boilers used throughout the year usually require the efficiency to be maintained over a large turn-down range and also a rapid dynamic response. Heating boilers are operated at a low load factor (i.e. they are used for only part of the year and rarely operate at their maximum output); it is necessary to minimize the capital cost while achieving an acceptable efficiency, particularly at reduced output, and an acceptable turn-down range.

4.1 Factors affecting overall boiler efficiency

There is an economic incentive to incorporate the smallest possible size of combustion chamber in a boiler, and to do this requires that the air/fuel mixing and the combustion reactions must be carried out in the minimum time. The speed of these processes can be increased by using more air than is required theoretically for complete combustion (the stoichiometric air requirement), but using additional air gives the disadvantage that extra heat losses occur in the warm flue gases. The design of any boiler is therefore a compromise between its size, and hence its cost, and the overall efficiency which is achieved.

The overall efficiency is the percentage of the fuel's calorific value which is recovered in the steam or hot water being produced. The sources of inefficiency are as follows:

(i)   Incomplete combustion—The combustion efficiency, defined as the fraction of the fuel's calorific value which is released by combustion, improves with increasing excess air above the stoichiometric requirement. The excess air level, X, is defined as:

$$X = \frac{\text{Actual air supply} - \text{Stoichiometric requirement}}{\text{Stoichiometric requirement}} \quad (1)$$

(ii)   Heat losses due to the sensible heat content of the warm flue gas components other than water vapour—This is determined primarily by the flue gas temperature, typically 50 K above the water/saturated steam temperature in the boiler, and by the excess air level.

(iii) Heat losses due to the latent heat and sensible heat in the water vapour in the flue gas, arising from the hydrogen component of the fuel and any moisture in it—Because of the relatively high hydrogen content of oil and natural gas in comparison with coal, and consequently higher latent heat loss, the calorific value of these fuels is often quoted on a net basis after subtracting the latent heat of the water vapour from the gross calorific value.

(iv)   Heat losses from the boiler casing—This is typically about 1 % of the rated boiler output, irrespective of the operating output.

Figure 5 shows how the principal losses vary with excess air level for a typical coal-fired boiler and demonstrates that there is an optimum excess air level giving the best efficiency. The actual efficiency obtained will depend on the particular design; to obtain the highest efficiency at the optimum requires near-complete combustion with minimum excess air and a low flue gas temperature, and hence a boiler with a large combustion chamber and extensive heat transfer surface. Such boilers are expensive and are used only for sites with a high load factor such as continuous process plant where the fuel savings justify the high capital cost. In contrast, at sites where the boiler is only operated for part of the year, and rarely at maximum output, capital cost is a more important

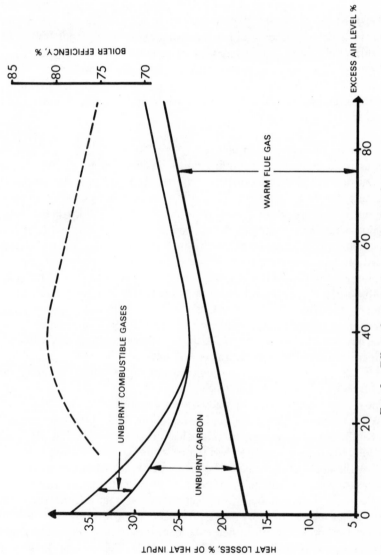

Fig. 5.   Effect of excess air on boiler heat losses.

consideration than fuel saving and the boiler will be more compact with a higher optimum excess air level. In practice, excess air levels range from 2 % in a large gas-fired boiler for process steam, to 60 % in a coal stoker-fired heating boiler. Increasing fuel prices provide an incentive to improve boiler efficiencies and it is reasonable to expect that boilers of the future will need to be somewhat more efficient. The fluidized combustion system provides a coal-burning technique which offers more promise for improvement than conventional stokers, particularly when control of emissions such as sulphur dioxide and nitrogen oxides is also required.

4.2 Features of fluidized bed boilers

A schematic diagram of a fluidized bed boiler is shown in Fig. 6. The combustion chamber contains the fluidized bed and provides sufficient freeboard height to avoid excessive elutriation of bed material or unburnt carbon at the operating fluidizing velocity. The gases leave the combustion chamber at approximately bed temperature, the actual value depending on the amount of combustion and heat transfer in the freeboard, and heat is recovered by passing the gas through a convective section, as with conventional firing. The main special feature of fluidized combustion is the constraint imposed by the relatively narrow temperature range within which the bed must be operated. With coal there is a risk of sintering the bed if the temperature exceeds 950 °C and combustion efficiency declines below 800 °C. Also, for efficient sulphur retention the temperature should be in the range 800 to 850 °C. In order to enable a bed to be operated within this temperature range, while obtaining the low excess level required for efficient operation, it is usually necessary to provide an in-bed bank of water/steam tubes.

When burning a high-grade coal or oil, about 50 % of the heat released by combustion is required to raise the fuel and combustion air (with 25 % excess) from ambient to bed temperature, and is removed from the bed by the hot dust-laden gas. The actual percentage depends on the bed temperature, and is reduced by air pre-heat. In order to maintain a constant temperature, the remainder of the heat must be extracted from the bed by some other means. In most fluidized bed boilers a low excess air level is achieved by providing an in-bed tube bank designed to extract the heat released in excess of that removed by the off-gas. The bulk of the heat in the off-gas is subsequently recovered in above-bed convective tube banks. Typically, with good quality coal, the tubes (and combustion chamber walls) extract about 50 % of the heat release to permit operation at 25 %

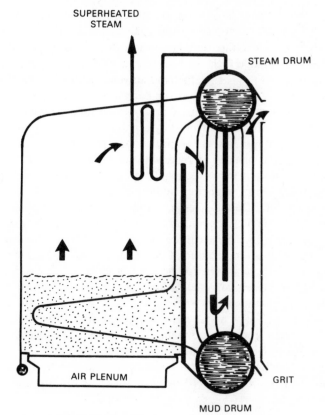

FIG. 6. Schematic diagram of fluidized bed boiler.

excess air. This represents more than 50 % of the boiler output because the heat in the off-gas is not fully recovered. Significant savings in the total amount of heat transfer surface in a boiler are given by the high heat flux to the immersed tubes, and this represents an important advantage of fluidized combustion. Extensive trials have demonstrated that, contrary to early fears, corrosion and erosion of these tubes are minimal.

It is necessary to design the boiler in such a way that the tubes in the bed remove 50 % of the heat released throughout the output range. If, at any output, there is too little heat transfer to the tubes, it is necessary to provide air in addition to combustion requirements in order to remove heat. The excess air level will then rise above the optimum, lowering the maximum output because the air is not fully utilized and reducing boiler efficiency through stack heat loss. If there is too much heat transfer, the excess air level must be allowed to reduce or the bed temperature allowed to fall, both of which lead to less efficient combustion. Techniques for varying heat flux to the tubes for turn-down are discussed in Section 5.2.

With a low-grade fuel, the heat release may be only slightly in excess of that required to heat the fuel and air required for efficient combustion to bed temperature. This is particularly the case for fuels with a high moisture content. In this situation, immersed tubes are not required and the bed temperature is controlled by supplying sufficient additional air as a heat carrier, to remove heat from the bed for subsequent recovery in the convective tube bank. For fuels with an intermediate calorific value a decision must be made on economic grounds as to whether to maintain bed temperature by supplying additional air or to install a few tubes in the bed to eliminate the stack heat loss due to the additional excess air. With high grade fuels, combustion in a bed without cooling surfaces (e.g. in a refractory-lined furnace) requires about 2·5 times the stoichiometric air to maintain a temperature of 900 °C, i.e. about 150 % excess air. This forms the basis of design for hot gas producers (see Section 8.2), but operation of boilers with such a high excess air level is inefficient.

There are circumstances where it is preferable to avoid the use of tubes in the bed, such as in a system for a conventional horizontal shell boiler, and alternative concepts are under development. These generally involve the use of a heat carrier, usually either recirculated flue gas or recycled fine particles, which is heated in the bed and subsequently transfers heat to the boiler. Boilers without in-bed tubes are discussed in Section 5.3.

5   FBC BOILER DESIGN CONSIDERATIONS

5.1 Heat transfer calculations
The first stage in designing a fluidized bed boiler is to calculate the fuel feed rate on the basis of the required thermal output of the boiler and a target efficiency. Subsequently, the air supply is calculated from a knowledge of the stoichiometric air requirement and a selected excess air level, as follows:

$$Q_i = \frac{Q_o}{\eta} \tag{2}$$

$$F_f = \frac{Q_i}{C} \tag{3}$$

$$F_a = F_f S_a (1 + X) \tag{4}$$

Of the heat in the fuel Q_i, a proportion remains unburnt, and the remainder is released by combustion, mainly in the bed but also in the freeboard. Subsequently, some of the heat released in the freeboard is transferred back to the bed by particle splashing. If the total fraction of the heat input available in the bed is f (typically 0.9–0.95) then the rate at which heat is supplied to the bed is,

$$Q_b = fQ_i \qquad (5)$$

Some of the heat (Q_g) is required to bring the air and fuel to bed temperature and leaves the bed surface in the form of hot combustion gases containing carry-over solids:

$$Q_g = (F_f + F_a)H_g - (F_f H_f + F_a H_a) \qquad (6)$$

If limestone or dolomite is added for sulphur retention, heat is required for heating to bed temperature and calcination and some heat is released by sulphation. There will be a net heat requirement, Q_s. The surplus heat release, which must be removed from the bed to maintain the bed temperature constant at T_b, usually by transfer to immersed surfaces, is therefore:

$$Q_e = Q_b - Q_g - Q_s \qquad (7)$$

The heat flux to surfaces immersed in the bed is usually correlated by the equation:

$$Q_t = hA(T_b - T_t) + \varepsilon\sigma A(T_b^4 - T_t^4) \qquad (8)$$

The first term is the flux due to convective heat transfer and the second that due to radiation. In a boiler for saturated steam, convection represents about 75 % of the heat transfer; radiation becomes increasingly important at the higher metal surface temperatures of steam superheater and air-heater tubes. The value of the convective heat transfer coefficient, h, is determined primarily by the particle size of the bed material and tube arrangement and metal temperature have only a secondary effect. For a given size of bed particle, the value of the coefficient is not usually changed significantly by variations in fluidizing velocity within the operating range of a boiler (although the velocity range determines the particle size selected).

For a boiler in which the combustion chamber has cooled walls, eqn. (8) can be used to calculate the heat flux to the wall area in contact with the bed, Q_w, using a value for h appropriate to the wall. Allowance should be made for heat transfer to the wall in the splash zone, through which the coefficient progressively reduces to that for the dust-laden gas in the freeboard. Radiation from the bed surface to a cooled freeboard, Q_r, should also be estimated. If $Q_w + Q_r$ is found to exceed Q_e it is necessary to reduce the wall area contacting the bed, either by reducing the bed depth or by installing insulation.

Using eqn. (8), the surface area of in-bed tubes required to extract the remainder of the heat is:

$$A_t = \frac{Q_e - Q_w - Q_r}{h(T_b - T_t) + \varepsilon\sigma(T_b^4 - T_t^4)} \tag{9}$$

The length of tubing to provide this surface area can then be calculated knowing the surface area per unit length and a suitable tube arrangement can be selected (see Section 5.6.7).

If, instead of installing a tube bank, the remaining heat is extracted by a heat carrier, such as additional air, recirculated flue gas or recycled fine particles, the required feed rate is:

$$F_h = \frac{Q_e - Q_w - Q_r}{(H_{hb} - H_{hi})} \tag{10}$$

Combustion will usually continue through the freeboard, and as mentioned previously, part of the heat released will be transferred back to the bed, by particle splashing and radiation. The net heat release in the freeboard will tend to increase the gas temperature, but this will be offset by heat transfer to the freeboard walls if these are boiler surfaces. It is necessary to estimate the net heat release and any heat transfer from the dusty gas in order to calculate the temperature of the gas as it passes from the combustion chamber to the conventional convective heat transfer tube bank. Subsequent heat transfer by gas convection is calculated as for the gases from conventionally-fired boilers.

5.2 Load control with in-bed tubes

For optimum efficiency throughout the turn-down range of a boiler it is necessary to maintain the excess level approximately constant as the air supply and coal feed rate are changed to vary output. This requires the heat flux to heat transfer surfaces in the bed to be varied approximately in proportion to output. From eqn. (8) it can be seen that it would have been particularly convenient if the heat transfer coefficient were to reduce in proportion to a decrease in fluidizing velocity, but unfortunately it is virtually independent of velocity for the particle sizes used. Since the temperature of the tube metal in the bed is usually constant (close to the steam saturation temperature at boiler pressure), it is clear from eqn. (8) that the only means of providing the required variation in heat flux are: (i) varying the bed temperature within its restricted operating range, and (ii) making provision to reduce the area of tubing in the active bed as output is reduced.

5.2.1 Variation of Bed Temperature with Multiple Compartments

For a bed without sulphur retention, the bed temperature can be varied between an upper limit just sufficient to avoid the risk of ash sintering (typically 950 °C) and a lower limit (about 750 °C) at which combustion efficiency becomes unacceptably low. The variation in heat flux to immersed tubes between these two temperatures can be calculated from eqn. (8), and will depend primarily on the tube metal temperature. For a low pressure hot water boiler with tubes at 80 °C, the bed-to-tube temperature difference can be reduced from 870 to 670 K giving a reduction in heat flux of only about 23 %. With evaporator tubes at 200 °C, the reduction in heat flux is still only about 27 %. For superheater tubes at 550 °C, it is possible to reduce the heat flux to 50 % by varying bed temperature.

The first stage of turn-down from maximum output is thus to reduce the air and coal feed rates in proportion, to maintain constant excess air level, and allow the bed temperature to reduce to its minimum value (Fig. 7). This will enable the boiler output to be reduced approximately in proportion to the above heat flux to the tube, i.e. by about 27 % for a boiler containing

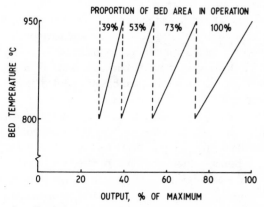

FIG. 7. Turn-down with fully-immersed tubes at 200 °C.

water evaporating at 200 °C. The percentage reduction in boiler output is not exactly the same as that of heat flux to tube because the heat content of the off-gas is also reduced at a lower bed temperature (eqn. (6)). It is not possible to reduce the air supply (fluidizing velocity) further at constant excess air without causing the bed temperature to fall below the minimum acceptable level, giving rise to inefficient combustion and smoke emission. In the case of beds with limestone addition for sulphur retention it is usually necessary to restrict the bed temperature to between 800 and 850 °C, which severely restricts the turn-down attainable.

The degree of turn-down which can be achieved simply by reducing air and coal feed rates and allowing the bed temperature to fall is insufficient for most commercial applications. Consequently in the design concepts which incorporate variation of temperature, turn-down is usually extended by sub-dividing the bed into compartments and operating the appropriate number of compartments to give a bed temperature within the operating range. The compartments need not be of equal area; for example, if 73 % turn-down can be achieved (by temperature variation) in a single compartment (Fig. 7) the output range 100 to 73 % is achieved with all of the bed, 73 to 53 % MCR (maximum continuous rating) with 73 % of the bed area, and 53 to 39 % MCR with 53 % of the area.

A disadvantage of achieving turn-down by varying bed temperature with multiple compartments is that, because of the thermal inertia of the beds, the dynamic response is somewhat slow. This is particularly the case with the deep beds necessary if limestone is used to retain sulphur. However, control techniques are being developed which should give acceptable boiler response, particularly with the use of crushed coal to provide rapid combustion response.[5] There is also a requirement to maintain stable boiler operation when reducing the number of compartments in operation, because of the concurrent need to increase the temperature of the beds remaining in operation from minimum to maximum. The reverse situation occurs when re-starting a compartment. Re-start of a compartment is straightforward providing that the bed has not cooled below 700 °C since it is possible simply to refluidize and commence coal feeding. However, if the bed has cooled below 700 °C it is necessary to use the auxiliary heating system, with a consequent delay in increasing output. It follows that turn-down by varying bed temperature with multiple compartments is most appropriate for boilers which are to be operated for long periods at steady output. This will usually be the case for very large boilers, and these will also usually comprise multiple beds to facilitate construction by assembling factory built modules.

5.2.2 Varying the Area of Heat Transfer Surface in the Bed

An alternative approach to load control is to provide some method of varying the area of heat transfer surface in the fluidized bed so that heat transfer to tubes can be modulated while maintaining a near-constant bed temperature. Variation in effective heat transfer area is most readily achieved by positioning the heat transfer tubes such that the surface area immersed in the bed can be changed by varying the bed level (Fig. 6). Some expansion of the bed occurs when the fluidizing velocity is increased and the most elegant scheme is to design the tube bank so that at a given velocity the bed is expanded to immerse sufficient heat transfer surface to extract the required heat flux. There are, however, two disadvantages to this concept. The first is that particles are splashed above the bed surface and some heat is transferred to the tubes above the nominal bed level. Because of this it is not possible to achieve the required output variation and also fully immerse all of the tubes at maximum output, and consequently somewhat more tubing is required than in designs in which tubes are fully immersed. The second disadvantage is that, for designs operated at high velocities the tube bank required to extract the heat might be up to 0·5 m in height, and to achieve this variation in bed level within the operating fluidizing velocity range necessitates the use of a relatively deep bed, giving a penalty in fan power cost. In order to overcome this second disadvantage various schemes have been proposed for varying the bed level by changing the amount of bed material in the fluidized bed. This is readily achieved in units with sulphur retention by limestone. The bed level can be raised by temporarily stopping the removal of spent stone, while continuing to feed crushed limestone, and lowered by temporarily stopping the limestone feed and removing spent stone at the maximum rate.

5.2.3 Circulation of Bed Particles Between Combustion Zone and Heat Transfer Zone

In this load control concept the bed is divided into two zones, an uncooled combustion zone which is operated at constant temperature and a heat transfer zone with immersed tubes. Bed material is interchanged between the two zones at the rate necessary to maintain the required temperature in the combustion zone. The temperature of the heat transfer zone 'floats', approaching that of the combustion zone at maximum output. The rate of particle interchange is usually controlled by the air supply to the heat transfer zone. which is either modulated or on/off.

In one form of the concept the heat transfer zone is at the side of the combustion zone, preferably consisting of a shallow bed to give a low mass of solids in order to allow rapid dynamic response. This system is used in the water-tube boilers manufactured by Stone International Fluidfire, and is described in Section 6.7 (see Fig. 14). An alternative is to have the heat transfer zone at the bottom of the bed and the combustion zone above with a second, upper, air distribution for the combustion zone.[6]

5.3 Boilers without in-bed tubes
5.3.1 Heat Removal by Fluidizing Gas
The simplest fluidized bed boiler is one in which the bed is in a refractory-lined combustion chamber, operated as a hot gas producer, and the steam or hot water is generated by passing the hot gas through a waste heat boiler. This design is appropriate for low-grade fuels with which the equilibrium excess air is not excessive. With higher grade fuels the excess air can be reduced by recirculating flue gas through the bed as a heat carrier; the gas flow is still higher than it would be if tubes were in the bed, necessitating the use of a boiler larger than the equivalent convective section of a unit with immersed tubes. The main justification for adopting this approach is to avoid the higher quality of water treatment needed for the high heat fluxes to tubes immersed in the bed, while providing a simple and reliable control system.

Similar considerations apply to combustion chambers with water-cooled walls where the excess air level could be reduced by installing tubes, but it is considered preferable not to do this. An example is the system being developed for conventional horizontal shell boilers by NEI Cochran (Section 7.1.2).

5.3.2 Removal of Heat by Recycle of Fine Inert Particles
An alternative method of heat removal to overcome the complication of load control with immersed tubes is to have a continuous flow of fine inert solids through the combustion bed of larger particles, forming a 'fast' fluidized bed. The recycle rate of fines is controlled to maintain the required bed temperature. Heat is recovered from the fines as they pass in dilute phase flow through the waste boiler. Alternatively, the fines can be separated before the waste heat boiler using high temperature cyclones, and passed to a separate fluidized bed of the fine material, operated at low velocity, with immersed boiler tubes. The main problems to be overcome in developing this concept are to devise a mechanical design which will withstand erosion and to control the fines recirculating rate. The concept is being developed by Lurgi GmbH[7] in Germany, the Battelle Institute[8] in the USA and the Ahlstrom Co.[9] in Finland. The Battelle system uses a finely

pulverized limestone as the recirculating material and this is claimed to give the advantage of very efficient sulphur retention.

5.3.3 Use of the Bed as a Devolatilizer/Partial Gasifier
A further approach to eliminating tubes in the bed is to enhance the proportion of the combustion which occurs above the bed. This concept has been adopted by Energy Equipment Ltd in the UK, initially as a means of firing conventional horizontal shell boilers (Section 7.1.3) and sub-sequently to generate high temperature hot gas for driers (Section 8.4). The bed is fluidized using a mixture of air and recirculated exhaust gas, with just sufficient air to maintain the required operating temperature and prevent excessive carbon accumulation. The remainder of the air, up to 30%, is supplied above the bed to burn the volatiles, fine carbon and products of gasification reactions in the bed.

A similar method of operation has been evolved by the Esso UK Research Centre.[10] This follows the development of a system to gasify high-sulphur oil in a fluidized bed of limestone, which is being demonstrated with Foster Wheeler in converting a boiler from natural gas in San Benito, Texas. Attempts to apply the desulphurizing gasification process to coal have inevitably led to the production of a low calorific value fuel gas with considerable fine carbon, as in the Energy Equipment system.

Both systems have been developed primarily as methods of converting existing boilers to fluidized bed firing, but can also be applied in new installations.

5.4 Start-up
Before fuel can be burnt in a fluidized bed it is necessary to heat the inert fluidized bed material to about 600 °C using an auxiliary heating system. Two methods of heating the bed have been used: (i) combustion of the auxiliary fuel as a flame above the bed fluidized with cold air, and (ii) passing hot gas through the bed.

In the case of a coal-fired unit, coal-feeding can commence at a temperature of about 500 °C, and the volatiles will burn above the bed. However, at this temperature, the rate of combustion of the residual char (carbon) particles is slow and the char accumulates to a high concentration in the bed. As the temperature rises, the char combustion rate increases and the equilibrium char/carbon concentration decreases towards the oper-ating level. If the char concentration exceeds the equilibrium level when the bed temperature reaches the operating level, all of the oxygen in the air will be used for combustion, rather than 80% (for combustion with 25% excess air) and the heat release will be above the design value. Consequently the bed temperature will continue to rise, with the risk of clinkering. For this reason, careful control of the coal feed rate is necessary during start-up to avoid excessive accumulation of char.

5.4.1 Above-bed Combustion Systems

The simplest above-bed start-up combustion system consists of an oil or gas burner directed at the surface of the bed, which is fluidized by cold air. This method has the advantage of a relatively low capital cost, but it is somewhat inefficient since only part of the heat released is transferred to the bed. The efficiency of heat transfer to the bed is improved in a variant in which gas is premixed with the fluidizing air. The gas burns as a flame above the bed surface until the bed temperature rises to about 650 °C; thereafter the combustion begins to take place within the bed and by 800 °C there is no longer an above-bed flame (see Section 5.6.3).

Since in these systems the bed is fluidized, part of the heat transferred to the bed is removed in heating the fluidizing air to the bed temperature and part is transferred to the immersed heat transfer surfaces. Consequently the auxiliary heat input needed to bring the bed to 600 °C is only slightly less than the rated heat release of the fluidized bed combustion system.

5.4.2 Hot-gas Start-up

Heating the bed by passing hot gas through it eliminates the inefficiency of transfer from the above-bed flame. The main disadvantage of hot-gas start-up is that the gas duct must be refractory lined and the distributor either manufactured of expensive heat-resistant material, with allowance for thermal expansion, or cooled by water or air.

If the bed is fluidized, heat is still removed by the gas leaving at bed temperature and by transfer to immersed tubes. However, these losses can be minimized by maintaining the bed static during the first phase of heating, taking advantage of the fact that the gas mass flow rate which can be passed through a cold bed without fluidizing is typically about six times that for a hot bed. (In most commercial fluidized bed combustion units, an air flow rate corresponding to maximum rated output, i.e. the maximum air supply from the fans, will just fluidize the bed when it is cold.) Thus, if hot gas at 900 °C is supplied to the cold bed at, say, 50 % maximum fan rating, the bed does not fluidize immediately and a hot zone at 900 °C forms at the bottom, gradually extending through the bed. However, the gas flow rate would be sufficient to fluidize the hot zone were it not for the restraining effect of the cold bed above. As the hot zone extends upwards, a situation is reached where there is insufficient cold bed to restrain the upward force of the hot zone, and the bed suddenly fluidizes. The hot gas flow rate is usually selected such that this occurs at a mean temperature at which coal feeding can commence, typically 500 °C. In principle, the onset of fluidization could be delayed by gradually reducing the gas flow rate, but this would increase the start-up time and complicate automatic control of start-up.

5.5 Selection of fluidizing velocity

Once the design concept for a boiler is selected, i.e. the start-up system, the method of achieving a low excess air level and the method of load following, the main design parameter to be selected is the fluidizing velocity. The optimum velocity represents a compromise between capital cost, boiler efficiency and fan power requirement because most other design parameters are dependent on velocity, including bed area, bed particle size, immersed tube surface area and hence tube layout, bed depth and freeboard height. As the design velocity is increased, the amount of elutriated unburnt carbon will be increased unless other parameters are changed to compensate. The excess air level can be increased, but this also increases stack gas heat losses; there is an optimum excess air level for any combination of other design parameters. To maintain combustion efficiency at constant excess air, the bed depth and freeboard height must be increased. A deeper bed will also be necessary to accommodate the tubes (slightly more because of a larger particle size) in the smaller bed area. A deeper bed increases the fan power consumption. Refiring of fines can be incorporated to maintain efficiency but fine inerts are also refired and remove heat from the bed, reducing the cost saving due to the high in-bed heat transfer rate.

Increasing the velocity also lowers the efficiency of sulphur retention, necessitating the use of more limestone. Design changes made to improve combustion efficiency will also improve sulphur retention. The importance of sulphur retention in determining the optimum velocity will be dependent on the sulphur content of the fuel and the emission level to be met.

Reducing the fluidizing velocity improves combustion efficiency and in-bed heat transfer and allows the use of a shallower bed but requires a greater bed area with more feed points. However, in general, it is more expensive to manufacture a short combustion chamber with a large cross-sectional area than a tall combustion chamber with a small cross-sectional area. Also, the air distributor is usually an expensive item and the cost is raised as the bed area is increased.

There is therefore an optimum fluidizing velocity above which the

savings in capital cost by reducing bed area are more than offset by
increases in operating costs due to lower efficiency and higher fan power.
For most applications this velocity is within the range 1 to 3 m/s, but lower
and higher velocities are occasionally justified.

5.6 Mechanical design

5.6.1 Air Distributor

The function of the air distributor is to admit the air with sufficient
uniformity over the cross-sections to ensure good fluidization, and to
support the weight of the static bed without allowing bed material to pass
into the air plenum. The 'nozzle stand pipe' type of distributor (Fig. 8; see
also Figs. 17 and 26, Sections 7.1 and 7.3.4, respectively) has been found to
be particularly suitable for fluidized bed combustion. The main advantage
is that, since the air enters the bed from holes or slots at the top of the
nozzle, static bed material forms an insulating layer between the hot
fluidized zone and the base plate. Consequently, the base plate may be
manufactured using mild steel without problems of thermal expansion, and
only the nozzles need be of expensive heat resistant material.

A porous tile of sintered metal or ceramic is an excellent distributor for a
small-scale combustor, but porous tiles present constructional difficulties
for large bed areas since many tiles are required and it is necessary to
seal between them while providing allowance for thermal expansion.

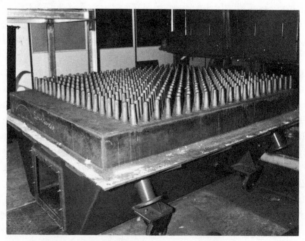

FIG. 8. Air distributor with nozzle stand pipes. (Reproduced with permission of
Babcock Power Ltd.)

Perforated plate distributors have a tendency to allow fall-through of bed material. Also, since the plate operates at or near bed temperature it must be manufactured from heat-resistant steel and the design must allow for the poor mechanical strength at operating temperature.

For units in which the bed is heated by fluidization with hot gases from oil or gas combustion, the distributor must be designed to accept the hot gas. One approach is to retain the nozzle stand pipe design and water cool the base plate, with insulation on the underside. The plate can either be of sandwich construction, forming a water space, or of water tube construction. An alternative approach is to use heat-resistant steel and allow for the thermal expansion. One method of doing this is to use a grid of pipes with multiple air outlets (sparge pipes) to introduce the air above the solid base of the combustion chamber. The sparge pipes may have drilled holes in the underside or be fitted with nozzle stand pipes. The advantages of sparge pipes is that they can expand from the fixed end to accommodate thermal expansion while maintaining a gas-tight seal. However, careful design is required to avoid circumferential temperature gradients which can give rise to distortion.

The nozzles are usually arranged on a uniform pitch of 75 to 100 mm over the base plate and have a diameter of 12 to 25 mm and a height of 50 to 100 mm. The hole size is a compromise between having an excessive number and allowing particle fall-through, typically 2 to 5 mm. Larger holes or slots are sometimes associated with the use of a 'bubble cap' to prevent fall-through.

A distributor pressure drop of about 12% of that across the bed (typically 1·5 mm water gauge per mm static depth) is required to provide uniform distribution for the beds used in fluidized combustion. However, the distributor pressure drop increases as the square of air flow, while that of the bed is constant (or decreases if bed level is maintained constant), and it is necessary to ensure good distribution of the minimum air flow rate. On this basis, the distributor pressure drop at the maximum air flow rate must be 48% that across the bed for a 2/1 turn-down and 108% that across the bed for a 3/1 turn-down. This requires a distributor with an open area in the range of 0·5 to 2%.

5.6.2 Coal Feeding

The early development of fluidized bed combustion was carried out using crushed coal and it was found necessary to feed it immediately above the distributor to ensure good combustion. This was achieved by pneumatic injection using pipes with a diameter of six times that of the largest coal particle, to prevent blockage. Typically, between one and two nozzles are required per square metre of bed area.

More recently, many units have been designed to utilize the coal as-supplied, without crushing and drying. Usually the coal is metered by a rotary valve or screw on to a chute, down which it falls to the bed surface. This method seeks to minimize entrainment of fines and relies entirely on the bed to distribute the coal. Alternatively a spreader, as used in a spreader stoker, is used to distribute coal over a wide bed area. This increases the possibility of rapid entrainment of fines, but it appears that the method is satisfactory providing that the coal is damp. Feeding uncrushed coal below the bed surface by screw feeder has also been applied successfully. Because the uncrushed coal has a lower rate of volatile release and elutriation, fewer feed points are necessary.

5.6.3 Gas Firing

Gas firing is usually incorporated so that it can be used as the start-up method, although it also provides a dual fuel firing capability. Gas firing is also used in some fluidized bed heat treatment furnaces. If the gas and fluidizing air are introduced separately to the bed, it is generally found that there is an above-bed flame above each gas entry and this does not heat the bed to a temperature at which combustion occurs below the surface. Consequently, the gas is always pre-mixed with the fluidizing air. With a cold bed, there is a sheet of flame at the surface, giving rapid heating of the particles, and combustion moves into the bed at a temperature of about 650 °C. It is essential that the fluidizing velocity during start-up should not exceed the flame speed of the gas being used, and this prevents the use of gas start-up in units designed to operate at high fluidizing velocity.

In small units it is acceptable to pre-mix the air and gas in the plenum but considerable care is required in adopting this approach in large units. The velocity of the mixture through the holes in the distributor must be greater than the flame speed and, preferably, a large plenum should be sub-divided into multiple units. A preferred alternative is to inject the gas into the standpipes, usually from a gas grid in or below the plenum.

All gas firing systems require an above-bed pilot burner and adequate safety interlocks.

5.6.4 Oil Firing

Small quantities of oil (which might, for example, be required in incinerators to maintain bed temperature) can be fed directly into the bed through multiple inlet tubes with air assistance. In this situation, a

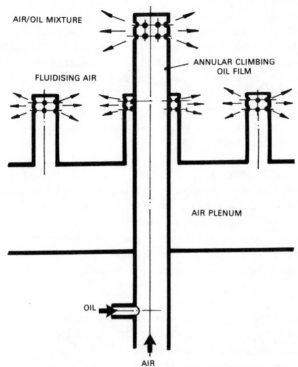

FIG. 9. Climbing film oil injection nozzle.

significant proportion of the combustion is likely to occur above the bed because of the rapid devolatilization and relatively poor in-bed lateral gas mixing. In order to burn oil efficiently as the main fuel, without excessive freeboard combustion, it is essential to disperse the oil very rapidly and uniformly throughout the bed. The oil is therefore introduced at a comparatively large number of feed points, located at the bottom of the bed in a zone of vigorous fluidization to avoid agglomeration.

Oil firing has been developed primarily by British Petroleum,[11] and they have evolved designs for injectors based on the principle shown in Fig. 9. Air is supplied to each nozzle, by separate pipes or from an air plenum, and oil is injected, from a pipe or oil plenum. The air causes the oil to flow as a climbing film up the nozzle walls and an air/oil mixture enters the bed through the holes. The nozzle is surrounded by air nozzles to ensure vigorous fluidization at the injection points. This is sometimes enhanced by having a tapered bed, with inclined containment walls to reduce the cross-sectional area at the distributor.

5.6.5 Firing Other Fuels and Wastes

Most other fuels and waste materials can be fired using the techniques described for coal, oil or gas. Semi-solid sludges are usually fed by a screw feeder at a point slightly above the bed surface. Slurries with a high water content are sprayed from above the bed in order to achieve some evaporation by the off-gas in the freeboard before the droplets fall to the bed. The main application for firing of slurries will be as a means of facilitating the disposal of waste products, with heat recovery as a secondary consideration.

5.6.6 Additive Feeding

Limestone or dolomite for sulphur retention are crushed so that the larger particles remaining after elutriation of fines are of the correct size to form the fluidized bed. The additive can be blended with coal before firing, but it is more usual to feed it separately by a pneumatic conveying system. Since the residence time of additives is long compared with that of fuel, one feed point is usually sufficient and there is no preferred location.

5.6.7 Controlling Bed Inventory

Irrespective of the bed material used, abrasion occurs because of particle impacts. This continuously reduces the size of particles within the bed with the resulting fine particles being elutriated. However, this loss of feed material is usually more than offset by the accumulation of spent limestone, coal ash or other inerts in the bed. With limestone addition, the stone is crushed such that the larger particles are the correct size to form the bed, the finer particles (typically 80–90%) being elutriated. As stone particles accumulate in the bed, it is necessary to withdraw material to maintain the required inventory of appropriately sized particles. Similarly, with crushed coal systems, the feed size is selected such that the larger ash particles are of the correct size to form the bed.

When commercial size grades of coal are used or other inert-containing fuels are fired, there is a gradual accumulation of over-size inerts in the bed. If the inerts are of relatively low density, as is the case with much of the coal ash, the particles will be mixed throughout the bed, increasing the mean particle size. This has the effect of increasing the minimum fluidizing velocity and therefore, for a given operating velocity, less gas will pass through the bed as bubbles, reducing the turbulence. With an excessive accumulation of over-size inerts, typically more than 20%, the reduced mixing rate will give rise to lateral temperature gradients, ultimately leading to clinker formation around fuel feed points. When this might occur, it is necessary to provide a bed regrading system which periodically withdraws the bed material, separates the oversize, e.g. by screening, and returns the correct size.

Large, dense particles, such as stones in coal, will segregate and gradually accumulate as a layer on the distributor. These segregated particles have some lateral movement and it is therefore possible to remove them from a single off-take point provided that the feed material contains only a low concentration of dense inerts. When this is not the case it is necessary to make special provision to facilitate withdrawal of segregated materials. This usually takes the form of incorporating an inclined distributor with the off-take at the lowest point. An alternative is to use a sparge pipe distributor with a hopper base below, allowing segregated inerts to be slowly withdrawn over the entire bed area.

In units with a high rate of off-take of bed material, e.g. large boilers with a high additive feed rate, it will usually be necessary to provide some method of cooling so that the material is safe for disposal.

5.6.8 In-bed Heat Transfer Tubes

Heat transfer tubes for steam raising will normally be horizontal if the water is circulated by a pump. For natural circulation they must be inclined at an angle of at least 10 to 20° to the horizontal, depending on the particular design (see Figs 12 and 16, Sections 6.3 and 6.9, respectively). Superheater tubes and air-cooled tubes will usually be horizontal. The tube diameter is typically in the range 25 to 120 mm and the centre to centre pitch is selected such that the gap between tubes is at least 25 mm. If tubes are installed vertically, gas bubbles tend to flow preferentially up them. This reduces the heat transfer coefficient somewhat, but the main disadvantage is that erosion by bed material following the bubbles can occur at bends from vertical to horizontal within the bed. Similarly, bubbles tend to flow preferentially up the line of horizontal tubes arranged above each other, and for this reason a triangular rather than a square tube pitch is preferred.

Extensive tests have shown that evaporator tubes of standard carbon steel can be installed in the bed without problems of fouling, corrosion or erosion.[12] However, careful selection of materials is necessary for in-bed superheater tubes and the tube banks of air-heating units.[12] The low chromium ferritic steels can be used up to about 500 °C. Austenitic steels such as types 304, 316, 321 and 347 are satisfactory up to 650 °C, with or without limestone addition. For metal temperatures up to 900 °C, high strength nickel-base alloys appear to be suitable for beds without limestone or dolomite addition. However, with limestone addition, nickel-base alloys can suffer severe grain boundary corrosion. For metal temperatures above 650 °C in beds of limestone it is therefore necessary to use an iron base austenitic steel, particularly type 347, up to 800 °C; above this temperature, it is necessary to use a low-strength super alloy such as GE 2541 as a cladding to protect a high-strength alloy.

6   FLUIDIZED BED WATER-TUBE BOILERS

The initial research on fluidized bed combustion of coal was carried out with the objective of developing very large water-tube boilers for power generation. Following pilot-scale test programmes, the next stage was to demonstrate the operation of boilers from which scale-up to power plant units could be carried out with confidence. The first two major boiler demonstrations were the conversion by Babcock Power Ltd of a 13·5 MW boiler at Renfrew, Scotland, and the purpose-designed 100 MW boiler at Rivesville, West Virginia, manufactured by Foster Wheeler in collaboration with Pope, Evans and Robbins for the US Department of Energy. However, by the time these were in operation in 1976 it was apparent that there would also be a substantial market for industrial coalburning boilers. Subsequently, the main development effort has been towards establishing commercially acceptable designs of boiler for industry, recognizing that this would also provide a basis for larger units. As with oil and gas firing, water-tubes are essential for outputs above about 20 MW and for small units to operate at high pressure. However, the ability to incorporate a tall combustion chamber also makes water-tubes potentially attractive for fluidized bed firing at lower outputs, providing that they can be manufactured at a commercially viable cost.

Some of the existing and planned water-tube boilers will now be described to illustrate the design considerations discussed in Section 5. It should be noted that this is by no means a comprehensive list of all the fluidized bed boilers and that, in particular, there are units operating in China and Russia about which few details are available.

6.1 Babcock Power Ltd (BPL) and Fluidized Combustion Contractors Ltd (FCCL)

The original FBC technology base from which BPL and FCCL have developed their commercial design expertise was the pilot-scale development carried out by the NCB[13] and BP[11] in the period to 1973. The data

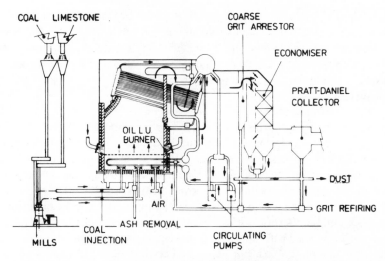

FIG. 10. Babcock boiler converted to fluidized bed combustion. (Reproduced
with permission of Babcock Power Ltd.)

acquired using several pilot plants (including a 0·5 MW unit, 0·9 m square
and 5·5 m high, at the Coal Research Establishment, supplied by BPL) were
assembled into a design package by NCB and BP, and licenced through
their jointly owned Combustion Systems Ltd (CSL) to BPL in 1973. In
order to demonstrate the operation of a FBC-fired boiler at a scale from
which scale-up could be made to large industrial boilers BPL converted
an existing boiler at their Renfrew Works in a joint venture with CSL.[14]
The boiler had a rated output of 13·5 MW (18 000 kg/h of superheated
steam at 28 bar and 294 °C) when fired with a spreader stoker.

A diagram of the Cross Type boiler and fluidized bed firing system is
given in Fig. 10. The bed is 3·1 m by 3·1 m and is installed in a refractory
combustion chamber. The air box below the distributor is compartmented
to allow independent operation of various sections of the bed. The coal,
dried and crushed below 6 mm, is supplied by a pneumatic conveying
system to feed points in the lower part of the fluidized bed. The combustion
gases, with elutriated ash and some unburnt carbon, pass through the
convection section and then to a grit arrester which separates particles
larger than 300 μm. All or part of the collected grits can be refired to the bed
to improve the combustion efficiency. The gases pass from the grit arrester
through an economizer to a Pratt-Daniel multicyclone grit arrester, and
then to the atmosphere via an induced draught fan.

Horizontal hairpin tubes are installed in the bed, designed to remove about half of the heat released at maximum output. Water from the boiler drum is circulated through the tubes such that a water/steam mixture is returned to the drum. The fluidized bed of coal ash or crushed limestone is heated to operating temperature by freeboard oil burners directed at the bed surface.

The boiler was commissioned in 1975 and has operated satisfactorily for more than 5000 h, confirming and extending the knowledge gained by the NCB using pilot-scale combustors. Both coal and oil have been fired and the studies have covered such factors as light-up, load control, uniformity of bed temperature, effect of fuel size, number of feed points and fluidizing velocity on unburned carbon carry-over, heat transfer to the immersed tubes, corrosion and erosion of the tubes, sulphur retention by limestone, and NO_x emission levels.[14] It has been demonstrated that a wide range of coals can be burned with optimum boiler efficiency at least as good as with the original stoker. When burning a coal containing 3·5 % sulphur it was possible to retain 90 % of the sulphur by feeding crushed limestone to the bed and the emission of NO_x was well within present limits for new coal-fired boilers in the USA. The boiler output could be modulated from 25 to 100 % of maximum by varying bed temperature and stopping or reactivating sections of the bed.

The first commercial application of the FBC system demonstrated at Renfrew was the conversion of an existing 20 MW (27 200 kg/h steam) boiler at the Central Ohio Psychiatric Hospital, Columbus, by Babcock Contractors, Inc. (BCI) of Pittsburg, a BPL (UK) subsidiary, in collaboration with the Riley Stoker Corporation. This conversion was a consequence of studies commissioned in 1977 by the State of Ohio Energy Resource and Development Agency (OERDA) with the aim of demonstrating that FBC provides a viable alternative to conventional stokers with stack gas scrubbers. The Columbus boiler was designed to burn high-sulphur Ohio coal with local dolomitic limestone to absorb the sulphur dioxide. The fluidized bed system was essentially a larger version of the Renfrew design, firing crushed coal and having start-up by above-bed burners and load following by modulating bed temperature and compartmenting the bed into three sections. The boiler conversion was completed in 1981, but funding was withdrawn before it was fully commissioned.

To exploit its design expertise in fluidized combustion, BPL has formed Fluidized Combustion Contractors Ltd (FCCL). This company provides FBC designs and project management to BPL in the UK, Riley Stoker Corporation in the US, Stork Boilers in Holland and other boiler makers world-wide.

6.2 Pope, Evans and Robbins—100 MW Boiler at Rivesville

Pope, Evans and Robbins (PER) became involved in fluidized bed combustion in 1965, following a visit by one of their staff to the UK. In the period to 1972 they operated a series of pilot-scale rigs at Alexandria Va., with funds from the US Office of Coal Research (OCR). Following design studies for power generation boilers (carried out jointly with Foster Wheeler Corporation) a contract was awarded for the supply of a demonstration boiler for the Monongahela Power Company's site at Rivesville, West Virginia, sponsored by OCR and others. The unit was designed to produce 136 000 kg/h of superheated steam at 92 bar and 496 °C, i.e. approximately 100 MW.[15] Installation commenced in 1974 and the boiler was first operated in 1976.

The fluidized bed occupies a total plan area of 3·6 m by 11·4 m and is divided along its length into four independent cells (Fig. 11). One cell does not have in-bed tubes and is used initially to establish coal combustion and to provide hot bed material for the other cells; subsequently, it acts as a 'carbon burn-up' cell for fines elutriated from the other beds and collected by cyclones. The remaining cells are larger and each contain an in-bed tube bank. The hot gases from the beds pass upwards through a bank of convective evaporator tubes arranged horizontally and then through an economizer. The total boiler height is about 7·5 m. Evaporation in the four convective sections is supplemented by an in-bed convective tube bank in one of the cells and the in-bed tubes in the remaining two cells are used for superheat. All of the in-bed tubes are horizontal and on a triangular pitch. Two pumps provide water circulation.

The static bed depth is 600 mm and this expands to about 1200 mm at the design air velocity of 3·6 m/s. The air distributor was originally an uncooled punched hole plate, designed to accommodate thermal expansion at the high metal temperatures. Some redesign has been necessary to reduce distortion and cracking. Coal (nominal size range 6–12 mm) is mixed with crushed limestone and fed to the bed pneumatically. The beds each have a system to remove oversize ash from the bed. The bed material flows by gravity to a classifying screen and the separated particles of the correct size are conveyed to storage bunkers from which they can be returned to the bed.

Preliminary tests[16] confirmed that start-up could be achieved by heating bed material in the carbon burn-up cell and then allowing it to flow into the other beds in succession. Full load performance tests have not yet been reported, although preliminary data indicate that the combustion efficiency was somewhat below the design value. A fly ash re-injection system has been installed to improve combustion efficiency.

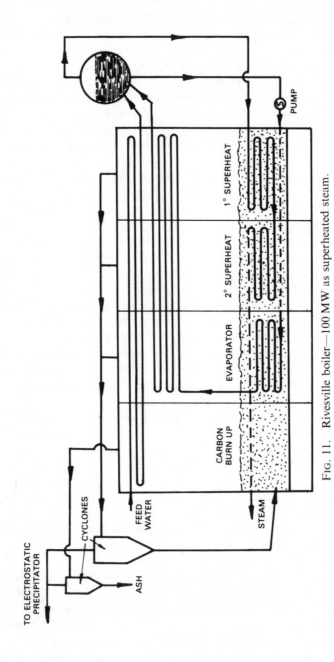

FIG. 11.   Rivesville boiler—100 MW as superheated steam.

A major continuing problem has been the provision of a reliable system for feeding limestone and coal to the bed. The use of coal of high moisture content (5 %) and containing fine material (below 3 mm size) caused arching in the storage bunkers and blockages in the rotary feeders and the fuel entry pipes to the bed. Modifications were made to the geometry of the entry pipes, their diameter was increased and the rotary feeders were replaced by belt conveyors. These modifications, combined with the installation of vibratory feeders for conveying the coal/limestone mix to the entry pipes have provided a workable system.[16] The performance of alternative fuel feed systems has been evaluated in order to provide design data for use in other FBC systems. Problems have also been encountered with the bed material drain and storage system; mainly blockages caused by clinker formation and moisture pick-up during cooling.

Ultrasonic testing and metallurgical examination of heat exchanger surface have been made at intervals during some 4000 h of coal firing and results have shown that there was no significant reduction in the wall thickness, i.e. erosion and corrosion did not appear to be problems. However, failure of the original in-bed surface support bars (made from high-nickel alloy) necessitated their reconstruction in 310 stainless steel alloy. Some redesign of the grid plates which frame the air distributor was also necessary in attempts to reduce distortion and cracking.

The Rivesville installation demonstrated that coal could be fired in a large industrial FBC boiler while maintaining proper operating conditions and stack emissions, and it provided valuable design information. However, following the successful operation of subsequent FBC boilers, the US Government discontinued funding for the project in early 1981 and the Rivesville boiler has now been permanently shut down.

6.3 Foster Wheeler Corporation

Following the design and initial operation of the Rivesville boiler, PER and Foster Wheeler formed the jointly owned Fluidised Combustion Company. This company has supplied a 45 000 kg/h boiler to Georgetown University, financed by the US Department of Energy, to demonstrate industrial-scale fluidized combustion of high-sulphur coal with strict environmental controls. The boiler, designed and constructed by Foster Wheeler Corporation, supplies steam for heating in winter and operation of air conditioning equipment in summer. Because it is intended to add a power generating turbine into the system the boiler is designed for operation at a steam pressure of 43 bar (saturated). However, current normal operation is at 19 bar (saturated).

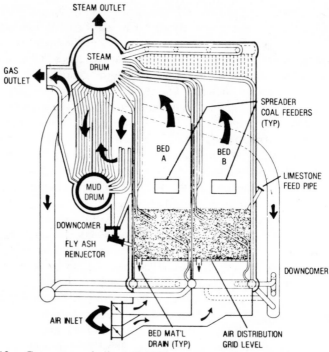

FIG. 12.   Georgetown boiler—30 MW as saturated steam. (Reproduced with
permission of Foster Wheeler Power Products Ltd.)

As shown in Fig. 12 the boiler is of the bi-drum water-tube construction
and has two independent beds, each designed to give a turn-down of 2/1,
thereby allowing a 4/1 range of boiler output.[17] The dimensions of each bed
are 5·09 m by 1·7 m and the bed depth at maximum output is 1·4 m. The
deep bed is required to achieve high sulphur retention using limestone sized
below 3 mm which forms the bed. The fluidizing velocity at maximum
output is 2·4 m/s. The coal is bituminous (sized below 32 mm) and is fed by
one spreader above each bed, positioned on the end wall. The feed rate of
crushed limestone is approximately 30 % of the coal feed rate. Elutriated
fines are refired from a drop-out hopper after the first convection pass and
from cyclones. The gas from the cyclones flows through a water preheat
economizer and then to a bag house. The stack terminates at roof level and
there are no visible emissions.

The distributor is a multi-plate assembly with stand pipes, the stand
pipes being stainless steel bolts, drilled straight through, on a pitch of about
40 mm. Start-up is by heating a start-up zone with an above-bed burner and

then fluidizing the remainder of the bed. To minimize the start-up time, about 5 % of coal is added to the cold bed. Once in operation, output is modulated by varying the air and coal feed rates and adjusting the bed level to cover sufficient area of tubing to maintain a near-constant bed temperature (see Section 5.2.2). An increase in bed level is achieved by stopping the off-take of spent limestone bed material and feeding limestone at maximum rate, and a decrease is obtained by stopping limestone feed and withdrawing bed material at maximum rate. Small fluctuations in output are accommodated by variations in bed temperature and excess air level.

The boiler was commissioned in July 1979. By 1980, start-up and operation had become routine procedures, carried out completely by the University's boiler house personnel. The unit is normally operated on a continuous basis around the clock, under automatic control. The sulphur content of the coal is around 2 % and a retention in excess of 80 % has been consistently achieved by feeding three times the theoretical requirement of crushed limestone. The initial design of the system for recycling cyclone fines to the bed was inadequate and, as a consequence, the combustion efficiency was below the design value. It is understood that this feature has now been improved. A further difficulty encountered has been local erosion of some of the wall tubes and in-bed tubes. The reason for this is considered to be a bulk circulation of bed material induced by the high air flow through the original fines recycling system. Despite these difficulties the boiler had operated for over 6000 h by mid-1981, demonstrating the validity of the design features and control system.

Following the success of the Georgetown boiler, the Foster Wheeler Corporation has received contracts for a further six FBC boiler installations; these are listed in Table 3.[18]

The Foster Wheeler technology has been licenced to Generator Industri A.B. of Sweden, who supplied their first FBC boilers for district heating at Eksjo, Sweden. Two hot water boilers, rated at 5 MW and 10 MW have been in operation there since 1979, burning wood and refuse. More recently a 16 MW boiler for bituminous coal was commissioned at Chalmers University, Sweden, in 1982. Generator Industri A.B. have orders for two 15 MW boilers for Sandviken and two 10 MW boilers for Landskrona, scheduled for commissioning in 1984. They will burn wood or coal and supply hot water for district heating.

6.4 Deutsche Babcock 35 MW boiler conversion at Dusseldorf
The Deutsche Babcock FBC technology is based on development work by

TABLE 3
Foster Wheeler Corporation boiler projects

Customer	Steam output (tonne/h)	Pressure (bar)	Temperature (°C)	Fuel	Start-up date
Royal Dutch Shell, Europoort, Netherlands	50	81	495	Bituminous coal	1982
Mitsui Toatsu Chemical Co., Japan	25	24	250	Bituminous coal	1982
Kentucky Energy Corporation, Franklin, Kentucky, USA	Two at 27	41	Saturation	Bituminous coal	1982
Canadian Dept. of Defence, Summerside, Prince Edward Island, Canada	Two at 18	9	Saturation	Bituminous coal and wood chips	1983
Ashland Petroleum Co., Catlettsburg, Kentucky, USA	Two at 145	31	370	Gas (CO, SO_2)	1983
Idaho National Energy Lab., Idaho Falls, USA	Two at 30	10	Saturation	Sub-bituminous coal	1984

the Bergbau Forschung Laboratory and by Ruhrkohle AG in 1977. The first large-scale demonstration was the conversion of a 35 MW boiler (50 000 kg/h of superheated steam at 17 bar and 400 °C) at Flingern Power Station, Dusseldorf. The boiler was one of several old coal-fired units which have been superseded by more recent oil-fired boilers, but are retained as stand-by although rarely operated. The conversion has been sponsored by Ruhrkohle AG with a government grant, and engineered by Vereinigte Kesselwerge AG (VKW), a subsidiary of Deutsche Babcock. (The Babcock companies in the UK, US and Germany are completely independent.)

The boiler conversion was very similar in principle to that at Babcock Power Ltd, Renfrew, in that the fluidized bed unit has been installed at the bottom of a refractory combustion chamber.[19,20] The bed unit was 5 m square with water-cooled walls 1·5 m high, a water-cooled distributor and two rows of evaporator tubes near the top of the unit, with flow by forced circulation. The bed was divided into four independent sections, each with an area of 6 m², by refractory dividing walls 1·5 m high. The distributor was a plate with stand pipes and 30 % of these had gas injection for start-up. During start-up the bed level was below the tube bank and the ignition burner was below the tubes, immersed in the bed in normal operation. There was a second burner above the tubes as a safety precaution and also gas detectors in the plenum. Heating time to 500 °C was 30 min; coal feed was then commenced and steam could be supplied to the main after a further 30 min.

Coal from an existing bunker was crushed in a hammer mill swept by hot gases extracted from the boiler. The gas conveyed the finer coal particles to an intermediate bunker, to which the coarser particles from the mill were conveyed mechanically. The crushed coal was then blended with limestone in a belt weighing system and fed by screw to four feed hoppers, one for each bed compartment. Each feed hopper had six rotary valves, on a common drive system, which metered the coal into six pneumatic feed systems through the distributor. Fluidizing velocities were in the range 1·5 to 2·5 m/s.

Each bed compartment had two water-cooled ash off-takes through the distributor from which the ash flowed by gravity through a cooler with a cooling-water tube bank (one cooler per compartment). The ash was conveyed pneumatically from the coolers to a storage bunker which had sufficient capacity to hold the material from all four beds if necessary. Fly ash was collected from the bottom of the convection section and from two cyclones in parallel and conveyed pneumatically to a second bunker from which it could either be blended with the coal for refiring or disposed with

J. Highley and W. G. Kaye

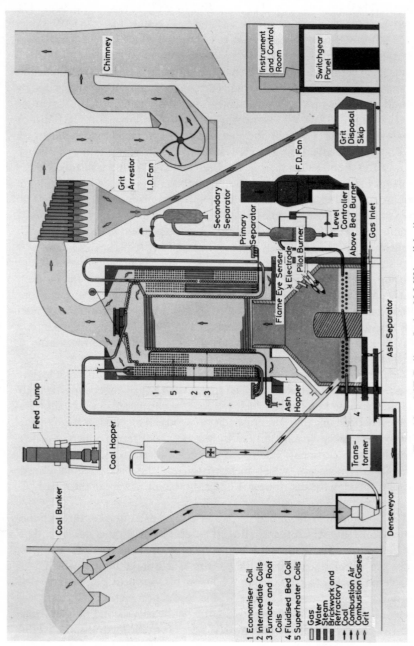

FIG. 13. ME Boilers Ltd—30 MW coil boiler.

the bed ash. Fly ash from the bag filter was conditioned with water in a screw conveyor and removed from site as a thick sludge.

The old stoker was removed from the boiler in July 1978 and the installation was completed in April, 1979. Cold testing commenced in May, the first coal was burned in July and by the end of 1979 the boiler had been in operation for 1100 h. The operation of the fluidized bed unit was generally reliable, although some early problems were experienced with the coal storage and handling system necessitating modifications which have proved successful. The unit was operated primarily as a large test facility to provide data on combustion, heat transfer and methods of minimizing the emission of sulphur dioxide and nitrogen oxides.[20] The test programme was to be completed by November 1981 and it is understood that operation of the boiler has now been terminated.

The data and expertise gained from operation of the Flingern boiler are being applied by VKW in the design of a 124 MW boiler (140 tonne/h of superheated steam at 120 bar and 530 °C) to be installed in a combined power and district heating scheme at Hameln (West Germany).[21]

6.5 ME Boilers Ltd—30 MW boiler at Sheffield

To demonstrate the operation of a large fluidized bed fired water-tube boiler in the UK, the NCB has sponsored the design and manufacture of a 30 MW unit, by ME Boilers Ltd of Peterborough, for installation at the British Steel Corporation's River Don Works, Sheffield.[22]

As shown in Fig. 13 the boiler is of the once-through coil type, and produces superheated steam at 45 bar and 440 °C. The fluidized bed is circular, 5·3 m in diameter, and contains the evaporator, comprising a single tube arranged as a double pancake coil. The gases from the bed flow up a central combustion chamber to two annular passes containing the superheater coils and, finally, the water preheat. This boiler type was selected on the basis that the load was planned to be steady at, or close to, maximum rated output; the design gives significant cost saving due to the elimination of the steam drum, although it is usually suitable only for applications where the load is steady at high output. However, although this was the original requirement, following changes in the site load pattern the boiler will be required to provide a 6/1 turn-down and follow rapid load changes caused by the operation of steam hammers. This facility has been provided by dividing the bed into six equal compartments, separated by refractory walls, each with its own fan, coal feed, start-up system and modulating controls. Start-up is by pre-mixing coke oven gas with the fluidizing air, and igniting with above-bed burners. (The boiler design facilitates operation of any or all beds on coke oven gas if it is available.)

The output of each compartment can be modulated by varying the air supply, and hence the bed expansion into the tube bank, and controlling the coal feed to maintain a pre-set temperature. In normal boiler operation, the output of one compartment is modulated and the others are either at maximum output or slumped. To facilitate a rapid increase in output, slumped beds are maintained at a temperature above 700°C so that they can be reactivated immediately. This requires the sequential operation of all the beds. To extend the 'slump' time, the static bed is entirely below the tube bank, thus minimizing the rate of cooling.

Installation of the boiler and ancillaries was completed at the end of 1980. All six beds had been operated satisfactorily burning coke oven gas by April 1981 and full rating was achieved burning coal in July 1981. However, various problems have been experienced with the ancillaries, particularly the start-up gas ignition burners and associated safety interlocks, and these have restricted the availability of the boiler for supplying steam to the site and test programmes.

6.6 Standard Kessel/Thyssen Energie 6 MW boiler

A consortium consisting of Standard Kessel and Thyssen Energie has been established in Germany to supply fluidized bed boilers based on technology developed by Thyssen Energie for incineration of sewage sludge. They have supplied a 6 MW boiler for a district heating scheme at König Ludwig, Recklinghausen, under the sponsorship of Ruhrkohle AG with a government grant. The FBC unit has been installed in a boiler house with several existing oil-fired shell boilers.

The boiler is a new, purpose-designed water-tube unit consisting of a vertical combustion chamber and two convection passes.[19] The combustion chamber is square, with a cross-section of 5 m², and contains a bed about 1 m deep with an immersed tube bank through which the flow is by natural circulation. The plenum is divided into three compartments but there are no divisions in the bed and the facility to subdivide the bed is not being used initially. Start-up is by an oil burner directed at the bed surface. Coal and limestone handling and blending is by a complex mechanical system, which supplies three feed hoppers at one side of the boiler. The coal supplied is closely sized, 8 to 13 mm, but undergoes considerable degradation in passing through the handling system. Coal from each hopper is fed directly into the bed by a horizontal screw, positioned immediately above the distributor. Provision is made to fill these screws with sand before they are stopped, in order to avoid the potential problem of the coal caking and blocking the screw. The flue gas is cleaned by passing through primary and secondary cyclones and a bag filter before discharge to atmosphere. Ash from these units is conveyed to a receiver outside the building by a drag-link and screw system. The unit is controlled manually and has comprehensive monitoring equipment.

It was commissioned in 1979 and operation has been generally very satisfactory. The expertise and performance data obtained have been applied by SK/TE in the design of a 8 MW boiler supplied to De Hazelaar, Netherlands, in 1982. This produces 10 tonnes/h of saturated steam at 20 bar and consists of a combustion chamber of water-tube construction connected to a shell and tube unit, similar to the boiler shown in Fig. 25. SK/TE are also manufacturing two boilers, 7·5 MW and 3·5 MW, to be installed at Luneburg for district heating.[21]

6.7 Stone International Fluidfire Ltd

Fluidfire Ltd was formed by the late Professor Douglas Elliott in 1972, initially to manufacture FBC heat treatment furnaces based on his expertise in gas combustion in fluidized beds (Chapter 9). Subsequently, Fluidfire began a boiler development programme in 1974 with the aim of manufacturing a product range for both domestic and industrial applications.[23] Following trials using gas-fired prototype boilers, a coal-fired prototype was commissioned in March 1978.[24] This was a water-tube boiler producing hot water, with a rated output of 300 kW. A special feature of the Fluidfire design was the use of a concave distributor, made of separate porous ceramic tiles, designed to create circulation of the bed material down at the centre and up at the sides (similar to that of Fig. 14). After a programme of trials at the factory, this unit was supplied to the Virginia Polytechnic and Institute, where it has been used for test and demonstration burning of a wide range of fuels.

The first industrial boiler is a 3 MW (4500 kg/h of steam) unit which was installed at Hayward Tyler Ltd, Keighley, Yorkshire, in 1980. As shown in Fig. 14, this is a two-drum D-type boiler, using the circulating bed concept.[24] The coal is screw-fed just below the surface of the bed at the centre, so that the circulating solids cause the fresh coal to move down into the bed and then to circulate. This technique is intended to ensure that most of the volatiles are released within the bed and to increase the residence time of finer particles before elutriation. Start-up is by pre-mixed air and gas supplied through the porous ceramic tile distributor. Load following is achieved by dividing the bed into an uncooled combustion zone and two

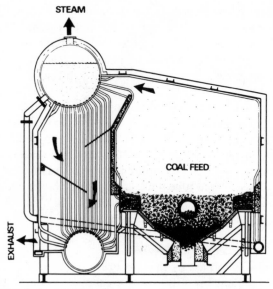

FIG. 14.   Fluidfire Ltd—3 MW boiler. (Reproduced with permission of Fluidfire Ltd.)

heat transfer zones (Section 5.2.3). The heat transfer zones are at each side of the combustion zone (Fig. 14) and contain closely packed tubes in a relatively shallow bed. In operation, the steam pressure is maintained constant by modulating the air supply to the heat transfer zones, in order to vary the heat flux to the tubes. As heat transfer is increased, the temperature in the combustion zone will tend to decrease and the controller then increases the coal and air supply to maintain a constant temperature. This technique is said to give a turn-down of 3/1 with rapid response; turn-down can be extended to 4/1 by varying the combustion bed temperature.

By December 1981 the 3 MW unit had operated for about 2500 h, with good reliability. A range of packaged boilers from 4500 to 23 000 kg/h has been designed, for pressures up to 55 bar and superheated to 400 °C. These are commercially available and are expected to find application in industrial combined heat and power installations. The recirculating bed concept is particularly appropriate to low grade fuels with high ash content, and a system to extract stones from the bed has been developed by Fluidfire.

6.8 Deborah Ltd

Deborah Ltd have acquired the rights to a circulating fluidized bed design

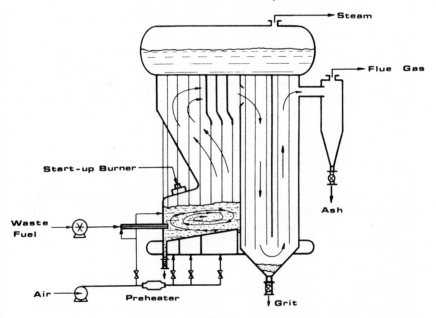

FIG. 15. Deborah Ltd—circulating bed boiler. (Reproduced with permission of Deborah Ltd.)

which has been under development since 1972. After work carried out using three small pilot plants, a 3200 kg/h boiler was supplied to a chemical works at Avonmouth (UK) in 1980.[25] It was designed primarily to burn waste acid tars and oil wastes, but can also burn fuel oil when these are not available. It had previously been necessary to pay for disposal of the tar because it had not been possible to burn it with conventional burners.

As shown in Fig. 15, the boiler is of 'A' type water-tube construction, with a longitudinal steam drum.[25] Circulation of the bed is obtained by using an inclined distributor plate (drilled hole plate covered by a mesh) with three separate air supplies which are adjusted to give a high fluidizing velocity in the deep zone and a low velocity in the shallow zone. Above the deep high-velocity zone, the boiler wall is inclined inwards to deflect the particles across the bed surface. The resulting circulation is said to be particularly appropriate for ensuring rapid horizontal distribution of fuels with a high volatile content, which would otherwise require multiple feed points. In the Avonmouth boiler, the oil is supplied to a single feed point. An important advantage of the system is that lead components in the oil are not vaporized but are deposited on the inert bed material, from which lead can subsequently be recovered.

Following commissioning, the boiler was handed over to the site staff and it has since operated on a continuous basis on 5 or 7 days per week. By December 1981 it had been operated for more than 7000 h. Following the successful performance of this boiler, Deborah Ltd have received an order from Technipetrol, Italy, for a near-identical unit to burn acid tar, to be commissioned in 1983.

The circulating bed concept is suitable for other fuels, particularly those with a high volatile content. A boiler has been supplied to Norfolk, Virginia, to burn used car tyres, wood chips and sawdust and provide 3 MW of hot water for district heating. This was commissioned in 1982. Two UK orders have also been received, for hot water boilers to burn coal; a 0·5 MW unit at Barnsley was commissioned in 1982 and a 1·2 MW unit at Knaresborough is scheduled for commissioning in 1983.

6.9 Gibson Wells Ltd

Gibson Wells Ltd have developed a design for a modular packaged boiler for conventional firing and, since 1980, have been collaborating with the NCB to adapt it for FBC. Gibson Wells Ltd became a subsidiary of Foster Wheeler Power Products Ltd in 1982.

The first FBC boiler has been installed at Stevenson Dryers, Ambergate, Derbyshire, to supply 22 tonne/h of steam at 16·5 bar, 257 °C for a turbine in a combined heat and power scheme. As shown in Fig. 16, it has two bed compartments, a gas pass containing a superheater, and a convective evaporator bank of straight tubes. The distributor is water-cooled for start-up by hot gas and it is an integral part of the boiler construction. The in-bed tubes are inclined for natural water circulation and the geometry is designed to achieve a turn-down of 2/1 per compartment by the natural variation in bed level and splashing with velocity (Section 5.2.3). The coal, sized 18 mm to zero is metered through rotary valves and falls down chutes to the bed surface. Start-up and load following are controlled automatically by a microprocessor. The boiler was successfully commissioned in 1982.

A second boiler has been supplied to Woolcombers Ltd, Bradford, Yorkshire, to burn coal and a proportion of waste wool grease. This has an output of 7·1 MW as saturated steam at 12·4 bar. The design is similar in principle to that of Fig. 16, but there is only one bed compartment (and the superheater is omitted from the down-pass). This boiler was also commissioned in 1982.

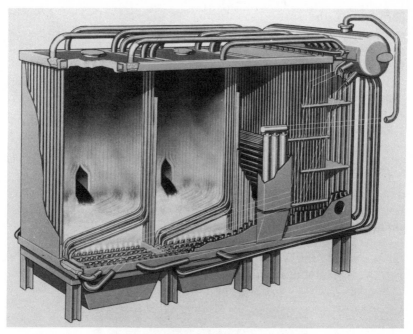

Fig. 16.   Gibson Wells Ltd—15 MW boiler.

6.10 Combustion Engineering, Inc.

Combustion Engineering began work on fluidized bed combustion in 1976, when they received a contract from the US Department of Energy to carry out pilot-scale trials and design a 19 MW boiler. Subsequently, the DOE placed a contract for the boiler to be manufactured for the US Navy's Training Facility at Great Lakes, Illinois.

As shown in Fig. 2, the design is a symmetrical 'A-type' with a central bed extending the entire length of the boiler and a convection bank at each side. In-bed evaporation surface is provided by hairpin tubes at both sides.[27] Superheating is achieved by an in-bed tube bank, positioned centrally, rather than by a gas convective unit. For turn-down the bed is divided into four equal zones along its length, three with superheat tubes and one without which is used for start-up. Two of these zones are further subdivided into halves, giving a total of eight bed zones. Each zone has an independently controlled coal and air supply but there are no divisions in the bed. The fluidized bed area is 13 m^2 and it is operated at a fluidizing velocity of 2·1 m/s. The coal is dried, crushed and fed pneumatically; the limestone for sulphur retention is sized below 3·2 mm, giving a mean bed particle size in the range 0·7 to 1·0 mm. Fines collected in the cyclone can be re-injected to improve sulphur retention and combustion efficiency. The gases from the cyclone pass through a tubular air heater and a bag filter before discharge to atmosphere.

The boiler was delivered to the site at the end of 1980 and was commissioned in 1981.

7 SHELL BOILERS

7.1 Horizontal boilers with circular furnace tube

There is an incentive to develop a fluidized bed system which can be installed in the circular furnace tube of a horizontal shell boiler, since it would allow conventional boiler designs to be utilized. Ideally, the fluidized bed unit should be suitable for use in standard boilers as manufactured for oil or gas since this would give minimum cost for a new boiler and also permit conversion of existing installations.

If a bed is installed in a furnace tube, as shown in Fig. 17, there is

Fig. 17.   Fluidized bed unit in 1·2 MW GWB Vekos Powermaster boiler. (Reproduced with permission of GWB Parkinson Cowan Ltd.)

insufficient heat transfer surface at the sides to permit operation at low excess air at the combustion intensity required to meet boiler output. Rather than providing additional heat transfer surface by installing in-bed tubes, most development groups have, until recently, accepted the resulting high excess air level. This approach offers the advantage of relatively simple start-up and load-following, but the gas flow rate through the boiler is higher, causing an increased pressure drop through the tubes in a standard boiler, which can be offset by incorporating additional tubes into the convective passes. However, there is now increasing interest in developing designs with in-bed tubes.

7.1.1 NCB/GWB Parkinson Cowan Ltd

A collaborative programme to develop a FBC firing system for conventional horizontal shell boilers was initiated in 1975, using a 1·2 MW Vekos Powermaster boiler.[22,28] This unit is the UK market leader in conventional coal-fired shell boilers and features coal feed from above the boiler via a central drop tube and a simple spreader to a static grate (Section 3.3.3). The initial fluidized bed firing system, installed in 1975 and shown in Fig. 17 consisted of a three compartment plenum having a plate distributor with multiple stand pipes.[22,28] Start-up was by pre-mixing natural gas with the fluidizing air in the centre-plenum, using a gas grid with an injector into each stand pipe, with ignition by an above-bed torch. The boiler's standard coal feed system was utilized (Fig. 4). The bed, consisting of silica sand sized 0·5 to 1 mm, initially 150 mm deep, was retained by a refractory wall at the rear of the combustion chamber.

The fluidized bed was readily heated to operating temperature using pre-mixed gas, and coal combustion was established. However, because of the high horizontal gas velocity at the rear of the combustion chamber, there was rapid elutriation of bed material, and operation for extended periods was not possible. Various designs of elutriation-reducing baffles in the combustion chamber were then evaluated. With the most successful baffle the bed level could be stabilized at about 65 mm deep and it was possible to operate for extended periods. Because of the limited heat transfer from the very shallow bed, the air was not fully utilized for combustion and consequently the maximum output was somewhat below boiler rating when using the standard fan. For the output range which was obtained, up to a maximum fluidizing velocity of 1·6 m/s, the overall boiler efficiency was similar to that of a mechanical stoker.

It was recognized that, in order to retain a fluidized bed in the combustion chamber at the higher fluidizing velocity needed to achieve boiler rating, it would be necessary to incorporate a sand refiring system which would return elutriated material from the gas reversal chamber at the rear of the combustion chamber to the bed. To avoid this complication, and also to improve boiler efficiency, work was initiated to reduce the excess air level by utilizing the lower surface of the fire-tube (covered by the plenum chamber) for heat transfer from the fluidized bed. The initial approach to this involved the use of sparge pipes, with downward air jets, instead of the original plenum and plate distributor.[28] However, considerable difficulty was encountered in achieving the required sand movement against the bottom of the fire-tube and alternative methods are now being considered.

However, following the increased market interest in coal-burning boilers, particularly fluidized bed fired, in the past two years, attention has again been given to the original plenum concept. Alternative baffle designs and methods of sand refiring to maintain a bed of adequate depth are being assessed. Also, consideration is being given to incorporating in-bed tubes. Manufacture of a purpose-designed 2·4 MW prototype commercial version is nearing completion.

7.1.2 NEI Cochran Ltd
NEI Cochran Ltd supplied a novel vertical shell boiler for fluidized bed combustion to the NCB Coal Utilisation Research Laboratories at Leatherhead (then BCURA), UK, in 1968 (see Section 7.2.1). In 1975 NEI Cochran started a programme to develop a fluidized bed firing system for horizontal shell boilers. The first prototype boiler at their Annan Works had a maximum output of 1360 kg/h of steam when fired by mechanical stoker. With fluidized bed firing an output of up to 2000 kg/h was achieved, the limit being that of excessive water carry-over in the steam.

Thus, for a given size of boiler, fluidized bed firing was shown to be capable of achieving outputs comparable with those from the use of oil or gas firing, as opposed to the substantially lower outputs obtained with mechanical stokers.

With the data obtained, a commercial prototype boiler with an output of 5 MW was designed and commissioned in 1979, Fig. 18. It is being used to prove the control methods and the long-term reliability of the combustion technique and the complete system.[29] It is of the wet-back type and is designed to be operated on forced draught or on balanced draught, using an induced draught fan. A bed depth of about 100 mm is used and this rests upon a distributor plate with vertical stand pipes for air distribution, similar to that of Fig. 17. Start-up is by gas mixed with the fluidizing air in the stand pipes and coal firing is commenced about 15 min after light-up.

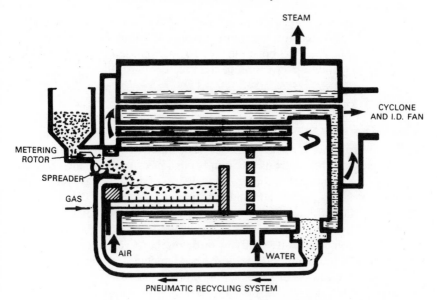

F<small>IG</small>. 18. NEI Cochran–horizontal shell boiler.

The coal handling system will depend largely upon customer preference, coal types to be burned and installation size. The coal feed rate is regulated by a novel feeder, controlled with the air feed rate, and the coal is thrown onto the bed by a sprinkler.

Large ash drops out in the wet-back combustion chamber, along with any bed material that is entrained. This mixture is recycled as it contains a high proportion of unburned carbon in addition to the bed material which needs to be returned. Fine ash passes through the convective tubes and is removed from the flue gases using a multicell cyclone.

There is little heat transfer from the fluidized bed and consequently it is operated with a high excess air level. To achieve outputs considerably in excess of stokers, the fluidizing velocity is above 3 m/s and the bed material is high-density alumina to minimize elutriation. As in all designs with little heat transfer from the bed, a turn-down of 2·5/1 is readily achieved from a single-bed unit. NEI Cochran intend to use a double bed to provide a 5/1 turn-down. There are no operator-dependent settings or adjustments and the commercial boilers will be fully automatic.

NEI Cochran have supplied two 1·2 MW hot water boilers to the NCB for a site on the Selby Coalfield commissioned in 1981. However, these are somewhat smaller than will normally be commercially competitive. The first boiler in the commercial range, for 4·5 MW, has been installed at a brewery in Tadcaster. It was commissioned in 1981 and is being operated on a daily basis. The other boilers in the new boiler house are stoker-fired but designed to facilitate conversion to fluidized combustion.

7.1.3 Energy Equipment Ltd

Energy Equipment (EE) Ltd have developed a firing system which provides a low excess air level and corresponding high boiler efficiency, without the use of in-bed tubes for heat transfer.[30] The fluidized bed is operated as a partial gasifier, the fluidizing medium being a mixture of air and recycled flue gas. The remainder of the air (up to 30 %) is supplied above the bed to burn the volatiles, fine carbon and products of gasification. This creates a high-temperature flame in the freeboard, possibly up to 1300 °C, which enhances radiation to the fire-tube. The recycled flue gas acts both as a heat carrier (Section 5.3.3) and a gasifying medium; the recycle rate is controlled automatically to maintain the required bed temperature. The excess air is lower than in other systems without in-bed tubes because: (i) the flue gas is used as a heat carrier; (ii) there is more heat transfer in the fire-tube because of the higher freeboard temperature; and (iii) the gases leave the fire-tube at a higher temperature. However, the high temperature precludes the use of limestone for sulphur retention and might possibly cause the distillation of volatile ash components with some coals, giving rise to deposits in the boiler. Start-up is by fluidization using hot gases from an external oil burner, and a multiple sparge pipe assembly is used as the distributor in order to accommodate thermal expansion.

The first trials were carried out at the EE factory and then, during 1976, a system was operated in a shell boiler at a factory in Stirling, Scotland.

In 1977 EE installed a fluidized bed unit into a 10 MW water-tube boiler at Cadbury Ltd, Bourneville, Birmingham, to convert it from a chain-grate stoker.[30] The bed area of 6·5 m² is divided into two compartments, each with a separate fan, oil-fired preheater and coal feeder. This unit has been operated for over 8000 h.

Energy Equipment are now applying this firing method to both shell and water-tube boilers. They have supplied three 1·5 MW shell boilers to the NCB for a site on the Selby Coalfield, commissioned in 1981, and a 6 MW shell boiler to Hungary.[30] A 8 MW water-tube boiler at the University of Nottingham was converted from oil firing to coal in 1982. A novel locomotive type boiler (see Section 7.3) has been designed and an order received from Ectona Fibres Ltd, Workington, for a unit to produce 27 tonne/h of steam, to be commissioned in 1983. In addition, the ability of the firing method to achieve high gas temperatures is being exploited in the EE hot gas generators (Section 8.5).

7.1.4 Danks of Netherton

Using NCB expertise (licensed through Combustion Systems Ltd) Danks have designed a prototype 6 MW horizontal shell steam boiler. The bed unit consists of a plenum with a distributor plate with stand pipes, incorporating start-up by gas pre-mixed with the fluidizing air. A novel feature of the boiler design is that additional in-bed heat transfer surface is provided by finned tubes across the circular furnace tube. The tubes are of large diameter and inclined to facilitate natural water circulation, although initially this will be assisted by forced circulation. The unit is installed at a Danks site and was commissioned in early 1981.

7.2 Vertical shell boilers

The development of new purpose-designed boilers for FBC offers the major advantage that the combustion chamber can be arranged for vertical gas flow and the correct amount of tubing can be installed in the bed in a configuration suited to achieving simple start-up and effective load control. Vertical shell boilers are potentially attractive for smaller outputs, but with the maximum permissible combustion chamber diameter of about 1·8 m, depending on steam pressure, the output is limited to about 4 MW at the usual maximum fluidizing velocity of 3 m/s. Consequently, for outputs above 4 MW, it is necessary to consider designs which provide larger bed areas than are available in a vertical cylindrical furnace. These designs are described in Section 7.3.

The potential of vertical shell boilers was recognized at an early stage in the development of fluidized combustion and a prototype was designed for the NCB by the British Coal Utilisation Research Association and has been used there (now NCB, CURL) since 1969 to supply process steam when required and also for test programmes. More recently, two prototype vertical shell boilers, one for hot water and the other for steam, have been designed by the NCB and operated at commercial sites. Both were installed in 1977, have operated five heating seasons with NCB supervision, and are expected to continue in commercial operation. A commercial design based on the hot water unit has been developed by EMS Thermplant, in collaboration with the NCB, and two 2·88 MW steam boilers have been in operation since 1981. Several boilers similar to the prototype steam boiler have been manufactured by various companies in the UK and South Africa. A novel design suitable for outputs up to 4 MW has been devised by

Wallsend Slipway Engineers (part of British Shipbuilders) in collaboration
with the NCB and a 3·8 MW prototype has been in operation since mid-
1980.

7.2.1 2·5 MW Boiler at CURL, Leatherhead

It was recognized at an early stage in the development of fluidized
combustion that a vertical shell was an appropriate concept for industrial
boilers with outputs up to about 4 MW. Preliminary pilot-scale studies of
fluidized combustion at high burning rates were carried out for the NCB at
the British Coal Utilisation Research Association (now the NCB's Coal
Utilisation Research Laboratory, CURL) in the period 1966 to 1968. Using
the data obtained, a prototype boiler, rated at 2·5 MW (3700 kg/h of steam)
was designed in collaboration with John Thompson Ltd (now NEI
Cochran), manufactured by them in 1969 and installed at BCURA.

The boiler shell was 2·4 m in diameter and 2·5 m high, containing a
combustion chamber 1·2 m diameter and 1·85 m high, and three passes of
convective tubes (Fig. 19). It was designed for a bed 0·75 m deep and three

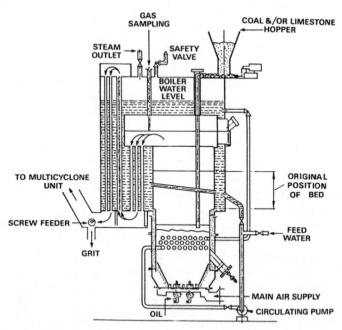

Fig. 19. Vertical-shell boiler at CURL, Leatherhead, UK. (Reproduced with
permission of BP and CURL.)

rows of tubes were installed across the bed, inclined at 10° for natural circulation. Crushed coal was fed pneumatically at four positions immediately above the distributor (the standard technique at that time) and start-up was by pre-mixed air/gas supplied to a central nozzle.

The boiler readily achieved the design output but, even with fines refiring, the combustion efficiency was only about 90 %. This was because the height of the combustion chamber was insufficient for the combustion of crushed coal at the design fluidizing velocity of 3·6 m/s. However, the excess air level was low in comparison with mechanical stokers, and the overall boiler efficiency was similar to that of stoker firing. A single test was carried out in which coal sized 12 mm to zero was fed above the bed and gave the improved combustion efficiency of 95 %. The development programme was suspended in March 1970, because of the adverse market for coal in industry, by which time the boiler had operated for 400 h.

Subsequently the boiler was modified to investigate fluidized combustion of oil with sulphur retention, under the sponsorship of British Petroleum.[11] A refractory-lined combustion chamber 1·1 m in diameter, with three rows of horizontal in-bed tubes, was positioned below the boiler and the original combustion chamber formed the freeboard; most of the inclined cross-tubes were removed. More recently, facilities for in-bed and above-bed coal feeding have been added. The fluidizing velocity is now usually in the range 1·8–2·7 m/s, with a static bed depth of 0·5–0·7 m, the bed material being sand or limestone sized 700–2000 μm. Some 4000 h of operation on oil has been achieved, including the development and demonstration of the climbing film oil nozzle (Section 5.6.4). The boiler has also been used to investigate the corrosion resistance of various metals for in-bed tubes.

7.2.2 National Coal Board Hot Water Boiler

This unit, the first prototype boiler to be designed for a fluidised bed burning commercial size grades of coal without crushing, was installed by the NCB at the CWS Market Garden site at Marden, Herefordshire, in 1977. It was manufactured as two separate components (Fig. 20), the shell unit containing the freeboard combustion chamber and four passes of convection tubes, and the bed unit with water jacket and tube bank.[22,31] The shell unit was designed and supplied by Clonsast Ltd. The bed unit, manufactured by G. P. Worsley Ltd, was assembled from four ring modules, two having a row of horizontal tubes, to facilitate modification of the in-bed tube position. The nominal rated output was 3 MW.

The internal diameter of the bed unit was 1·52 m, as was that of the combustion chamber which formed a freeboard of 0·9 m above the bed unit. Air was supplied by a forced draught fan capable of supplying 5000 standard m³/h of air against a pressure of 760 mm water gauge. Washed singles coal (12 to 25 mm) was fed via a variable speed screw feeder to a central drop tube through the top of the boiler. Elutriated solids from the bed were carried by the flue gas through the two convection passes and then through interconnecting ductwork to twin cyclones in parallel, one being used at low gas flow and both at high flow. From the cyclones, the flue gases passed to the chimney via an induced draught fan.

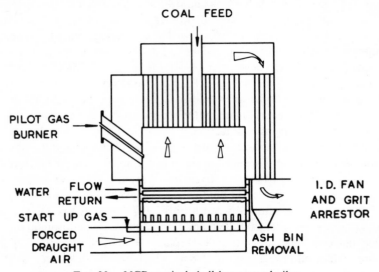

Fig. 20.   NCB vertical-shell hot water boiler.

As originally installed, start-up was by pre-mixed air and gas supplied through a drilled-hole sparge pipe distributor. The unit was commissioned during the summer of 1977 and operated through the 1977/78 winter, supplying hot water for site heating. During this mild winter the heat demand rarely exceeded 1·2 MW and consequently the boiler was considerably over-rated, giving rise to load-following problems. Also, some distortion of the sparge pipes occurred.

During the summer of 1978 the fluidized bed unit was re-designed to eliminate the use of sparge pipes and to make it more suitable for operation primarily at low output, while retaining the ability to obtain maximum output if necessary. As shown in Fig. 20 the modules with tubes were positioned at the top of the unit, to permit load control by varying the bed level. Start-up was by pre-mixed gas introduced through a grid to the stand pipes in the centre compartment of the plenum. Following these modifications, the boiler operated reliably throughout the 1978/79 winter.

Tests were carried out to assess the possibility of obtaining load control by the variation in bed level with fluidizing velocity.[22,31] With a static depth of 200 mm, the bed expanded to cover the lower tube row, but not the upper, at a fluidizing velocity of 1·7 m/s. This resulted in an excess air level of 25 % when the coal feed was controlled to maintain a bed temperature of 900 °C at this velocity. However, with this bed depth, there was insufficient variation in heat transfer to the tubes with changes in velocity to maintain a constant excess air level; the excess air level increased with increasing velocity. In contrast, with a static bed depth of 130 mm, the excess air level was approximately constant, albeit at a high value, demonstrating that the bed expansion with fluidizing velocity was well matched to the tube geometry.

Further major modifications were made to the installation for the 1979/80 winter in order to assess start-up by hot gas from an external burner supplied to the bed through a water-cooled distributor plate, together with output control by on–off operation of the bed. Other improvements included a new coal feed system, new high-efficiency cyclones and a fully-automatic control system. In this form the boiler operated reliably through the 1979/80, 1980/81 and 1981/82 heating seasons under NCB supervision. During this period, trials were carried out burning various commercial coal types to assess the effect of rank and size distribution on combustion performance.[31] The boiler is expected to continue in commercial operation, meeting site heating requirements.

7.2.3 National Coal Board Steam Boiler

This prototype vertical-shell steam boiler, designed for burning commercial size grades of coal, was installed by the NCB at Antler Ltd, Bury, Lancashire, in 1977.[22,31] The boiler, designed by Clonsast Ltd and manufactured by J. S. Forster Ltd, has a nominal rating of 2·3 MW. The fluidized bed unit (made by G. P. Worsley Ltd) is in the form of a flat plate distributor with vertical stand pipes (spaced on a 75 mm pitch) and is fitted directly to the base of the furnace tube.

Within the combustion chamber (1·34 m diameter) are ten thermosyphon cooling tubes (89 mm diameter) for in-bed heat transfer (Fig. 21). These tubes enter one side of the bed near the base and then turn vertically upwards to pass through the bed and freeboard to the overhead steam space. The combustion gases leaving the bed flow through two convection passes in series at the opposite side of the bed. These gases are cleaned by

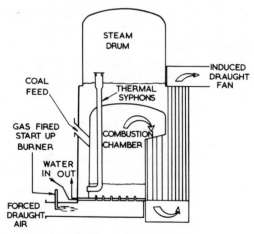

FIG. 21.   NCB vertical-shell steam boiler.

two 0·6 m diameter cyclones in parallel and pass to an ID fan at the base of the stack. Air is supplied to the bed by a forced draught fan rated for 4000 standard m³/h at 760 mm wg. Washed singles coal (12 to 25 mm) is fed to the combustion chamber via a screw feeder and an inclined tube situated in the same side as the thermosyphon tubes. Four 50 mm ID tubes provided in the distributor allow removal of bed material by gravity flow. One tube is connected to a vibrating screen system for separating large size accumulated ash, cleaned sand of the correct size being returned to the bed by a pneumatic system.

The original start-up method was by pre-mixed air and propane gas fed to the centre section of the distributor, each stand pipe in this section being fitted with an internal concentric gas tube fed from a separate gas header. Air was supplied to the whole of the bed at a rate just sufficient for fluidization and the gas mixture ignited by a 73 kW above-bed gas burner.

The unit was commissioned in September 1977 and successfully met the site heating requirements from January to April 1978 under manual control.[22] Start-up and load following were automated for the 1978/79 heating season and the boiler performed reliably during this period. Load following was by modulation of air supply in response to steam pressure and control of coal feed to maintain a constant bed temperature. On the basis of operation under manual control, this system was considered preferable to varying bed temperature, even though the disposition of heat transfer surfaces was not well matched to the bed expansion characteristics. The bed level was limited to 180 mm static in order to avoid sub-stoichiometric conditions at low output and excessive sand carry-over at high output, but this resulted in a high excess air level at high output. Consequently the output was restricted to 1·95 MW at the maximum fluidizing velocity of 3·1 m/s because the air could not be fully utilized for combustion.

By 1979 it was considered that start-up by pre-mixed air and hot gas was adequately developed and demonstrated and that, as at the CWS Marden site (Section 7.2.2), it was appropriate to test and demonstrate an alternative. Since start-up by supplying hot gas to the entire bed was being incorporated at Marden, it was decided to investigate a variant which had the objective of reducing the amount of premium fuel required. This involved passing hot gas through a sparge pipe distributor supplying part of the bed and then, having established combustion in this zone, gradually fluidizing the remainder of the bed using a flat plate/stand pipe distributor.[22] To incorporate this concept, the Antler boiler was modified by removing the stand pipes from about 25 % of the distributor and fitting four sparge pipes into this area. In view of previous experience of thermal distortion of sparge pipes with drilled holes, the sparge pipes were fitted with stand pipes in an attempt to minimize temperature gradients. However, these were short in order to accommodate the assembly within the height of the stand pipes on the base plate. Distortion of the sparge pipes again occurred, causing non-uniform depth of bed above the distributor and consequent mal-distribution of the air supply giving rise to hot spots where clinkers formed occasionally. Despite these problems, the boiler was operated to meet the site heating requirements for most of the 1979/80 heating season and the feasibility of start-up by supplying hot gas to part of the bed was demonstrated.

In order to ensure reliable operation during the 1980/81 heating season, a new bed unit was designed and installed, incorporating a water-cooled distributor for whole-bed hot gas start-up (Fig. 21). The start-up sequence and subsequent load following were controlled by a microprocessor. For load following, the boiler was operated at one of two preset outputs, 'high fire' and 'low fire', or the bed was slumped, according to the steam pressure. At each firing rate the air was maintained constant and the coal feed was switched between two preset rates, one above and the other below the required average rate to maintain a near-constant bed temperature. This system was intended to eliminate the occasional excessive coal feed rates with the previous system, which modulated the air and coal supplies independently. Once the various air and coal feed rates had been selected and sequences established for output change, slump and restart, the boiler met the site heating requirements with a high level of reliability. It is expected to continue to operate in this form on a commercial basis.

Two boilers based closely on the original installation at Antler Ltd were manufactured by Whites Combustion Ltd in 1979 and 1980. The rights to this design have recently been acquired by Robert Jenkins & Co. Ltd and one of these boilers is to be installed at their Rotherham (UK) factory as a demonstration unit. They intend to market vertical shell boilers under licence from the NCB. The second Whites boiler (manufactured by Worsley Fabrications Ltd, Manchester) has been installed at Peirpoint and Bryant's factory at Warrington. A near-identical boiler has also been manufactured in South Africa by World Energy Resources Consultancy Service (Pty) Ltd (WERCS) and was commissioned at Zululand in 1980; a similar unit has been supplied to a site in Natal. Two boilers of this design have also been manufactured in the UK by Allied Boilers Ltd.

7.2.4 Wallsend Slipway Engineers Ltd

Wallsend Slipway Engineers Ltd, in conjunction with the NCB, have designed a range of vertical shell boilers with outputs up to 4 MW. The first unit, a 3·8 MW hot water boiler[22] has been installed for a district heating scheme at Edmonton, London, and was commissioned during mid-1980.

The boiler is 4·8 m high and 3·4 m in diameter and incorporates a central combustion chamber 2·8 m high and 1·85 m in diameter (Fig. 22). In-bed cooling to transfer about 50% of the heat release is provided by thermosyphon tubes. The fluidized bed is supported on a stand pipe distributor base and air is fed to the distributor through a single plenum chamber. The central section of the base has no stand pipes and fluidizing air is introduced into this section via stainless steel sparge pipes. The bed material is sand, sized 620 to 1400 μm, and the operating temperature is nominally 900 °C. Coal (washed singles) enters the combustion chamber from two screw feeders via inclined tubes located above the bed and passing through the shell. The diameter of combustion chamber required for this output, 1·85 m, was close to the maximum which can be used. Consequently, it was not possible to provide a larger-diameter gas reversal chamber at the top, as in the CURL and Antler boilers. Instead, the combustion gases flow directly into the tubes of the first convection pass, which are connected horizontally to the chamber (Fig. 22). The combustion chamber is central and the three convection passes are located in the annular space between it and the shell. The gases are cleaned by twin cyclones in parallel prior to entry to the stack. There is no induced draught fan.

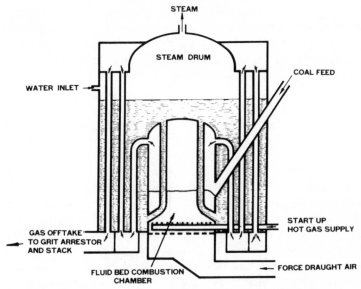

FIG. 22. Wallsend Slipway Engineers Ltd—4·3 MW boiler.

Start-up is provided by hot gas (at 800–850 °C) from a 350 kW external
oil burner. The hot gas is introduced to the initially static bed in the start-up
zone above the sparge pipes, and fluidization commences as the sand is
heated. Coal is supplied to the start-up zone at an initially low rate when the
bed temperature reaches 470 °C; at 700 °C the burner is switched off and the
remaining bed area is progressively fluidized. This system utilizing a burner
with a rating only 7 % of boiler output was designed following its use in the
Antler boiler (Section 7.2.3). During operation, the bed level (nominally
200 mm) is monitored and make-up sand is metered from an overhead
hopper through a rotary valve into the coal drop tube. The bed is cleaned
regularly (using a vibrating sieve system) in order to maintain a coarse ash
content of 5 % or below. Boiler control is effected by modulation of air flow
according to the outlet water temperature; the bed temperature is
maintained constant by adjustment of coal flow. By mid-1981 the boiler
had operated for some 1900 h. The start-up procedure for the partial bed
technique has been reliable and the improved design of sparge pipes has
overcome the problems of distortion encountered at the Antler site. The
results of a preliminary test programme indicated that the geometry of the
in-bed tubes was well matched to bed expansion, providing operation over
a 2/1 range at near-constant temperature and excess air level. The tests also

showed that the three passes of convective tubes provided excessive gas cooling and consequently the smoke boxes were rearranged to combine the second and third passes in parallel. This modification also reduced the gas pressure drop through the boiler and simplified the exhaust gas offtake.

7.2.5 EMS Thermplant Ltd

EMS Thermplant Ltd became involved in fluidized combustion through designing and installing the replacement bed units and other ancillaries for the NCB's boilers at CWS Marden, Antler Ltd and Rists Ltd (Section 7.3.2) and in providing staff for operating and maintenance at these sites. In collaboration with the NCB, they designed a range of hot water and steam boilers based on the unit at CWS Marden. Two of the 'Thermaster' steam boilers, rated at 2·88 MW, have been installed at the NCB's Tredomen Engineering Works, South Wales, and were commissioned in mid-1981.

The boilers, manufactured by Danks of Netherton have the combustion chamber at one side, a larger-diameter reversal section, and two passes of convective tubes (similar to Fig. 21). The bed units are similar to that at Marden (Section 7.2.2), with horizontal tubes and pumped water circulating through them. The bed system consists of a water-cooled plate with stand pipes, to facilitate start-up by hot gases from an external burner. The water flow (by forced circulation) is through the base plate to the in-bed tubes and then to the boiler shell. The bed material is sand sized 620 to 1400 μm and the static bed depth (nominally about 150 mm) is such as to cover the in-bed tubes when fluidized at maximum velocity. Bituminous coal, nominally sized 25 mm to 13 mm, is metered by a screw feeder and falls by gravity through a vertical tube at the top of the boiler. The gases are cleaned by a multi-cell collector before entry to the stack. Three bed drain ports are provided from the base plate, one of which is connected to a vibrating screen to separate oversize ash from the recycled sand.

The boilers met the site heating requirements throughout the 1981/82 winter, demonstrating reliable automatic operation. Load following was achieved by switching between preset high and low firing rates or slumping the bed, in response to steam pressure. This method of operation is analogous to the usual high fire/low fire/off control of an oil burner for a factory heating boiler.

In addition to firing the bituminous coal, trials were carried out using a local low volatile coal sized 13 mm to zero. This was fed to the bed using a novel reciprocating ram feeder. Because of the coal's relatively low reactivity, the elutriation of unburnt carbon was somewhat more than with bituminous coal. Also, it was necessary to heat the bed to the higher temperature of 600 °C before coal combustion was established.

7.3 Novel horizontal shell boilers

The vertical shell boilers, discussed in Section 7.2, provide an appropriate method for accommodating a vertical combustion chamber within a shell boiler in order to utilize the relatively low cost of shell construction. However, at the maximum combustion chamber diameter of about 1·8 m, the bed area available for combustion precludes outputs above 4 MW because of the excessive fluidizing velocity which would be necessary. Thus, for outputs above 4 MW, it is necessary to adopt a different method of construction. Most manufacturers have adopted a horizontal design and incorporated a vertical combustion chamber. Since a horizontal shell is essential for outputs above 10 MW, to provide sufficient water surface for steam release, a horizontal shell concept allows a standard design to be used from 4 MW upwards.

A similar requirement for a novel boiler design arose in the development of high pressure steam boilers for railway locomotives; the small-diameter shells to contain the high steam pressure did not have sufficient space to accommodate the combustion chamber. The solution adopted involved providing a separate combustion chamber adjacent to the shell, usually in the form of a rectangular box fabricated from steam-raising walls. This so-called 'locomotive boiler' concept has also been used for most of the novel fluidized bed shell boilers (Sections 7.3.1 to 7.3.3). A different approach is adopted in the Wallsend Slipway Engineers boiler (Section 7.3.4).

7.3.1 Johnston Locomotive-type Shell Boilers

Johnston Boiler Company, using technology provided by the National Coal Board and British Petroleum Ltd (licensed through Combustion Systems Ltd) have designed a range of locomotive-type fluidized bed shell boilers (1–15 MW capacity) for operation at steam pressures up to 20 bar. A prototype unit commenced operation at their factory in Michigan, USA, in September 1977 and, in addition to use as a test facility, provides 4600 kg/h steam at a pressure of 10 bar for space heating purposes.[32] The boiler incorporates a vertical rectangular combustion chamber and a shell containing three horizontal gas-tube passes (Fig. 23). The combustion chamber walls and roof are fabricated from flat plate, with stay bars to withstand boiler pressure. The bed is divided into three compartments by divisions formed by inclined tubes with connecting fins, although there is provision at the bottom for particle interchange to equalize bed

148   J. Highley and W. G. Kaye

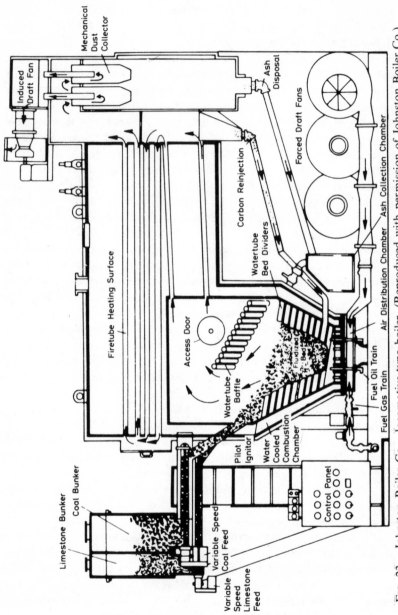

FIG. 23. Johnston Boiler Co.—locomotive-type boiler. (Reproduced with permission of Johnston Boiler Co.)

depths. Each bed compartment has its own air distributor, forced draught fan, pre-mixed gas start-up system and above-bed coal/limestone feeder. Oil firing is achieved by the use of climbing film nozzles (Section 5.6.4). There is an above-bed baffle, formed by inclined water-tubes, which increases the gas flow path between the bed and the first convection pass. Coarser entrained particles, which are disengaged in the gas reversal chamber after the first pass, are returned to the bed by a pneumatic system.

The bed material is silica sand (3·2 mm maximum size) or crushed limestone and the static bed depth is 500 mm. Excess bed material produced during coal firing or limestone addition is removed via overflow weirs (at the side of each bed) to a disposal system.

In the prototype unit, the fluidizing velocity is 1·8 m/s at maximum output and the bed temperature is 850 °C. For bituminous coals the combustion efficiency is 97 % at an excess air level of 20–25 %. Load following is achieved by variation of bed temperature and on–off cycling of the individual beds.

Commercial versions of the prototype boiler are currently in operation and incorporate some minor differences in design characteristics, such as additional in-bed tubes. One coal-fired unit in Ohio generates 18 200 kg/h steam at 13·8 bar. The bed temperature is 850 °C and a combustion efficiency of 95 % is achieved with refiring of carryover material into the centre bed. Three 7 MW units (generating 11 400 kg/h steam at 13·8 bar) have been successfully used in Brazil for burning coal of 52 % ash content and calorific value 12 500 kJ/kg. The removal of defluidized coarse solids from this system necessitated the design of a conical distributor terminating in a large bore tube at the base. The hot bed material is removed continuously by means of a water cooled screw conveyor. Orders for about twenty units had been received by mid-1981. These included seven boilers, each producing 22 300 kg/h steam (15 MW), supplied to Campbell Soup Co. and commissioned in 1982.

7.3.2 National Coal Board Locomotive-type Boiler

To demonstrate the suitability of locomotive-type design in the UK, the National Coal Board commissioned Clonsast Ltd to design and supply a 10 MW unit, which was installed in 1978 at Rists Wires and Cables Ltd, Newcastle under Lyme, a subsidiary of Joseph Lucas Ltd. The design selected incorporates a central combustion chamber with twin convective sections, giving a 'double-ended' boiler (Fig. 24).[22,31] The boiler, manufactured by J. S. Forster Ltd, produces saturated steam at 17 bar. The bed unit and start-up system were designed and supplied by Lucas

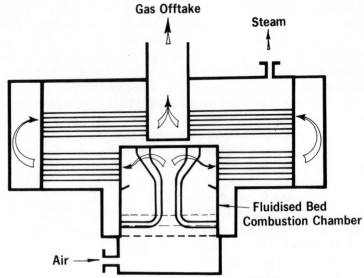

F<small>IG</small>. 24.   NCB locomotive-type boiler.

Aerospace Ltd and other ancillaries were designed and supplied by EMS Thermplant Ltd.

The combustion chamber walls are water-cooled flat plate, with stay bars, and additional in-bed heat transfer is provided by banks of thermosyphon tubes which project horizontally into the bed and turn vertically upwards to enter the steam space at the top centre of the shell. The combustion gases leaving the chamber divide and flow through the horizontal gas tubes in the connecting shell sections before recombining in a single flue gas duct.

The fluidized bed area is $6 \cdot 27 \, m^2$ and the fluidizing air enters from vertical capped stand pipes mounted on a distributor plate. The bed material is sand, sized 620 to 1400 μm, and the static bed depth is nominally 190 mm. Coal (washed singles or smalls) is admitted above the bed from two rotary feeders via inclined injector tubes; alternative systems to reduce the number of feed points are being investigated.

As originally installed, start-up was by hot gas from two oil burners, supplied to the plenum. As an alternative to the usual water-cooled plate, it was decided to use an uncooled plate of stainless steel, segmented to accommodate thermal expansion. However, it was found that distortion of these plates occurred, probably because of non-uniform gas temperature across the gas duct from the burners. An interim solution was to reduce the gas temperature, extending the start-up time considerably. A second unsatisfactory feature of the initial design was that, although the total amount of in-bed heat transfer surface was correct, there was too much surface in the lower part of the bed. This precluded operation at low output since a low excess air level and operating temperature were required for a heat balance, and this condition gave rise to smoke emission. To improve both start-up and turn-down, a refractory-lined spacer was installed between the boiler and the distributor plate, increasing the distance between the plate and the lower tubes. In this form the boiler performed reliably through the 1980/81 heating season, operating about 108 h/week. Typically the boiler efficiency was 80% at an excess air level of about 30%. A turn-down of 2·4/1 was achieved by a combination of: (i) bed expansion into the tube bank; and (ii) controlling the bed temperature within the range 800 to 950 °C.

During 1981 the fluidized bed system was almost entirely replaced in order to incorporate improvements in design and control developed at other NCB sites. The new start-up system consists of an in-duct oil-fired burner, which pre-heats the fluidizing air to 900 °C. The plenum chamber is lined with ceramic fibre and the distributor is water cooled and insulated on the underside. Start-up and load-following are fully automatic and based on preset air flow rates and corresponding fuel feed rates. The cyclone dust collectors have been replaced by a bag filter unit; although not essential at this site, the use of a bag filter will provide experience relevant to larger boilers. The installation was successfully re-commissioned in November 1981 and it operated through the 1981/82 winter with excellent reliability. It is expected to continue in commercial operation.

7.3.3 Babcock Composite Boiler

The Shell Boiler Division of Babcock Power Ltd has designed a fluid bed boiler which combines a combustion chamber of water-tube construction with a shell-type convection section in locomotive-type configuration. The objective is to provide a design for boilers in the range 6 to 30 MW, with pressures of 30 to 40 bar, without the cost of full water-tube construction.

A 4·3 MW unit (providing 7000 kg/h steam) was installed at the NCB's Coal Research Establishment in 1980 for test purposes and, during subsequent commissioning, successfully met the design output. The fluidized bed unit incorporates vertical water-tube membrane walls with inclined in-bed tubes for natural circulation. The off-gases from the combustion chamber pass directly into the convection banks of a two-pass horizontal shell (Fig. 25).

The bed area is 3·4 m² and the fluidizing air enters through stand pipes

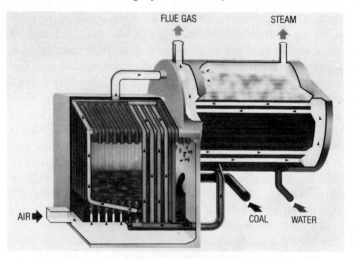

FIG. 25.   Babcock Power Ltd—composite boiler.

mounted on the distributor plate. The bed material is sand sized 500 to 1000 μm, and the static bed depth is nominally 150 mm. Coal (washed singles or smalls) is introduced onto the bed via two inclined drop tubes (150 mm diameter) each located near a corner of one wall of the combustion chamber and at a height of 450 mm above the stand pipe nozzles.

Start-up is provided by a 2·3 MW above-bed natural-gas burner directed onto the bed surface. The in-bed tubes immediately below the burner are fully immersed in the bed even when it is not fluidized in order to protect them from flame impingement. A supplementary 0·6 MW gas burner in the duct to the air distributor is used to preheat the fluidizing air to 400 °C, avoiding the expense of a water-cooled distributor associated with a higher gas temperature. The boiler can reach maximum steam output firing coal in approximately 60 min from a cold start.

Four outlet ports in the bed plate are provided for the removal of bed material. A vibrating sieve system separates oversize ash and facilities are provided for returning correctly graded material to the bed. The flue gases are cleaned by a multi-cell grit arrester.

The boiler is designed to operate under balanced draught conditions at 30 % excess air and a fluidizing velocity of 2·3 m/s at maximum output. A turn-down of 3/1 is anticipated by a combination of modulating the air and coal flow rates to vary the amount of tube area immersed and allowing the bed temperature to 'float' between 750 and 950 °C. The boiler is to be controlled by a microprocessor digital control unit which switches between four pre-set outputs within this range. Outputs below 33 % rating will be obtained by on–off operation. In order to reduce the frequency of switching on and off when operating in this latter mode, and also to accommodate rapid increases in steam supply found in some boiler applications, some thermal storage has been incorporated into the design by controlling the water level in the shell between two working levels.

The 4·3 MW unit will be moved to a commercial site near Basingstoke in 1983. Meanwhile, a range of composite water-tube/shell boilers with outputs up to 30 MW is being designed using the operating experience and data from the prototype. Orders have been received for a 22 tonne/h (15 MW) boiler for Capper Pass Ltd, Hull, and a 13 tonne/h (9 MW) boiler for Ciba Geigy, Cambridge, for commissioning in 1983. For boilers to provide outputs above 30 MW and to operate at higher pressures, it will be possible to utilize a similar combustion chamber integrated into a fully water-tube boiler.

7.3.4 *Wallsend Slipway Engineers Horizontal Shell Boiler*
Wallsend Slipway Engineers Ltd have designed a range of horizontal shell boilers for outputs up to 17 MW at 17 bar pressure. In collaboration with the National Coal Board, a 5 MW capacity unit has been installed at the Coal Research Establishment for test purposes and was commissioned in September 1980.[31] As shown in Fig. 26, the circular fire-tube of a

FIG. 26.   Wallsend Slipway Engineers Ltd—horizontal shell boiler. (Reproduced with permission of Wallsend Slipway Engineers Ltd.)

conventional boiler is extended by vertical walls (appropriately stayed) through to the bottom of the boiler. This feature increases the height of the combustion chamber and allows the plenum and distributor to be positioned below the boiler. In addition, vertical gas flow through the combustion chamber is achieved by incorporating multiple outlet ports, connected to two large-diameter ducts leading to the first reversal chamber.

The 5 MW boiler has an overall length of 3·1 m, an overall width of 3·1 m and an overall height of 4·1 m. The fluidized bed is 2·5 m by 1·5 m and the combustion chamber contains in-bed heat transfer surface in the form of 36 inclined thermosyphon tubes (76 mm OD) arranged in three rows entering each side of the chamber (Fig. 26). Water flow is by natural circulation. The fluidizing air enters the bed through vertical capped stand pipes mounted at 85 mm spacing on a water-cooled distributor plate. The bed material is sand, sized 620 to 1400 μm, and the static bed depth is nominally 170 mm. Coal (washed singles) is metered by a variable speed screw to a drop-tube at the top centre of the boiler shell. The flue gases enter the two large diameter ducts and subsequently flow through two conventional convection tube passes before leaving the boiler at about 250 °C. Gas cleaning is provided by two cyclones in series.

At maximum output the boiler is designed to operate at 900 °C bed temperature with 30 % excess air and a fluidization velocity of 2·45 m/s. Load control (3/1 turn-down) is achieved by the variation in bed expansion and splashing into the thermosyphon tube bank as the fluidizing velocity is changed.

Start-up is by hot gas supplied to the entire bed area. About 40 % of the combustion air needed for maximum output is heated to above 900 °C by a 1 MW capacity in-duct natural gas burner and passes through the insulated plenum and water-cooled distributor plate. This is capable of raising the bed temperature to about 500 °C within 25 min from cold. Coal is admitted at an initially low rate and the burner is extinguished when the bed temperature reaches about 700 °C. The coal feed rate is subsequently increased and the boiler can be stabilized at about 850 °C bed temperature.

By the end of 1981 the boiler had operated for over 1600 h. Start-up and operation had become routine and performance tests confirmed that the design was correct. In particular, the excess air level was near-constant over the turn-down range, indicating that the in-bed tube geometry is well matched to the bed expansion characteristics. A microprocessor control system was being developed and facilities for removing over-size bed material had been added.

It is expected that the prototype boiler will be moved to a commercial site during 1983. Following the successful operation of the prototype, two orders have been received. A 1·5 MW boiler at Sheffield was commissioned in 1982 and two 7 MW boilers for Brymbo Steel, North Wales, are scheduled for commissioning in 1983.

8 FLUIDIZED BED FURNACES

8.1 The development of fluidized bed |furnaces

Furnaces and kilns in the process industries consume substantial quantities of fossil fuel, predominantly oil and gas. As the price of these fuels increases, reflecting their diminishing availability, there will be an incentive to convert to coal. Wherever possible it will be most efficient to burn the coal directly, and it is likely to be worthwhile to consider modifying the process to facilitate this. The alternative will be to produce a clean, low calorific value gas from the coal, with the attendant energy loss in conversion. As a first stage towards increasing the use of coal in process furnaces in the UK, the National Coal Board has collaborated with G.P. Worsley & Co. Ltd (GPW) in the development of fluidized bed furnaces to supply hot gases for direct contact drying plants.

The furnace development programme was initiated by the NCB in 1974.[28,33] Market potential for coal-fired hot gas furnaces was identified in direct contact drying of many products in which a small degree of contamination by ash is acceptable, and this relatively simple application has enabled the furnace system to be developed and demonstrated commercially. The first product dried by a coal-fired fluidized bed furnace has been grass in the production of cattle feed pellets. This application proved to be particularly suitable for the development of the furnaces because, being seasonal, it was convenient to make furnace modifications during the winter shut-down. Also, during the eight month operating season there were both extended periods of continuous operation and periods of operation on a daily basis, giving the opportunity to demonstrate the reliability of the furnace and start-up system under a variety of load conditions.

Following the demonstration of reliability by the successful operation of five 5 MW furnaces at grass drying plants, installed in the period 1976 to 1978 (Section 8.3.1), units have been offered commercially by GPW for continuous industrial drying processes. The first industrial unit was a 15 MW furnace supplying gas at 1000 °C for clay drying in cement manufacture, installed in 1979. A second industrial unit of 7·5 MW was installed in 1981 for drying stone at a quarry (Section 8.3.2).

Another UK company, Energy Equipment (EE) Ltd, has also developed a fluidized bed hot gas furnace (Section 8.4). The first two units (11 MW and 7 MW) have been installed at grass drying plants in the UK, and a 22 MW unit has been installed at a grass drier in France. In these furnaces the beds are operated as partial gasifiers having substantial above-bed combustion with secondary air, as in the company's combustion system for boilers described in Section 7.1.3.

In order to extend the application of fluidized bed furnace to the majority of industrial drying processes it is necessary to develop units which produce clean heated air. For temperatures up to about 150 °C, this can be achieved by using a shell boiler and steam-to-air heat exchanger. In principle, a similar system can be used with thermal fluids to produce somewhat higher air temperatures (it is likely that the capital cost of a thermal fluid heater will be less than that of a water-tube boiler for high temperature steam) but these are yet to be demonstrated. For higher air temperatures, up to about 600 °C, fluidized bed furnaces with in-bed air-heating tubes are being developed (Section 8.5). These can also be used as an alternative to a boiler with a steam-to-air heat exchanger for lower air temperatures. A further design concept, giving an almost-zero dust content, is to heat the air by passing it through a fluidized bed of inert material which, in turn, is heated by the combustion bed (Section 8.5.3).

8.2 Furnace design considerations

A schematic diagram of a fluidized combustion furnace supplying hot combustion gas to a drier is shown in Fig. 27. The fluidized bed is normally operated at a temperature of 950 °C and combustion above the bed increases the off-gas temperature to about 1000 °C. Gas at this temperature is suitable for drying many products which are not damaged by exposure to high temperatures and for these it is standard practice to use inlet drier temperatures up to the maximum compatible with materials of construction, usually about 1000 °C. If a lower gas temperature is required, the hot gas from the furnace can be diluted with cold air or recycled drier exhaust gas. An alternative method of producing lower temperature gas is to operate the fluidized bed at a lower temperature, but this is limited to a minimum of approximately 800 °C by the need to ensure efficient combustion. In the application to grass drying a gas temperature of 1000 °C can be used with wet grass but temperatures down to 500 °C are used after extended periods of dry weather.

In the fluidized bed furnace virtually all of the heat generated by combustion is utilized in heating the fluidizing air (and the coal) to produce

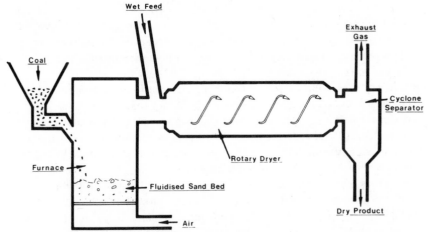

FIG. 27. Rotary drier with fluidized bed hot gas furnace.

hot gas, and the furnace is operated with approximately 150 % excess air to maintain the bed temperature at 950 °C. In principle the excess air level can be reduced to improve overall plant efficiency, by recycling exhaust gas through the bed in substitution for the air not utilized for combustion. Provision for this has been incorporated in some of the furnaces.

Most high temperature driers are designed so that the exhaust gas and dried product are at an equilibrium temperature of about 120 °C, and the input rates of hot gas and wet product are controlled to maintain this exhaust temperature. With fluidized bed hot gas furnaces it is preferable to operate with a constant gas flow rate and inlet temperature and to control the drier by automatically modulating the wet product feed rate to maintain the required exhaust temperature. All of the present GPW furnaces are operated in this manner.

For fluidized bed furnaces to be commercially acceptable they must be reliable in operation and easy to maintain. These requirements dictate that the design of the furnace and its start-up and output control system should be as simple as possible, and preferably make use of standard components familiar to operators and maintenance engineers. In particular, the selection of the start-up technique is a major design consideration because the requirement for an auxiliary start-up fuel and burner system involves a significant penalty in the cost of a fluidized bed coal furnace relative to oil-fired furnaces and introduces novel automatic control features. In two NCB prototype furnaces start-up was by above-bed oil burner; subsequent GPW and EE units have incorporated start-up by passing hot gas, from an external burner, through the bed. This change to hot gas start-up was made to ensure reliable fully automatic start-up with a simple control system, and to reduce auxiliary fuel requirements, but it has necessitated the development of an air distributor which is also suitable for high temperature gas during start-up.

Because less than half of the fluidizing gas is utilized for combustion, the coal burning rate per unit bed area is about half that in a bed with cooling surfaces, operated at the same velocity. Because all of the heat is removed by the gases, load control is simply a matter of varying the air supply and automatically controlling the coal feed rate to maintain the required bed temperature. The ultimate limitation on turn-down is in the range of velocities over which the bed can be operated with sufficient turbulence to ensure good mixing without excessive elutriation. For the beds of fairly wide size distribution used in combustion a velocity range (and turn-down range) of 2/1 is usually the maximum which can be achieved.

8.3 G.P. Worsley furnaces

8.3.1 Furnaces for Grass Drying

Following the operation of an NCB 6 MW prototype furnace of Mark I design,[28,33] GPW have manufactured two Mark II furnaces and three Mark III furnaces (Fig. 28).[33,34] All of these consist of two modules, each with an output of 2·5 MW as hot gas at 950 °C. The furnace modules are refractory lined, with bed dimensions of 2·3 m by 1·3 m and a height of 2·75 m. Start-up is by hot gas supplied from an oil-fired combustion chamber for each bed. In the Mark II design the two bed modules and two

FIG. 28.   G.P. Worsley Ltd—7·5 MW hot gas furnace.

burner modules are separate units, but in the Mark III design they are contained within a single integrated unit. The air distributor comprises multiple sparge pipes, designed to accommodate the thermal expansion when supplying hot gas during start-up. The Mark II furnaces initially had simple drilled-hole pipes, but these distorted, primarily as a result of non-uniform heating from above when the combustor had to be maintained on stand-by with a hot bed. To overcome this, sparge pipes with vertical stand pipes are now used in order to ensure that the horizontal pipes are insulated by static bed material.

The fluidized bed material is silica sand, normally sized 1 to 2 mm, with static bed depth of 150 mm above the stand pipe caps. The coal is washed bituminous singles (13 to 25 mm) or smalls (13 mm to zero), metered by rotary valves and fed to the bed surface by gravity down a chute. Typical steady-state operating parameters are as follows:

Fluidizing velocity	3 m/s
Bed temperature	900 °C
Coal burning rate	112 kg/h m^2
Combination efficiency	98 to 99 %
Excess air level	160 %

By the end of 1981 the five furnaces had operated for a combined total of more than 28 000 h. Based on this UK experience, G. P. Worsley supplied a 6 MW furnace for lucerne drying in Hungary in 1981.

8.3.2 Industrial Furnaces

Following the successful operation of the 5 MW furnaces on a seasonal basis, Mark IV furnaces were offered commercially for continuous process applications in 1979. The first unit supplied for industrial process drying was a 15 MW furnace installed at the Ketton Portland Cement Company.[33,34] The furnace, which supplies hot gas at 1000 °C to a clay drier/crusher, consists of two modules each with a bed area of 8·2 m^2. The hot gases flow from the bed modules to a central junction box, which acts as the ash drop-out zone, and through a refractory lined 2·2 m diameter duct to the plant. The bed units are elevated above ground level to facilitate drainage of bed material. Each module has 13 large diameter sparge pipes, each with four rows of stand pipes on a 75-mm-square pitch.

The fluidizing air for each bed is supplied by two forced draught centrifugal fans, one fan providing combustion air through the start-up burner and the other providing additional fluidizing air. The amount of air supplied by the second fan is regulated by a damper remotely operated from the control panel.

Each module has two hydraulically operated rotary coal valves suitable for singles or smalls grade coal. At high output the coal burning rate in each module is approximately 1 tonne/h.

The furnace was commissioned in September 1979 and handed over to the cement plant operators. It has since been operated in conjunction with the plant, usually on six days per week with over-night shut-down. The furnace performance has proved satisfactory, with the rated output readily achieved and gas temperatures up to 1080 °C produced when burning untreated smalls grade coal. Since commissioning, the furnace has successfully met the process heat requirements on almost every working day; the total operating time to the end of 1981 was over 3000 h.

In 1980, G. P. Worsley received orders for a further two industrial hot gas furnaces. One was for a twin module 10·8 MW unit for a chalk drier in Denmark, to burn coal with 20 % ash. The other was for a single 7·5 MW unit for a roadstone drying plant at Amey Roadstone Corporation's Blodwel Quarry, Shropshire. These were followed by an order in 1981 for a twin module 10·8 MW furnace for Ruhrgebeit, West Germany, for coal drying.

The design of these furnaces is significantly different to that of the Mark IV in order to improve refractory and air distributor integrity. The 7·5 MW furnace is shown in Fig. 28. Fluidizing air is supplied by two forced draught fans to a cylindrical steel plenum chamber beneath an air distributor of sparge pipes with bubble cap type stand pipes. The ceramic fibre lined plenum chamber has an air cooled extension at one end to house the start-up oil burner. The furnace casing is refractory lined and coal is delivered to the fluidized bed via two rotary valves or screw feeders. Start-up is fully automatic and the controller maintains the hot gas temperature at the required value. Hot gas from the furnace passes through a refractory lined connection chamber into the drier drum. The drier is operated with an induced draught fan and additional ambient air can be admitted at entry if temperatures below 750 °C are required.

The two 10·8 MW twin bed units and the 7·5 MW furnace were commissioned in 1981 and have operated satisfactorily, fully meeting the drying plant requirements. A second 10·8 MW furnace was supplied to Ruhrgebeit in 1982 and a second furnace, rated at 12·3 MW, has been supplied to Amey Roadstone. Also in 1982, Worsley supplied a 6 MW furnace to Cardiff and a 1·8 MW furnace to a site in Lancashire, both for coal drying.

8.4 Energy Equipment Ltd

The EE firing system for boilers, described in Section 7.1.3, is effectively a hot gas producer and EE are offering it for use in this application. The operating principle is the same as for EE boilers in that the bed is fluidized by a mixture of air and recycled exhaust gas and secondary air is supplied above the bed to enhance freeboard combustion.[35] This system facilitates the production of hot gas at temperatures somewhat in excess of 1200 °C, coupled with a low excess air level to minimize stack heat losses. However, a second fan is required to recycle the warm dust-laden exhaust gas, incurring both capital and operating cost penalties.

Following the demonstration of the EE system by the conversion of the 10 MW boiler, they supplied an 8 MW hot gas furnace to Dengie Crop Driers, Southminster, Essex.[35] This was commissioned in April 1980 and, once initial control problems were overcome, it operated satisfactorily during the 1981 drying season.

In May 1981 EE commissioned a 16 MW furnace for a grass drying plant at Suippe, France. The main fuel at this site is wallpaper waste from a local factory and the furnace design has proved to be appropriate for this. It has also successfully burned smalls coal, waste derived fuel and a mixture of straw and coal in the ratio 3/1. Further commercial orders are to be expected for similar EE furnaces for both agricultural and industrial drying in the UK, France and elsewhere.

8.5 Production of clean hot gas

Many products are at present dried by direct contact with the combustion gases from burning natural gas and distillate oil and substitution of dust-laden gases from coal would be unacceptable because of product contamination. Examples are detergents, dried milk, instant coffee, sugar, pigments, pottery, fertilizers and plasterboard. Consequently, coal can only be utilized by producing clean heated air for direct contact with the product. Inevitably, the capital cost of such systems will be substantially more than that of the hot gas furnaces described in the previous sections, but it is expected that their use will be justified because of the high price differential between coal and natural gas or distillate oil.

The only technique which allows the production of completely dust-free heated air is to pass the air through tubes immersed in the fluidized bed, with above-bed heat exchangers to recover heat from the off-gas. This concept is being developed in the USA by Fluidyne Engineering Corporation (Section 8.5.1) and in the UK by Encomech Engineering Services Ltd in collaboration with the NCB (Section 8.5.2 and Fig. 28).

However, an alternative which produces heated air that is almost dust-free is to pass the air through a fluidized bed of inert material, separated from the combustion bed by a division wall through which heat transfer occurs (Section 8.5.3). Also it would be possible to produce heated air with a low dust content by circulating the bed material between a combustion bed and an air-heating bed.

8.5.1 Fluidyne Air Heater

Fluidyne initiated design studies for FBC air heaters with in-bed tubes in 1971. These studies led to the construction of two test units at their facility near Minneapolis, Minnesota, both having a bed area of $0·21\,m^2$. These units have been used to obtain data on combustion efficiency, sulphur retention and heat transfer and also to perform corrosion tests on candidate in-bed heat exchanger tube materials.[36] Part of this work has been sponsored by the US Department of Energy.

The experience gained through operating these units for several thousand hours has been applied in the design of a larger pilot plant with a bed dimension of $1·6\,m$ by $1·0\,m$. This represents a 'vertical slice' of a full-scale industrial air heater. It was commissioned in April 1977 and by mid-1980 had operated for over $2500\,h$, including a continuous $500\,h$ test and over 150 start-ups and stops. Fluidizing velocity is in the range $0·9$ to $2·1\,m/s$, temperature 790 to $900\,°C$ and excess air level between 15 and $40\,\%$. Recycle of fly ash to the bed improved sorbent utilization and also increased combustion efficiency for fuels of low reactivity. Start-up is by pre-heated air at $760\,°C$ and takes about $3\,h$.

The in-bed tube bank is designed to produce heated air at $425\,°C$ and consists of straight 'bayonet' tubes of type 304 stainless steel. The maximum design tube wall temperature is $700\,°C$ and the bed temperature is controlled automatically to ensure that this metal temperature is not exceeded. The bayonet design was selected to accommodate thermal expansion without a sophisticated tube support system. After $2500\,h$ of operation they were straight and in excellent condition, with no visual evidence of corrosion or erosion.

It is understood that the information obtained from these pilot plants is to be used in designing air heaters for a variety of applications and fuels.

8.5.2 Encomech Air Heater

Encomech Engineering Services Ltd carried out design studies for a FBC air heater in 1978 and this led to the manufacture and testing of a $200\,kW$ pilot plant in collaboration with the NCB.[37] As in the Fluidyne concept,

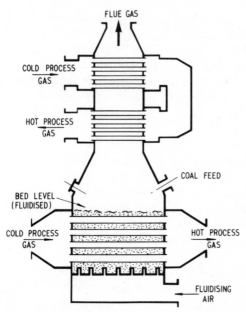

FIG. 29. Encomech air heater with ceramic tubes.

air is heated by passing through tubes immersed in the bed (Fig. 29), but instead of using tubes of stainless steel, Encomech use ceramic tubes. The design of the in-bed tube bank, and particularly the sealing arrangement which accommodates thermal expansion, is based on ceramic tube heat exchangers which Encomech manufacture for industrial applications involving dust laden, corrosive gases at temperatures up to 1350 °C. Ceramic recuperators are used to recover heat from the off-gas leaving the bed.

Because of the relatively low heat transfer coefficient between a tube and the air passing through it, the heat flux is considerably lower than that to a steam-raising tube in a boiler. Many more tubes are required for a given heat output, and a substantially deeper bed would be necessary if the fluidizing velocity was the same as for a boiler, giving rise to a higher pressure drop and the need for a more powerful fan. Consequently, it is economic to design air heaters for a lower fluidizing velocity than is used in boilers; the increased cost of the distributor and containment is more than offset by savings in electricity consumption.

The 200 kW pilot plant was installed at the NCB's Coal Research Establishment in 1980 and preliminary investigations showed that ceramic tubes of an appropriate material were able to withstand both steady operation and transient conditions without cracking. Combustion efficiencies up to 98 % were obtained without fines recycling when feeding coal sized 13 mm to 25 mm to the bed surface, using a screw feeder. Despite the close packing of tubes in the bed there were no significant vertical or lateral temperature gradients. Subsequent operation has been concerned primarily with obtaining heat transfer data and also demonstrating long-term mechanical stability of the tubes. A 1000 h endurance test as well as over 40 rapid starts from cold have been completed without any problems. Air temperatures of up to 600 °C have been obtained.

The first commercial air heater of this design was manufactured in 1982 and installed at a maltings. It has an output of 1·8 MW as air at 200 °C. This is diluted with recycled air to provide a temperature in the range 70 to 105 °C. A range of designs for outputs up to 7 MW has been prepared and further orders are expected for commissioning in 1983.

8.5.3 Designs with a Secondary Heat Transfer Bed

Applied Combustion Systems, Rangiora, New Zealand, have developed a design of FBC air heater based on research carried out at the University of Canterbury, New Zealand, in the period 1971–77.[38] Their first commercial unit commissioned in 1980, is rated at 5·5 MW and produces clean air at 300 °C from the combustion of dirty wet wood bark. As shown in Fig. 30, the unit consists of a central combustion bed and an annular bed through which air is passed to be heated. The division between the two beds, manufactured of stainless steel, is corrugated to increase the area for heat transfer and also for strengthening. About 1·2 MW is transferred across this division to the air passing through the annular bed and a further 0·9 MW is transferred across the division in the freeboard above the beds. The hot combustion gases from this unit are passed to a conventional air heat exchanger, where a further 3·4 MW is recovered.

The air distributor for the combustor consists of multiple sparge pipes with drilled holes, or in some cases bubble caps. To compensate for uneven distribution along the pipes, alternate pipes are fed from opposite ends. Start-up is by two 750 kW diesel oil burners firing into the bed and this takes 3 h from cold. However, it has been found that the slumped bed will retain enough heat to allow re-start without auxiliary fuel for 16 h. After several hours of operation it was found that large dense particles, such as stones, were segregating and causing uneven air distribution. Also, clay particles accumulated in the bed causing an increase in minimum fluidizing velocity and changing the fluidizing properties. Both of these problems

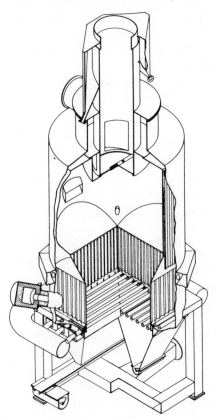

Fig. 30.   Air heater with annular heat transfer bed. (Reproduced with permission of Applied Combustion Systems.)

were overcome by drawing bed material down past the sparge pipe distributor, through a twin hopper arrangement and screw feeders, to a screening system. Material of the correct size is returned to the bed. This 'bed reconditioning' system is operated automatically on a time clock.

Preliminary operating data from the 5·5 MW unit indicate that its performance is close to the design parameters, even though the scale-up from the prototype was about 100/1.

In an independent development in Yugoslavia, a pilot-scale (1 MW) air heater of similar design has been manufactured by CER, Čačak. This unit has been designed on the basis of data from pilot plants at the Boris Kidric Institute, Belgrade,[39] and has a combustion bed 1 m square, with secondary beds on three sides. It is initially being developed to provide hot air for drying maize, using the cobs as fuel, but the design is also suitable for industrial use burning lignite. The prototype was commissioned at the manufacturer's factory in 1981 and a 3 MW commercial version is expected to be completed in 1982.

A disadvantage of these designs with completely separated beds for combustion and heat transfer is that a relatively deep bed is necessary to immerse sufficient area of the division wall for the required heat exchange. An alternative method of transferring heat between the beds would be to have interchange of the bed material at a controlled rate. A prototype to investigate this possibility is being manufactured by the NCB. However, it is inevitable that the heated air will be slightly contaminated by fine ash and combustion gases, somewhat restricting the product range with which the design could be utilized.

8.6 Process heating applications
Within the process industries there are very many applications where, at present, heat is supplied by an oil or gas burner which is an integral part of the unit design. Examples include tubular heaters for oil at refineries, calcination vessels and lime kilns. In principle, many of these applications could be converted to coal by the installation of a gasifier to produce a low-calorific-value fuel gas. However, gasification introduces a loss in efficiency of coal utilization, particularly if the fuel gas is cooled for cleaning. Also, the installation cost is high, the process is more complex than combustion, and there is a residue of unconverted carbon which must be utilized (e.g. in a FBC boiler). There is therefore an incentive to modify the process so that coal can be burned directly. In some cases it will be appropriate to fire pulverized coal using modified burners, accepting the cost penalty of the pulverizing plant installation, maintenance and power consumption. The alternative will be to modify the process to facilitate the use of FBC. However, it is likely that this approach will usually require a completely new design of plant and therefore it will not be followed until a new or replacement plant is required.

In the case of processes requiring heat at up to about 300 °C, an option is to use an organic thermal fluid instead of direct firing. Those processes which already use a thermal fluid would be converted to coal by installing a FBC thermal fluid 'boiler'. In addition it is to be expected that several processes, which are at present direct-fired by oil or gas, will be changed to use a thermal fluid heated by coal. However, although there has been some discussion on the potential for FBC thermal fluid units by potential suppliers and users, little progress has so far been made. The main concern appears to be the risk of coking of the fluid in the boiler tubes and it will be essential to ensure that this is avoided both under normal operation and fault conditions.

In order to exploit fully the advantage of FBC in a novel heater design it will be necessary to utilize the high heat transfer coefficient to immersed surfaces. Where the requirement is to heat a gas or a liquid, this can be done by immersing tubes in the bed and adopting the design principles used in boilers or clean air heaters. This approach has been followed by Exxon in their design studies for refinery oil heaters. An alternative possibility for gas heating would be to incorporate a secondary bed fluidized by the gas, separated from the combustion bed by a division wall for heat transfer, similar to the air heater design of Section 8.5.3 and Fig. 30. A similar secondary bed can also be used to heat solids which can be processed in particulate form. To demonstrate this concept, the NCB, G.P. Worsley & Co. Ltd and British Plaster Board Ltd have collaborated to design, manufacture and operate a prototype gypsum calciner with a central FBC combustor and a surrounding fluidized bed of powdered gypsum. The 5 tonne/h (0·9 MW) prototype was commissioned in 1980 and, following performance trials, began continuous production in mid-1981. The prototype has demonstrated high efficiency and good reliability and a full-scale 20 tonne/h calciner is being designed. It is expected that FBC units of similar design will find application in other industries in which a powder-heating process is involved.

ACKNOWLEDGEMENTS

The authors wish to acknowledge the co-operation of the companies involved in industrial fluidized bed combustion in providing up-to-date information on their developments. Recognition is made of their colleagues at the Coal Research Establishment, both for their efforts in progressing the commercial application of fluidized combustion, which are reflected in many of the projects described, and for assistance in preparing this chapter. Special thanks are due to Dr T. J. Peirce, who made valuable editorial suggestions and to Mr E. C. Smith, who prepared most of the diagrams and graphs.

The chapter is published with the approval of the National Coal Board but the views expressed are the authors' own and not necessarily those of the NCB.

REFERENCES

1. *Fluidised Combustion: Systems and Applications*, Inst. of Energy Symp. Ser. No. 4, London, 1980.
2. *Proc. of the Sixth International Conf. on Fluidized Bed Combustion*, Vols. I, II, III, US National Technical Information Service, 1980.
3. DRYDEN, I. G. C. (ed.), *The Efficient Use of Energy*, IPC Science and Technology Press, Guildford, 1975.
4. LOWREY, H. H. (ed.), *Chemistry of Coal Utilisation*, 1945; First Supplementary Volume, 1963; Second Supplementary Volume, Elliott, M. A., (ed.), 1981; John Wiley and Sons Inc, New York.
5. MACDONALD, D. M. and STOTON, J., ref. 1, Paper|IB-4.
6. BERANEK, J., KASPAR, M., BAZANT, V. and CHLADEK, A., British Patent 1 375 011.
7. REH, L., SCHMIDT, H. W., DARADIMOS, G. and PETERSEN, V., ref. 1, Paper VI-2.
8. NACK, H., ANSON, D. and DI NOVO, S. T., ref. 1, Paper VI-4.
9. ENGSTROM, F., ref. 2, pp. 616–21.
10. JOHNES, G. L., ref. 1, Paper I(b)-6.
11. BARKER, D. and BEACHAM, B., ref. 1, Paper IA-3.
12. MINCHENER, A. J. and COOKE, M. J., ref. 1, Paper IA-5.
13. HOY, H. R. and KAYE, W. G., *J. Inst. Energy*, **52**, 86–93 (1979).
14. BEACHAM, B. and MARSHALL, A. R., *J. Inst. Energy*, **52**, 59–64 (1979).
15. GAMBLE, R. L., *Proc. of Fourth International Conf. on Fluidized Bed Combustion*, MITRE Corp., 1975, pp. 133–51.
16. CLAYPOOLE, G. T. *et al.*, ref. 2, pp. 600–10.
17. GAMBLE, R. L., ref. 1, Paper IB-1.
18. GAMBLE, R. L., *Industrial Power Conference*, St. Louis, ASME-IPC/FU/2, 1981.
19. STOPPEL, K. G. and LANGHOFF, J., *Vereins Deutscher Igenieure Berichte*, **322**, 37–43 (1978).
20. LANGHOFF, J. and WIED, E. W. V., ref. 1, Paper IB-3.
21. JEFFS, E., *Energy International*, 26–8 (September 1980).
22. HIGHLEY, J., KAYE, W. G. and WILLIS, D. M., ref. 2, pp. 318–33.
23. VIRR, M. J., *Proceedings of Fourth International Conf. on Fluidized Bed Combustion*, MITRE Corp., 1975, pp. 631–47.
24. VIRR, M. J., ref. 1, Paper IA-1.
25. ORMSTON, D., ROBINSON, E. and BUCKLE, D., ref. 1, Paper IIA-1.
26. GIBSON, T., *Proc. of Steam Boiler Plant Technology Conference*, I. Mech. E., C 138/81, 1981, pp. 63–70.
27. ACCORTT, J. I., COMPARATO, J. R. and NORCROSS, W. R., ref. 1, Paper IB-6.
28. GIBSON, J. and HIGHLEY, J., *Vereins Deutscher Igenieure Berichte*, **322**, 17–28 (1978); reprinted in *J. Inst. Energy*, **52**, 51–8 (1979).
29. BREALEY, L. and WILSON, J. H., ref. 1, Paper IA-2.
30. CAPLIN, P. B., *Proc. of Fluidized Combustion Conference*, Energy Research Institute, University of Cape Town, 1981, pp. 1–20.
31. FISHER, M. J., PAYNE, R. C., TATE, P. and WILLIS, D. M., ref. 1, Paper IA-6.
32. MICHAELS, H. J. and BEACHAM, B., ref, 1, Paper IA-4.
33. HIGHLEY, J., KAYE, W. G. and WHEATLEY, P. C., ref. 2, pp. 584–95.

34. WHEATLEY, P. C. and HIGHLEY, J., ref. 1, Paper IIA-3.
35. PRITCHARD, A. B. and CAPLIN, P. B., ref. 1, Paper IIB-4.
36. HANSON, H. A., KINZLER, D. D. and NICHOLS, D. C., ref. 2, pp. 160–8.
37. LAWS, W. R. and REED, G. R., 1982 *Industrial Energy Conservation Technology Conference*, Houston, Texas.
38. DOBBS, R. M., GILMOUR, I. A. and HOULT, B. D., ref. 1, Paper IIA-4.
39. JOVANOVIĆ, L. and ARSIĆ, B., ref. 1, Paper IIA-2.

Chapter 4

PRESSURIZED FLUIDIZED COMBUSTION

A. G. ROBERTS, K. K. PILLAI and J. E. STANTAN

*National Coal Board Coal Utilisation Research Laboratory,
Leatherhead, UK*

1 INTRODUCTION

The ability to operate a gas turbine on products from the combustion of coal has long been a goal for combustion engineers and the advent of pressurized fluidized combustion (PFBC) may result at last in this goal being realized. The low temperature (750 to 950 °C) at which coal can be burned in fluidized beds minimizes the volatilization of alkali metals in the fuel and avoids any sintering of the coal ash; as a result the likelihood of corrosion or erosion of gas turbine blades is reduced.

Pressurized fluidized bed combustion has three potential advantages compared with atmospheric operation:

(i)   There is an increase in specific power output and hence a potential reduction in capital cost. This effect is shown in Fig. 1 where the bed area per unit power output is shown as a function of fluidizing velocity and operating pressure. It will be seen that at a typical fluidizing velocity of 2 m/s, the bed area at atmospheric pressure (100 kPa) is *c*. 2·0 m^2 for every MW of power output, reducing to *c*. 0·2 m^2 or less for operation under pressure.

(ii) By combining a gas turbine and a steam cycle, the efficiency of power generation can be increased, as will be explained later.

(iii) Emissions of oxides of nitrogen are substantially reduced.

To these advantages must be added that which is common to all applications of fluidized combustion, namely, the ability to reduce sulphur emissions at very little extra cost and virtually no loss of efficiency.

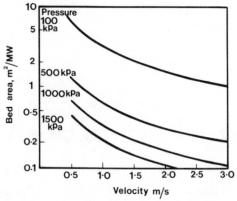

FIG. 1.   Effect of operating pressure and fluidizing velocity on bed area.

It has been appreciated for several years that if these potential advantages can be realised in practice, then electricity can be produced from coal more cheaply and with lower pollution levels.

2   THERMODYNAMIC CYCLES

It is axiomatic with PFBC that the hot gases must be expanded through a turbine. Thus, PFBC cannot be considered in isolation, but must be related to gas turbine operation and to the thermodynamic cycle into which the gas turbine is incorporated. The ideal gas turbine for PFBC probably does not exist, and is unlikely to do so, at least in the foreseeable future. For example, most modern gas turbines utilise high gas velocities with large changes of direction through the turbine in order to reduce the number of stages, in conjunction with ever-increasing turbine inlet temperatures. These are conditions which are not ideal for a coal combustion system, but the cost of re-designing and developing new compressor and turbine machinery for a specific application is generally stated to be prohibitive. All serious PFBC studies and proposals therefore, are built around modifications to existing gas turbines.

There are two forms of PFBC—'air' heaters and supercharged boilers. Both can be combined with gas and steam turbines to give combined cycles, although 'air' heaters can also be used in cycles in which the only prime mover is a gas turbine or a multiplicity of gas turbines. It is convenient to classify cycles using 'air' heaters as 'air cycles' and those using supercharged boilers as 'supercharged boiler combined cycles'.

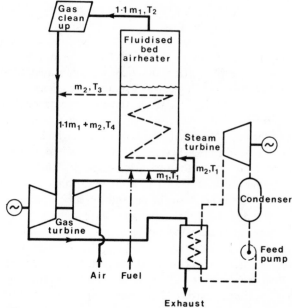

FIG. 2.   Basic 'air' cycle or air heatu combined cycle.

'Air' Cycles. A simple form of air cycle is shown diagrammatically in Fig. 2. Air from the compressor is split into two streams. Most of the air passes through tubes immersed in the fluidized bed, where it is heated to a temperature approaching the combustor bed temperature. The remaining air is used to fluidize the bed. Combustion products from the bed are cleaned before mixing with the first stream to pass through the turbine. Because the fluidized bed has a relatively slow response rate to changes in fuel feed, it is usual in all cycles to by-pass a small amount of compressed air directly to the turbine to control the turbine inlet temperature.

The amount of tubing in the bed can vary from zero (the so-called 'adiabatic' combustor, in which case all the air passes through the bed and the excess air is high—*c.* 350 %) to a maximum which leaves just enough fluidizing air for combustion of the fuel (i.e. about 20 % excess air). In this latter case, the amount of air which can be heated in the tubes is about twice the amount of fluidizing air.

Cycle efficiencies in simple 'air' cycles (i.e. where there is no additional steam turbine) are slightly lower than those for the same gas turbine operating with conventional oil-fired combustors. The lower efficiencies arise because of the higher pressure losses associated with a fluidized bed and a gas clean-up system, the effect being less significant for gas turbines operating at a high pressure ratio. The efficiency can be improved substantially by using the exhaust gases from the turbine to produce steam, either in a waste heat boiler as shown in Fig. 2, or in a separately-fired boiler. This would be the general practice. If a separately-fired boiler is used, it seems aesthetically logical to use an atmospheric fluidized bed, although the complications of providing feeding systems for both pressure and atmospheric pressure operation could be prohibitive.

A variant of the air cycle is that shown in Fig. 3 in which the air heated in the tubes forms part of a closed cycle. Generally, such units would operate at a lower pressure ratio than conventional gas turbines, but at a significantly higher pressure level. This leads to substantial increases in the heat transfer coefficient between the air and the tube wall and hence to a reduction in the amount of heat exchange surface required. The fluid in the closed cycle loop can also be one having thermodynamic advantages (e.g. helium) compared with air.

Another variant of the air cycle giving extra output at peak periods, is the 'airstore' system shown in Fig. 4. In a gas turbine, approximately two-thirds of the turbine output is used to drive the compressor. With the airstore system, the air is supplied from a pre-compressed source at periods of peak load and consequently the whole of the turbine output is available to drive the generator over the peak load period. The system requires an underground cavity into which air can be compressed during off-peak periods. During such times the generator acts as a motor driving the compressor for storing the air.

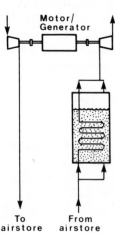

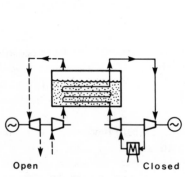

FIG. 3.   Closed 'air' cycle.   FIG. 4.   Airstore cycle.

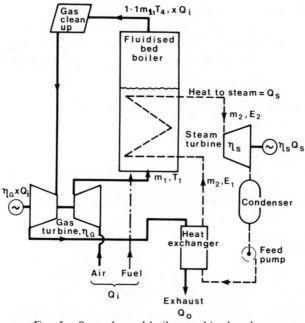

FIG. 5. Supercharged boiler combined cycle.

Supercharged Boiler Combined Steam/Gas Turbine Cycles. This type of cycle is illustrated in Fig. 5. Bed temperature is maintained at 850 to 950 °C by generating steam in tubes immersed in the bed. The steam is used to drive a steam turbine in a more or less conventional steam power generation loop.

Because some energy is recovered from the gas turbine cycle at a higher temperature than that recovered from the steam cycle, the overall efficiency of the combined cycle is typically 40 to 41 % compared with about 38 % for the basic steam cycle. The reason for this increase in efficiency is described below.

Consider a conventional steam turbine cycle, with a total thermal input Q_i, and heat rejection (in the gas and ashes) to the atmosphere Q_0. The boiler efficiency η_b is given by:

$$\eta_b = \frac{Q_i - Q_0}{Q_i}$$

If the steam cycle efficiency is η_s, then the plant efficiency η_p is given by:

$$\eta_p = \eta_b \eta_s$$

Now consider a combined steam–gas turbine cycle as shown in Fig. 5. Consider the same thermal input Q_i and heat rejection Q_0. If x is the fraction of the heat input Q_i, entering the gas turbine, and the efficiency of the gas turbine plant is η_g then the net power generated by the gas turbine is:

$$P_g = x\eta_g Q_i$$

If Q_s is the total heat absorbed in the steam cycle,

$$Q_s = Q_i(1 - \eta_g x) - Q_0$$

and the power generated by the steam turbine is

$$P_s = \eta_s[Q_i(1 - \eta_g x) - Q_0]$$

the efficiency of the combined cycle is therefore given by

$$\eta_{cc} = \frac{P_s + P_g}{Q_i}$$
$$= \eta_g x + \eta_s(\eta_b - \eta_g x)$$
$$= \eta_p + x\eta_g(1 - \eta_s)$$

since x, η_g and $(1 - \eta_s)$ are always positive, the combined cycle efficiency is always greater than the equivalent steam cycle plant efficiency η_p. The final efficiency of the combined cycle depends on the ingenuity employed in transferring as much heat as possible from the gas turbine exhaust back into the steam cycle.

In conventional steam generation systems there is every incentive to operate with low values of excess air. This is no longer the case in combined cycles where the cycle efficiency increases slightly as the excess air increases (up to about 100 % excess air, depending on the turbine inlet temperature). This increase arises basically because more of the heat is being incorporated into the cycle at the higher temperature (i.e. in the gas turbine). Operation at high excess air is beneficial to combustion, reduces the possibility of corrosion and provides a margin of safety for turn-down. On the other hand, the bed has to be larger for a given output (hence pressure vessel costs are increased) and the amount of gas to be cleaned increases.

One version of the supercharged boiler combined cycle which is currently attracting some interest is a variant of the old 'Velox boiler' cycle which was used successfully with conventional gas-fired units in the 1930s and 1940s. In this cycle the pressure level and gas turbine inlet temperatures are maintained fairly low (say 6 atm and 700 °C), so that there is little output from the gas turbine, i.e. the turbine generates only sufficient power to drive the compressor with little excess. Although it does not have the high efficiency of the true combined cycle the turbine conditions pose no operational problems and such a cycle could be used as a stepping stone to more advanced conditions. In its own right, this cycle offers a way of replacing oil by coal on existing oil-fired sites where there often is not sufficient space to accommodate coal handling equipment together with a conventional PF boiler.

2.1 Some characteristics of PFBC/thermodynamic cycles

Some of the ways in which requirements for fluidized combustion interact with those imposed by the choice of cycle and operating parameters are very complex. An attempt has been made in Table 1 to indicate a relative order of merit for different aspects of four cycles.

TABLE 1

Relative characteristics of some cycles

	Cycle—Air (Fig. 2)		Cycle—Combined (Fig. 5)	
		Excess air		
	c. 350%	20%	20%	100%
1. Bed area/MWe	4	1	1	3
2. Amount of tubing	None	3	2	1
3. Quantity of gas to be cleaned (this affects the gas cleaning costs)	4	1	1	3
4. Bed height (this affects the pressure loss and loss of cycle efficiency)	1	4	3	2
5. Dust concentration at turbine inlet	2	1	2	2
6. NO_x emissions (function of excess air)	4	1	1	3
7. Complexity of load changing and control	1	1	4	3

'1' indicates less than '2' which is less than '3', etc.

Bed Height and Amount of Tubing. The two cycles which are most likely to find commercial application are those illustrated by Figs. 2 and 5. The key feature of such schemes, and therefore of PFBC, is that the heat release per unit area (see Fig. 1) becomes so large that the height of the bed is determined by the volume required to accommodate the heating surface. This is illustrated by the following simplified analysis for the air heater and the supercharged boiler.

Referring to the 'air' cycle of Fig. 2, a heat balance around the turbine inlet mixing point yields the equation:

$$(1 \cdot 1 m_1 + m_2)T_4 = 1 \cdot 1 m_1 T_2 + m_2 T_3 \tag{1}$$

(Note: the coal mass flow is about $0 \cdot 1 m_1$, varying slightly with excess air.)

Taking into account that for stoichiometric combustion, the heat input (gross) is $3 \cdot 1$ MJ/kg air, then a heat balance around the combustor yields the equation:

$$m_2(T_3 - T_1)c_p + (1 \cdot 1 m_1 T_2 - m_1 T_1)c_p = 3100 m_1/\Omega \tag{2}$$

where c_p is the specific heat of the gas or air (kJ/kg K) and Ω is the excess air factor (ratio of total air to stoichiometric air).

A heat balance across the air heater tubes yields the equation:

$$m_2(T_3 - T_1)c_p = hA \, \Delta T \tag{3}$$

where h is the overall heat transfer coefficient across the tube walls, A is the surface area of the tubes and ΔT is the log mean temperature difference between the bed and the air in the tubes.

The tube geometry and the fluidizing conditions yield the equations:

$$A = 4\phi Ha/d \tag{4}$$

and

$$1 \cdot 1 m_1 = U\rho a \tag{5}$$

where ϕ is the fraction of bed volume occupied by the tubes, H is the bed height, a is the cross-sectional area of the bed, d is the tube diameter, U is the fluidizing velocity and ρ is the gas density at bed surface.

Rearranging eqns. (1) to (5) gives:

$$H = \frac{U\rho d}{4 \cdot 4 \phi h \, \Delta T} \left[\frac{3100}{\Omega} - (1 \cdot 1 T_2 - T_1)c_p \right] \tag{6}$$

Selecting some typical values, viz. $T_2 = 1173$ K, $T_1 = 573$ K, $T_4 = 1143$ K and excess air $= 35\%$ ($\Omega = 1 \cdot 35$) then from eqns. (1) and (2) it follows that $T_3 = 1130$ K.

Taking typical fluidizing conditions and tube bank conditions as:

$U = 1$ m/s
$\rho = 3 \cdot 0$ kg/m^3 (10 bar and 1173 K)
$d = 25 \cdot 4$ mm
$\phi = 0 \cdot 17$
$h = 255$ W/m^2 K and
$\Delta T = 90$ K (log mean temperature) then
$H = $ approx. 7 m

Following a similar analytical procedure for the supercharged boiler combined cycle to that given for the air heater cycle, and using similar nomenclature (see Fig. 5), the following equation can be written:

$$1 \cdot 1 m_1 T_4 c_p - m_1 T_1 c_p + m_2 (E_2 - E_1) = 3100 m_1 / \Omega$$

where E is the specific enthalpy of water or steam, $m_2(E_2 - E_1) = hA \, \Delta T$, $A = 4\phi Ha/d$ and $1 \cdot 1 m_1 = U\rho a$.

These equations can be combined to give:

$$H = \frac{U\rho d}{4 \cdot 4 \phi h \, \Delta T} \left[\frac{3100}{\Omega} - (1 \cdot 1 T_4 - T_1) c_p \right] \tag{7}$$

This is a similar formula to eqn. (6). However, the overall heat transfer coefficient h and temperature difference ΔT are considerably greater in the boiler than in the air heater, and the bed height is therefore considerably reduced. Using the same conditions as the air heater example, but with values of h and ΔT appropriate to the boiler (viz. $h = 300 \, \text{W/m}^2 \, \text{K}$; $\Delta T = 350 \, \text{K}$) gives a bed depth of about $1 \cdot 5$ m. In boilers it is more usual to use tubes of 38 or 50 mm diameter which would increase the bed depth to 2 or 3 m.

The bed depths calculated in the above examples are, of course, the depths necessary to accommodate the tube bank. In practice it is usual to allow an additional space of $0 \cdot 5$ to 1 m below the tube bank to allow adequate mixing of coal within the bed.

In very broad terms, designs for PFBC are usually based on fluidizing velocities of about 1 m/s for air heaters and 1 to $1 \cdot 5$ m/s for combined steam/gas turbine cycles, for the following reasons:

(1) These velocities result in bed depths of about 8 m and 3 to 4 m, respectively—values which are only slight extrapolations from existing pilot-scale plants.

(2) The resulting pressure drop (about 6 kPa/m of bed depth) is tolerable. Although pressure loss due to bed depth does not have a major effect on cycle efficiency at full load, it can become a problem in matching turbine and compressor characteristics at reduced loads (i.e. reduced pressure levels) since it is independent of the pressure level.

Equations (6) and (7) and the examples given, illustrate a second major feature of PFBC, namely that the residence time of the gases in the bed is high compared with AFBC units; typically, up to 8 s for air heaters and up to 4 s for supercharged boilers compared with a maximum of $0 \cdot 5$ s at atmospheric pressure. This has a significant effect on combustion efficiency, sulphur retention, etc., as is described later.

3 HISTORIC DEVELOPMENT AND PRESENT STATUS

The potential benefits of PFBC for power generation were first realized by Douglas Elliott and it was largely through his encouragement that pioneer pilot-plant investigations began in 1969 at the British Coal Utilisation Research Association, now the Coal Utilisation Research Laboratories (CURL) of the National Coal Board.

Subsequently a number of other pilot-scale combustors were operated and a list of the major activities is given in Table 2. In addition to the CURL plant, the units which have contributed most to PFBC in terms of combustion, heat transfer and emission behaviour have been the 'Miniplant' operated by Exxon (formerly Esso) Research and Engineering Co. of Linden, New Jersey, and the work at Argonne National Laboratory with small combustors. An important feature of the Miniplant was that it incorporated a unit for regeneration of sulphur dioxide sorbent, and most of the important work in this field has been carried out with this unit.

Both the CURL combustor and the 'Miniplant' incorporated cascades of gas turbine blades and target specimens in the exhaust gases downstream of hot gas clean-up systems. It is experiences with these cascades that have largely maintained optimism concerning future gas turbine operation.

Two companies have so far operated small gas turbines in conjunction with PFBC. The first was the unit of the Combustion Power, Inc. at Menlo Park, California. The combustor was adiabatic, i.e. it did not incorporate cooling tubes in the bed so that it operated with a high excess air level to control the bed temperature. As a demonstration, the venture was not entirely successful since malfunctioning of the gas cleaning system resulted in erosion damage to the turbine.

The second company to operate a small gas turbine was Curtiss-Wright of Wood-Ridge, New Jersey, who were awarded a contract in the mid-1970s by the then US Energy Research and Development Administration (now the Department of Energy) to design, build and operate a 13 MWe pilot-scale combined cycle power plant based on the 'air-heated cycle' of Fig. 2. This plant, which will incorporate a 7 MW gas turbine, is planned for the 1980s, but Curtiss–Wright have built a smaller plant to obtain design data and have successfully operated it in conjunction with a small gas turbine.

A further small combined cycle air-heater plant is planned in West

TABLE 2
Current PFBC activities

	Pilot plants without gas turbines				Pilot plants with gas turbines			Demonstration or commercial plants	
	NASA–Lewis	Exxon	NCB–CURL	IEA	Curtiss-Wright	Combustion power	Curtiss-Wright	AEP–Stal Laval	General Electric
Bed dimensions (m)	0·2 dia	0·3 dia	0·6 × 0·9 or 0·75 × 1·2	2 × 2	0·9 dia	2·1 dia	c.3 dia	—	3·7 × 3·7(4)
Bed area (m²)	0·03	0·07	—	4	0·66	3·66	c.7	—	53·5
Bed depth (m)	1·8	3–4·3	0·55 or 0·84	3·5–4·5	4·9	0·6	4·9	3·7	c.3
Combustor pressure (bar)	5·4 max	10 max	1·2–2·7	12 max	6·5	3·7	7·8	15·2	9
Bed temperature (°C)	870 max	950 max	950 max	950 max	900	950	900	840	950
Fluidizing velocity (m/s)	2·1 max	2·1 max	2·1 max	3 max	0·8	2·4	0·8	0·9	1·4
Coolant medium	Water	Water	Water	Steam	Air	(adiabatic)	Air	Steam	Steam
Steam temperature (°C)				440				496	540
Steam pressure (bar)				30				88	238
Steam turbine power (MWe)							6ª	110	238
Gas turbine power (MWe)					0·1	1	7	67	50
Start-up date or status as of 1982	Shut down	Shut down	1969	1980	1979	—	1983	Detail design stage	Conceptual design

ª Waste heat recovery boiler.

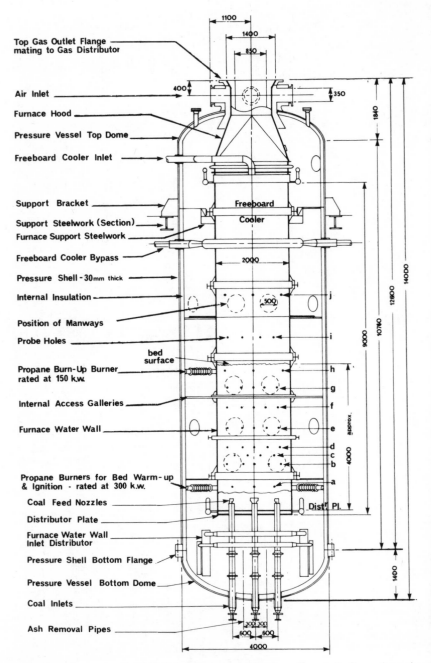

FIG. 6. Diagram of IEA Grimethorpe combustor. Key (all dimensions in millimetres): a, 505; b, 1095; c, 1295; d, 1680; e, 2265; f, 2850; g, 3270; h, 3770; i, 4880; j, 6000.

Germany by Bergbau-Forschung/Vereiningte Kesselwerke, although progress has been delayed.

The main technical step forward in the next decade will be through the Grimethorpe plant, commissioned in 1980. This is a large experimental steam generator jointly sponsored by three of the member countries (UK, USA and FRG) of the International Energy Agency which has been built at a UK National Coal Board Power station. The objective of the work is to obtain experimental data on a larger operating scale than hitherto, but with a plant having a greater degree of flexibility than would be possible in an integrated combined-cycle power plant. Investigations with cascades of turbine blades will be carried out and it is expected that a gas turbine will be connected to the plant at a later stage of the work. A diagram of this combustor is shown in Fig. 6. Economic studies of PFBC have been legion. Apart from the Curtiss-Wright proposals they have been dominated by two US companies—Westinghouse and, latterly, General Electric. It is beyond the scope of this chapter to discuss the economics in any detail and the reader is referred to the ECAS studies (see Bibliography). Because of the cost of power stations and the risks inherent in any new technology, commercial application of PFBC is still a few years away. The nearest approach is the design study currently being made by two private companies—American Electric Power in the USA in conjunction with Stal-Laval of Sweden. This study contemplates a 70 MW Stal-Laval gas turbine (which has a pressure ratio of 16/1) in conjunction with a supercharged boiler and a steam turbine at an existing AEP site. If the potential continues to be satisfactory, this plant should be built in the mid-1980s.

4 EFFECT OF PRESSURE ON FLUIDIZATION

There is very little published information about the effect of pressure on fluidization. This probably reflects the difficulties and cost of experimentation under these conditions. Nevertheless, many workers have commented on the 'improvement in fluidization quality' as operating pressure is increased and on the 'smoothness' of fluidization compared with atmospheric pressure. The usual explanation for this 'improvement' is that pressure brings about a reduction in bubble size. This is certainly true for beds composed of fine particles (up to, possibly, $200\,\mu m$), but direct evidence is conflicting for operation with beds of particles of 500 to $1000\,\mu m$—the size appropriate to PFBC.

Operation of a fluidized bed must lie somewhere between a lower limit defined by 'minimum' fluidization and an upper limit defined by the terminal velocity of the bed particles. Both these criteria depend to some extent on pressure. The generally accepted relationship for minimum fluidization conditions, which is based on Ergun's equation for the pressure drop through static beds, can be written:

$$\text{Ar} = 150 \frac{(1 - e_{\text{mf}})}{e_{\text{mf}}^3} \text{Re}_{\text{mf}} + \frac{1 \cdot 75}{e_{\text{mf}}^3} \text{Re}_{\text{mf}}^2$$

where $\text{Ar} = \rho_f(\rho_s - \rho_f)gd^3/\mu^2$ (Archimedes or Galileo number), $\text{Re} = \rho_f dU/\mu$ (particle Reynolds number), e_{mf} is the voidage fraction, ρ_s is the particle density, ρ_f is the fluid density, d is the particle diameter, μ is the fluid viscosity, U_{mf} is the minimum fluidizing velocity and U is the superficial fluidizing velocity.

For PFBC operation, a more useful way of expressing the relationship is to use a parameter called the fluidization number,

$$F_n = \text{Ar}^{1/3} = (3C_D\text{Re}^2/4)^{1/3}$$

and a parameter called the velocity number,

$$V_n = (4\text{Re}/3C_D)^{1/3}$$

where

$$C_D = \text{drag coefficient} = \frac{4g(\rho_s - \rho_f)d}{3\rho_f U^2}$$

These numbers have the merit that F_n is proportional to the particle diameter and independent of velocity, while V_n is proportional to velocity and independent of the particle size.

This relationship between F_n and V_n is plotted in Fig. 7 for a voidage fraction e_{mf} of 0·5, which is typical for fluidized bed combustion systems. Also plotted in Fig. 7 is the Pettyjohn & Christiansen relationship for the terminal velocity of cubes (sphericity of 0·8) which is considered to be appropriate for particles typical of those in fluidized beds. It will be seen that this is of a similar form to the minimum fluidization relationship.

Experience has shown that satisfactory operation of FBC's occurs within a band having the same general shape as the minimum fluidization and terminal velocity lines. This is depicted by the 'good fluidization' line in Fig. 7. It is possible that this line may also be an equilibrium line in that if conditions alter to move the operating point away from the line the size distribution of the bed may re-adjust itself (i.e. more fines are elutriated or retained) so as to move towards the line.

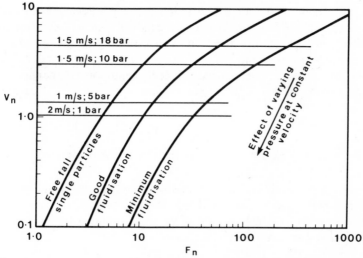

FIG. 7. Relationship between V_n and F_n and some operating conditions.

Shown on Fig. 7 are values of V_n corresponding to a number of operating combinations of fluidizing velocity and pressure: (i) 1·5 m/s and 18 bar which represent about the upper limit of velocity and gas turbine pressure ratio, (ii) 1·5 m/s and 10 bar—a pressure typical of many gas turbines, (iii) 1 m/s and 5 bar which corresponds to a typical minimum load or start-up condition, (iv) 2 m/s and atmospheric pressure which corresponds to a typical atmospheric pressure boiler. For each combination there is a value of F_n (and therefore d_p) which will produce good fluidization; the values of d_p corresponding to the above conditions are 1700, 1000, 450 and 700 μm, respectively. The range of Reynolds numbers is considerable—from 5 to 100. It is not surprising, therefore, that changes in some operating parameters are more important at high than at low velocity numbers.

The effect of changing different operating parameters would appear as vectors on Fig. 7. Since velocity, pressure, mass flow and temperature are not independent, there are a number of possible vectors. For the sake of clarity only one is shown in Fig. 7, that of pressure at a constant velocity. It will be seen that at low values of V_n the effect of change in pressure is not significant (the vector is parallel to the 'good fluidization' line), but at high values of V_n a rapid 3/1 reduction in pressure at constant velocity would move operating conditions close to the minimum fluidization line. If this reduction in pressure (which corresponds to a reduction in load in many gas turbines) is carried out gradually, the particle size in the bed will probably adjust to remain on the 'good fluidization' line. If a pressurized unit is shut down at a full load condition (i.e. on the 'good fluidization' line) it has to be re-started at low pressure and from a condition which may be near the minimum fluidization line.

The above remarks are intended only to highlight some of the effects that combining a PFBC with a gas turbine can have on fluidization characteristics.

5 THE EFFECT OF PRESSURE ON COMBUSTION RATE

As with the physical behaviour of fluidized beds (Section 4), and probably for similar reasons, there is very little published information on the effect of pressure on the combustion rate of carbon (r_c) in a fluidized bed. The combustion rate can be expressed as:

$$r_c = K_{eff} A_s p \tag{8}$$

where K_{eff} is the effective reaction rate coefficient (kg/m^2 atm s), A_s is the active carbon surface area (m^2) and p is the partial pressure of oxygen in the particulate phase of the fluidized bed (atm).

The value of p will depend on the oxygen partial pressure at the air inlet and upon the fluidization conditions which determine the gas exchange rate between bubbles and particulate phase. In general, deeper beds, smaller bubbles and larger particles promote gas exchange and will cause p to increase. In PFBC deep beds are used and closely spaced tubes limit the growth of bubbles, although particles are generally smaller than in AFBC.

The effective reaction rate coefficient, K_{eff}, is given by:

$$\frac{1}{K_{eff}} = \frac{1}{K_{diff}} + \frac{1}{K_s}$$

where K_{diff} is the external diffusion coefficient and K_s is the surface reaction rate coefficient. The consensus now is that in fluidized beds both K_s and K_{diff} are significant, with K_{diff} becoming more significant as particle size and pressure increase.

K_s is a function only of particle temperature and is essentially independent of pressure.

K_{diff} in a fluidized bed is given by the expression:

$$K_{diff} = M \phi_m Sh \varepsilon_b D_g / dRT_m$$

where M is the molecular weight of carbon, ϕ_m is the mechanism factor (depending on whether carbon dioxide or carbon monoxide is the primary combustion product), Sh is the Sherwood number ($= 2 + 0.68(\mu/\rho D_g)^{1/3}(\rho d U/\mu)^{1/2}$), ε_b is the tortuousity factor to account for particles surrounding a burning particle ($-$), D_g is the molecular diffusion coefficient (m^2/s), R is the Universal gas constant (atm m^3/kg—mol K), d is the particle diameter (m) and T_m is the mean temperature around the particle (K).

Since $D_g \propto 1/P$, where P is the plant operating pressure it will be seen that the product $(Sh - 2)D_g$ is inversely proportional to the square root of the pressure, and hence $K_{diff} \propto 1/P^n$ where n is less than 0·5.

Thus, eqn. (8) can be re-written:

$$r_c = \frac{P}{aP^n + b}$$

here a and b are constants for given fluidization and temperature conditions. For kinetic control $aP^n \ll b$ and for diffusion control $aP^n \gg b$. In either case, it will be seen that the burning rate should be enhanced at elevated pressures.

6 PERFORMANCE CHARACTERISTICS OF PFBC

Although the fundamental effects of pressure on fluidization and combustion have received very little attention, the performance of PFBC's has been adequately studied in the various experimental rigs listed in Section 3. Certain well-defined characteristics have become apparent; these are discussed below.

(1) *Combustion Efficiency.* The main variables influencing combustion efficiency are combustion temperature, excess air and superficial gas residence time. However, in commercial designs the superficial gas residence time is so large (in the air heater or supercharged boiler cycles) or the excess air so great (in the adiabatic combustor) that combustion efficiency, in practice, will always be high—99 % or greater. Factors which also assist in attaining these high efficiencies are the better gas-solids contacting (smaller bubbles) probably brought about by pressure and the fact that closely-spaced tube banks restrict the growth of bubbles in the deep beds.

Unlike atmospheric operation, practically all the combustion takes place within the bed, i.e. freeboard combustion is negligible. This is fortunate, since it is also a necessity where gas turbine operation is concerned. It is doubtful whether turbine inlet temperature could be adequately controlled if freeboard combustion were significant, and any extension of combustion into the gas-cleaning cyclones would, sooner or later, lead to sintering of the dust and problems of dust removal.

(2) *Heat Transfer.* Heat transfer rates have been measured in PFBC's at pressures up to about 10 bar and fluidizing velocities up to about 2·5 m/s. Over these ranges, no significant effect of pressure *per se* has been detected. Fluidization studies have suggested that at very high Reynolds numbers there may be a change to a new fluidization regime, and that the accompanying change in heat transfer mechanism will result in heat transfer rates that are pressure-dependent. It is doubtful whether this regime will be reached in any foreseeable PFBC design.

Because, however, fluidizing velocities (and therefore bed particle sizes) are appreciably lower in PFBC than in atmospheric units, the heat transfer coefficients are significantly higher. For example, at 1 m/s and 850 °C the heat transfer coefficient is about 350 W/m² °C compared with 270 W/m² °C at 2·5 m/s.

(3) *Sulphur Retention.* Pressurized operation brings about radical changes in the sulphur retention characteristics of sorbents. The most important of these changes, which are described and discussed more fully in Chapter 5, can be deduced from the curves shown in Fig. 3 of that chapter. They concern the effects of temperature, and the effectiveness of the two main types of sorbent—limestone and dolomite.

It can be seen that at atmospheric pressure (full curves) the sulphur retention efficiency with limestones, and to a smaller extent with dolomites, increases with temperature up to 800–850 °C and then falls sharply, but at high pressure (broken curves) the efficiency increases with temperature over the whole range.

A pre-requisite of good sulphur retention efficiency is the development of porosity in the sorbent particle. With limestones, porosity is developed by calcination, and this occurs readily at atmospheric pressure, but only with greater difficulty and at higher temperatures in pressurized plant. The curves in Fig. 4 of Chapter 5 show how the equilibrium temperature for calcination of calcium carbonate depends on the partial pressure of carbon dioxide, and hence on the total combustor pressure and level of excess air. As a consequence of the difficulty in achieving calcination of calcium carbonate at high pressure, the effectiveness of limestones is reduced.

Dolomites undergo two stages of calcination:

(a) Half-calcination $(CaCO_3 . MgCO_3 = (CaCO_3 + MgO) + CO_2$
(b) Full calcination $(CaCO_3 + MgO) = (CaO + MgO) + CO_2$

The 'half-calcination' reaction occurs rapidly at all temperatures above about 600 °C in both AFBC and PFBC and results in the development of considerable porosity in the dolomite particle. However, at atmospheric pressure the rapid evolution of carbon dioxide, enhanced by that from the calcium carbonate calcination (full calcination) reaction, can cause severe decrepitation of the particles, forming fines that are elutriated from the combustor before absorbing much sulphur. On the other hand, at high pressure, decrepitation is curtailed, and sufficient porosity is developed by half-calcination to give high utilization even under conditions preventing calcination of the calcium carbonate component. This explains the reversal of the relative effectiveness of limestone and dolomite.

(4) *NO_x Emissions.* The NO_x emissions from fluidized bed combustors are discussed fully in Chapter 6. Experimental data have shown that increasing pressure reduces the NO_x emissions (approximately in proportion to the square root of pressure), although the reasons for this are not fully understood.

(5) *Alkali Emissions.* The emission of alkalis is, in principle, of considerable importance because of their influence on hot corrosion of gas turbine blades. This type of attack is initiated when alkali salts react with sulphur dioxide and oxygen to form alkali sulphates which condense on the hot blade surfaces. Alkali metals are transported through the system as fine particulates, and as vapour species which are in equilibrium with solid alkali metal species in the bed or in suspension.

Equilibrium calculations,† reproduced in Fig. 8, indicate that bed temperature and chlorine content of the coal are the most important factors in determining the amount of alkali in the combustion products. Also shown in Fig. 8 are alkali measurements made on two experimental plants which, bearing in mind the difficulties in making such measurements, are in good agreement with the theoretical values.

The alkali metal content of the gas phase is not in itself a sufficient measure of the tendency to condense alkali sulphate in the gas turbine, even though it is at least an order of magnitude higher than the present gas turbine limit for liquid fuels. This is because: (i) the chlorine/alkali metal ratio in the PFBC efflux is higher than with oil products and this might be expected to suppress condensate (alkali chlorides are more volatile than sulphates), and (ii) reaction with the alumino-silicates in the coal ash (known as 'gettering') will tie up some of the alkali metal in a harmless form.

† Report to the Electric Power Research Institute by General Electric Co., Report No. CS-1469.

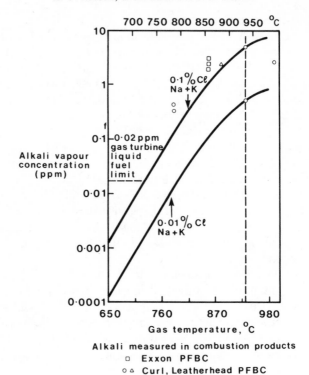

FIG. 8. Calculated alkali vapour concentrations as a function of gas temperature in a coal-fired PFBC.

Thermochemical calculations (see footnote, p. 189) indicate, however, that with typical chlorine/alkali metal ratios the alkali metal sulphate condensation on turbine blades will be many times that of a conventional liquid fuel-fired gas turbine, even with 90% gettering.

Thus, although only a small proportion of the input alkali (perhaps 1 or 2%) is vaporized from a PFBC, the efflux must be considered to be potentially corrosive. Whether corrosion of metals occurs depends, of course, on the temperature and on the corrosion resistance of the alloy. Present experience on the various rigs suggests that if the metal temperature is not higher than about 800 °C, then existing turbine alloys with existing corrosion-resistant coatings should give adequate life. For higher gas turbine inlet temperatures (up to 950 °C) a solution may lie in the use of cooled blades and/or in the metallurgical development of more corrosion-resistant coatings and claddings.

7 PROBLEM AREAS

There are possibly three areas of development where pressurized operation introduces particular problems. These are associated with coal feeding, gas cleaning and load control.

7.1 Coal feeding

In all PFBC rigs and schemes to date (and, incidentally, in many gasification processes), coal is dried and crushed to a size consistent with the fluidizing velocity (e.g. $3 \cdot 2$ mm × 0 for a design velocity of $\simeq 1$ m/s). It is then fed pneumatically into the base of the bed from a lock hopper system. Such a system consists of a feeder vessel which always operates at a pressure higher ($0 \cdot 3$ to 2 bar) than the fluidized bed, coal being fed from a number of offtakes into air-conveying lines, one line to each coal nozzle. In this way a number of offtakes can be accommodated from one feed vessel. The latter is connected to a lock hopper which is filled at atmospheric pressure, pressurized, and the coal dropped through a valve into the feeder vessel. Coal feed rate to the combustor is controlled by varying the pressure in the feeder vessel, or by varying the speed of a rotary (metering) valve, or by the use of an L-valve. The latter should be commercially feasible, although it has so far been used only on a very small rig.

Such a system can only be utilized if the coal is completely dried (i.e. contains virtually no surface moisture). It is technically proven, and can be scaled to any required size. It is, however, costly both in capital since it requires crushing and drying equipment as well as large pressure vessels, and in energy since the cost of compressing an inert gas for pressurizing the lock hopper and for maintaining the pressure in the feeder hopper is a debit to the cycle efficiency. The energy required to boost the pressure of the coal conveying air (air can safely be used for conveying, but inert gas will probably be a safety requirement for pressurizing the hoppers), however, has an insignificant effect on cycle efficiency.

Anything which can reduce the cost of coal preparation and feeding will further enhance the attraction of PFBC. Three developments being investigated, but which are technically and/or economically unproven, are:

(i) Development of a coal 'pump' which would replace the lock hopper system, and the drier—This is mainly a mechanical engineering problem, and past attempts at a solution have generally failed on grounds of reliability.

(ii) Feeding run-of-mine coal into the bed, thus eliminating coal crushing and drying—Simply feeding large coal onto the bed surface (as is an established method in AFBC) is not a solution since it is unlikely that large coal can be circulated through the deep, tightly-packed tube banks inherent in PFBC. An exception might be the adiabatic combustor of Fig. 2. Screw feeding run-of-mine coal below the tube bank is technically feasible, but whether it is economic will depend on whether the cost of a multiplicity of screw feeders and attendant pre-distribution systems is cheaper than that of drying and crushing. In any event, means for removing debris (i.e. lumps of 'stone') from the bed would be needed.

(iii)   Pumping a coal/water slurry into the bed, which would eliminate the lock hopper system and the drier—This would be technically feasible, but the economics are uncertain. Thermodynamic efficiency would be lowered due to the latent heat of the added water, although some of this energy would be recovered in the gas turbine as additional gas obtained at a negligible compression cost.

7.2 Gas cleaning

The crucial question controlling the successful application of PFBC is whether or not the gases from the combustor can be cleaned sufficiently to avoid (or reduce to an acceptable level) erosion of the gas turbine. This means reducing the dust concentration from a value of 10 000 to 40 000 ppm (grams of dust per 10^6 grams of gas)—about 8 to 30 g/m^3—at the combustor exit to something probably less than 300 ppm (220 mg/m^3) at the turbine inlet, with a negligible proportion of particles greater than 10 μm in size. This represents a collection efficiency of more than 97 %. The actual amount of dust elutriated depends on the excess air, ash content of the coal, nature of the ash, sulphur content of the coal (and therefore the amount of sorbent which is added), and the fluidizing conditions. The range 10 000 to 40 000 ppm probably covers most combinations.

In the absence of experience with a large modern turbine, the likelihood of erosion is difficult to assess. Previous experience (in the 1950s and 1960s) with coal-fired gas turbines using conventional (i.e. pulverized-fuel-fired) combustion systems can be little more than a guide since the abrasiveness of PFBC dust is considerably less than that produced by high-temperature suspension systems.

The most comprehensive investigations into PFBC turbine erosion have been made by General Electric, using earlier (proprietary) data from suspension systems together with information obtained by exposing cascades of blades in high velocity gas streams on the CURL and Exxon pilot plants. General Electric engineers have come to the conclusion that

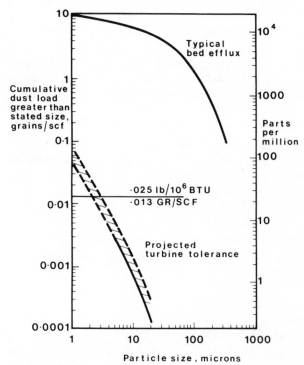

FIG. 9. Projected turbine dust tolerance. (Adapted from R. R. Boericke *et al.*, *6th International Conference on Fluidised Bed Combustion*, Atlanta, 1980, p. 724.)

erosion should not occur if the concentration of dust does not exceed that given in Fig. 9, i.e. if not more than 2 ppm is greater than 10 μm and not more than 8 ppm is greater than 5 μm. The quantity less than about 2 μm is not important since particles of this size do not present an erosion hazard.

The exposure of cascades of turbine blades for periods of 1000 h on the CURL plant at Leatherhead and on the Exxon Miniplant, together with the operation of a small turbine for 1000 h on the Curtiss-Wright plant have given confidence that by using cyclone dust collectors to clean the gas, erosion can be avoided. The cyclone is a well-established, reliable gas cleaning device having a high efficiency in the 5 to 10 μm range. However, because it has a low efficiency in sizes below about 5 μm it is unlikely to clean the gases sufficiently to comply with environmental requirements, which are generally strict. For example, the current emission limits in the US are 0·2 lb particulates per million Btu gross heat input to the furnace (about 90 mg/MJ), but are likely to be reduced to 0·03 lb/10⁶ Btu in the future. As presently conceived a combined cycle PFBC plant would therefore contain a number of cyclones in series prior to the gas turbine, followed by a more or less conventional bag filter unit or electrostatic precipitator at the stack. A modern 70 MWe gas turbine with a pressure ratio of 16/1 would probably require nine sets of cyclones in parallel, each set consisting of three cyclones in series, and each cyclone being 1·5 to 2 m in diameter.

Such a gas clean-up system, together with associated lock hoppers and control equipment for depressurizing the collected dust is, of course, expensive, and a number of alternative ways of cleaning the gas are being investigated. Small-scale experiments have shown that ceramic filters, moving-bed filters and electrostatic precipitators operating at high temperature and pressure can clean the gases sufficiently to meet environmental requirements as well as gas turbine requirements. However, it will require a major development programme occupying many years, to demonstrate that they can be reliably scaled-up to the huge units which would be necessary in practice.

7.3 Load changing and start-up

The combination of a PFBC and a gas turbine leads to complications in changing load and in start-up which will be resolved only when large-scale plant is in operation. The exact method of varying load will depend on the type of cycle and on the type of gas turbine but some generalizations can be made; these are discussed below.

(a) One of the problems arises because the gas turbine is a quick-acting device whereas the PFBC has a relatively slow response rate and constitutes an enormous storage of energy compared with a conventional gas turbine combustor. This has particular importance in the event of a generator trip when it becomes necessary to disconnect the PFBC from the turbine in a fraction of a second in order to prevent overspeeding of the turbine.

Large, quick-acting, valves for the combustor exhaust stream have still to be developed for this purpose. In this respect, single-shaft gas turbines (where the generator, turbine and compressor are mechanically connected) have an advantage in that the compressor acts as a brake in the event of a generator trip.

(b) In the supercharged boiler (Fig. 5), there is the added complication of reducing the steam load at the same time as the gas turbine load is reduced. This can best be illustrated by considering the stages in a load reduction.

(i) The first step in reducing load is to reduce the inlet temperature to the gas turbine. This can be accomplished by reducing the bed temperature and by-passing some of the compressor air directly to the turbine. The effect is to reduce the gas turbine output and the pressure in the system.

(ii) The bed temperature can probably be reduced to a minimum of 750 °C in this way before the combustion efficiency becomes unacceptably low. This, together with by-passing of compressor air can reduce the gas turbine output to almost idling. The output of the steam cycle is reduced, because of the reduction in the temperature differential across the tubes, but only to about two-thirds of the full power output (the temperature difference between the bed and the tubes at full load is *c.* 350 °C falling to 200/250 °C at minimum bed temperature.

In order to reduce the steam cycle output further, additional measures have to be taken; two methods have been proposed.

(iiia) Arranging the boiler as a number of identical modules with, say, two modules per gas turbine and two, three or four gas turbines per steam turbine (this may be desirable for other reasons). When minimum bed temperature (and/or maximum by-pass) has been reached, further load reduction is accomplished by shutting down a gas turbine and slumping the associated bed modules. The steam output in these beds is thus reduced almost to zero. Such a procedure represents a large step change in output, however, and in order to produce a smooth reduction in load it would be necessary to make a rapid increase in the output from the remaining active beds and gas turbines.

(iiib) Reducing the bed depth by transferring bed material to a separate container and thus reducing the amount of tube surface which is immersed in the bed. This would take advantage of the fact that heat transfer coefficients to tube surfaces above the bed are much lower than to those in the bed. The tubing above the bed would also cool the gas so that the turbine entry temperature is also reduced.

Start-up. A PFBC has a large thermal capacity and cannot be heated up at a rate compatible with that at which a gas turbine unit has to be speeded up and synchronized to avoid prolonged operation close to compressor surge conditions. In most instances, start-up of the gas turbine unit will have to be with conventional (oil- or gas-fired) combustors or with an electric motor, and will be independent of the initial stages of the PFBC start-up for which combustion air may need to be supplied independently.

In the combined cycle, a full depth bed with all the cooling surface immersed can be operated only at about the full working pressure. It is therefore necessary to start up with an uncooled bed, and this is usually arranged by providing a space below the tube bank, of depth 0·5 to 1 m to accommodate a start-up bed. This can be heated initially by burning gas or oil in the fluidizing air or in the bed. Once coal combustion is initiated, the bed depth can be increased as the air flow and pressure increase. There is a delicate balance at this stage between maintaining fluidizing velocity, maintaining the correct turbine inlet temperature without overheating the conventional start-up combustor and minimizing the amount of independently-supplied air.

8   CONCLUSION

Combined cycle power generation using pressurized fluidized-bed combustion offers the possibility of significantly cheaper electric power from coal, in an environmentally-acceptable plant. This can be achieved without the development of more advanced gas turbine technology although it still has to be demonstrated that prolonged gas turbine life is possible with the hot gas clean-up systems currently available.

It is not possible to review all aspects of PFBC in a single chapter. The authors' main objective has been to illustrate the close relationship between the PFBC and the gas turbine. The two cannot be studied in isolation.

ACKNOWLEDGEMENT

The chapter is published with the approval of the National Coal Board but the views expressed are the authors' own and not necessarily those of the NCB.

BIBLIOGRAPHY

Comparative evaluation of Phase 1 results from: *Energy Conversion Alternative Study—ECAS*, NASA TMX-7855, 1976.
Comparative evaluation of Phase 2 results from: *Energy Conversion Alternative Study—ECAS*, NASA TMX-73515, 1977.
Energy Conversion Alternative Study—ECAS, General Electric Phase II Final Report, NASA CR134949, 1977.
Energy Conversion Alternative Study—ECAS, Westinghouse Phase II Final Report, NASA CR134942, 1977.
Energy Conversion Alternative Study—ECAS, United Technologies Phase II Final Report, NASA CR134955, 1977.
Evaluation of a PFBC Combined Cycle Power Plant Conceptual Design, Vols. 1–4. Burns & Roe Industrial Services Corporation Report, FE-2371-36, to US Department of Energy.

Various papers in: *Proc. 5th International Conference on Fluidised Bed Combustion*, Vols. 1 to 3, Washington, 1977.

Various papers in: *Proc. 6th International Conference on Fluidised Bed Combustion*, Vols. 1 to 3, Atlanta, 1980.

Pressurised Fluidised-bed Combustion Technology Exchange Workshop, US Department of Energy, Conf. 7906157, Secaucus, New Jersey, 1979.

References 1 to 6 contain information on economic and cycle evaluations; References 7 to 9 contain papers by most of the groups active in the PFBC and give a fair cross-section of present-day effort.

Chapter 5

SULPHUR RETENTION IN FLUIDIZED BED COMBUSTION

J. E. STANTAN

*National Coal Board Coal Utilisation Research Laboratory,
Leatherhead, UK*

1 INTRODUCTION

All fossil fuels contain sulphur. In coals the sulphur occurs mainly as pyrites (FeS_2) associated with mineral matter, and as organic sulphur compounds in the coal substance. The sulphur content of coals can vary over wide limits, from less than 1% up to 10%. For example, most British coals have between 1 and 2% of sulphur. Eastern US coals often contain up to 5% sulphur, but the western US coals generally have low sulphur contents (down to <1%).

Similar variability in sulphur content is exhibited by petroleum, and the variability is even greater in the products because of the tendency for sulphur to be concentrated in the heavier oil fractions when the petroleum is refined.

When fuels are burned the sulphur is released as sulphur dioxide and, to a smaller extent, as sulphur trioxide, both of which are air pollutants. Because fuels can vary widely in calorific value, the polluting effect from combustion is best expressed in terms of the quantity of pollutant released per unit of combustion heat liberated. For example, when burning a typical British coal of 1·5% sulphur content and gross calorific value (higher heating value) of about 28 MJ/kg, the sulphur dioxide released is 1·1 g/MJ.

Most countries now impose limits on emissions of sulphur oxides from combustion plants. In some instances the limits are severe; for example, in the United States all new thermal power plants must employ flue gas desulphurization to remove 90% of the sulphur dioxide, or to limit

199

emission to less than $0.52\,g/MJ$ ($1.2\,lb/10^6$ Btu), whichever is the more severe; the limit is relaxed to 70% removal for emissions below $0.26\,g/MJ$ ($0.6\,lb/10^6$ Btu). These strictures are mandatory even for combustion of the low-sulphur western coals.

The ash of coal often contains some free lime which, under suitable conditions, can act as a sorbent to 'fix' sulphur as calcium sulphate. Oil has only a negligible ash value, so that all of its sulphur content is liberated on combustion. The quantity of sulphur that can be retained by coal ash is usually too small to satisfy air pollution limits, and means for additional reduction in emission are therefore required. In this chapter, 'reduction in emission' and 'retention of sulphur' will be regarded as synonymous terms referring to the 'free' sulphur content of the coal, not to the total (free and fixed) sulphur.

One way of bringing about further reductions in sulphur oxides emissions is to provide additional lime for the furnace. Fluidized bed combustors provide ideal environments for this approach, and allow the potential for 'fixing' sulphur with lime to be realized to the full. The combustion temperature is about the optimum for absorption of sulphur oxides by lime, and the lime residence time is prolonged, usually several hours.

The lime can be fed to the combustor as limestone (calcium carbonate, $CaCO_3$) or as dolomite (calcium magnesium carbonate, $CaCO_3 . MgCO_3$) both of which occur widely and are extensively quarried. The stone is crushed to a suitable particle size (top size as little as $1.5\,mm$ for use at low fluidizing velocities, or up to $5–10\,mm$ for high fluidizing velocities) and is fed pneumatically to the base of the bed, or by gravity to the top of the bed, either separately or mixed with the fuel.

Compared with wet gas-scrubbing processes, which are the principal means for flue gas desulphurization in conventional plant, the retention of sulphur by lime in fluidized bed combustors has the advantage of not requiring additional scrubbing plant (including flue gas cooling and reheating equipment), of easier handling and disposal of spent sorbent, and of greater overall thermal efficiency. The process therefore has considerable technical and economic attractions, and these have provided the major motivation for the research and development of fluidized bed combustion during the past decade, particularly in the light of diminishing oil and natural gas reserves in the United States, the consequent need to turn to high-sulphur coals and the increasingly stringent air pollution legislation.

In this chapter we shall examine how sulphur retention in fluidized bed combustion is influenced by plant operating conditions, attempt to explain sorbent behaviour in terms of the chemistry and physics of the process, and indicate possible routes to more efficient and economical sorbent utilization. A selection of literature for further reading is listed at the end of the chapter.

2   THE EFFECTS OF OPERATING VARIABLES

All of the studies of the effects of operating variables on sulphur retention have been made with bench-scale and pilot-scale fluidized bed combustion plant. The principal centres of activity have been:

(a)   For atmospheric pressure:
Argonne National Laboratory (Argonne, Illinois, USA);
National Coal Board (Coal Research Establishment, Stoke Orchard, Gloucestershire, and Coal Utilisation Research Laboratory, formerly the British Coal Utilisation Research Association, Leatherhead, Surrey, UK);
Pope, Evans and Robbins, Inc. (Alexandria, Virginia, USA).

(b)   For pressurized combustion:
Argonne National Laboratory;
Exxon Research and Engineering Co. (Linden, New Jersey, USA);
National Coal Board (Leatherhead).

2.1 The calcium/sulphur mol ratio (Ca/S ratio)
Of all the plant operating parameters the most significant from the viewpoint of sulphur dioxide removal is the rate of supply of sorbent (limestone or dolomite). It is the only operating variable that allows sulphur retention to be controlled independently of other plant performance factors, and it is also the only variable allowing the sulphur dioxide emission to be reduced by any amount up to virtually complete removal. The most usual way of expressing the sorbent feed rate, and the most satisfactory for comparing performance, is in terms of the mol ratio of calcium in the sorbent fed to sulphur in the fuel burnt, referred to as the calcium/sulphur mol ratio, or more shortly as the Ca/S ratio.

For pure sorbents, a Ca/S ratio of 1 is equivalent to a stone/sulphur mass ratio of 3·12 for limestone, or 5·75 for dolomite. These figures show that for a given Ca/S ratio the mass of dolomite needed is nearly twice that of limestone. They also indicate that for fuels of high sulphur content and for operating conditions and emission limitations requiring high Ca/S ratios, the sorbent forms a significant proportion of the total feedstock to the

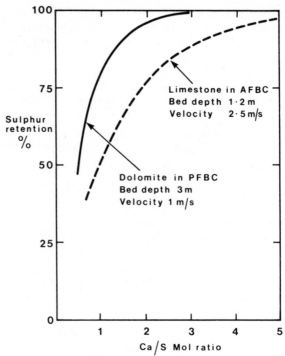

FIG. 1.   Variation of sulphur retention with Ca/S mol ratio: curves for typical
operating conditions.

combustor, and the spent sorbent becomes a major proportion of the
effluent solids.

The effect of varying Ca/S mol ratio on sulphur retention is illustrated by
the curves in Fig. 1. These curves are based on experimental data obtained
with dolomites in pressurized fluidized bed combustors (PFBC) and with
reactive limestones in atmospheric pressure combustors (AFBC) and they
relate to operating conditions likely to be employed in commercial plant. It
can be seen that excess sorbent is required to achieve high sulphur
retention and, in practice, the requirements for 90% sulphur retention
would be:

(a)   For limestone in atmospheric pressure fluidized bed combustors
      having shallow beds operated at high fluidizing velocities—Ca/S
      ratios between 3 and 5.
(b)   For dolomite in pressurized combustors with deep beds and low
      velocities—Ca/S ratios in the range 1·5 to 2·5.

The effect of Ca/S ratio, c, on the fraction of sulphur retained, R, can be approximated reasonably well by an empirical relationship of the form:

$$R = 1 - e^{-mc} \tag{1}$$

in which the parameter m is a function of the other main operating variables: bed depth, fluidizing velocity, sorbent particle size, sorbent type, bed temperature and pressure.

2.2 Bed depth and fluidizing velocity

Not surprisingly, sulphur retention is improved by an increase in bed depth or a decrease in fluidizing velocity. The effects of these two parameters are roughly equal but opposite, and they can be combined to define a 'superficial gas residence time', t_s, which is the ratio of bed depth to superficial fluidizing velocity. An empirical relationship between R and t_s is:

$$R/(1 - R) = t_s R_1/(1 - R_1) \tag{2}$$

where R_1 is the value of R when t_s is 1 s, all other operating conditions being unchanged.

A less empirical approach is based on the assumption that the rate of sulphur dioxide absorption, $-dS/dt$, is first-order with respect to sulphur dioxide concentration, S:

$$-dS/dt = kS \tag{3}$$

where k is the rate constant, which is found to be a function of sorbent particle size, sorbent type, temperature, pressure and the fraction α of sorbent sulphated. If we assume that the fuel burns and liberates sulphur dioxide uniformly throughout the bed—a reasonable assumption for atmospheric pressure combustors—we may integrate eqn. (3) to give:

$$R = 1 - (1 - e^{-k't_s})/(k't_s) \tag{4}$$

where k' is the value of k corrected for the actual (rather than superficial) gas residence time in the bed and for the inert solids (ash) content. For the lower velocities and deeper beds of pressurized combustors, a more valid assumption would be that all of the sulphur dioxide is liberated from the fuel at the base of the bed; integration of eqn. (3) would then result in:

$$R = 1 - e^{-k't_s} \tag{5}$$

For given values of k' and t_s, the value of R calculated by eqn. (5) for pressurized combustors is higher than that calculated by eqn. (4) for atmospheric pressure. The forms of both equations show that R is independent of the sulphur content of the fuel, provided that there is no change in the value of k' and hence of k and α. The value of k falls with an increase in α and may become negligibly small before the sorbent is fully sulphated. The relationship between k and α can be deduced from plant data or determined by the laboratory techniques discussed later, making allowances for differences between laboratory and fluidized bed conditions.

The fraction of sorbent sulphated, α, is itself a function of the residence time of the sorbent in the bed, and this in turn is related to fluidizing velocity (and hence sorbent throughput) and to bed depth. Thus, it may be an oversimplification to consider the effects of these variables only in terms of superficial gas residence time.

2.3 Sorbent particle size

The rate of sulphation of sorbent is directly related to the specific surface area of the particle which, for a spherical particle, is inversely proportional to its diameter. It might be anticipated therefore, that an increase in particle size would have the same effect as a decrease in Ca/S ratio or in superficial gas residence time. There are, however, several complicating factors.

A simple dependence on particle size may be clouded by the fact that sorbent particles are not spherical and often have a significant, readily accessible, internal surface. They are present in the bed as particles having a wide range of sizes which, because of attrition, decrepitation and elutriation, will differ from that of the sorbent fed. The finer particles, which would be expected to attain a high degree of sulphation, may have their residence in the bed curtailed by elutriation, so that it is often the coarse particles that achieve the greatest percentage sulphation.

Some of these effects are illustrated by the experimental results, obtained by Pope, Evans and Robbins, Inc., shown in Fig. 2. Starting with the coarsest sorbent particle size, there is a slight improvement in sulphur retention with decrease in particle size until a point is reached at which the sorbent residence time is curtailed by increases in elutriation, and sulphur retention suffers. A second critical stage is reached with further reductions in particle size when, with all of the sorbent being elutriated, the residence time of the particle is little more than that of the gas. Finer grinding does not then greatly affect particle residence time but again brings the benefits of increased specific surface.

The maximum and minimum in sulphur retention occur at particle sizes depending on sorbent reactivity and fluidizing velocity. The results in Fig. 2 were obtained with a limestone at a fluidizing velocity of 4 m/s and a bed

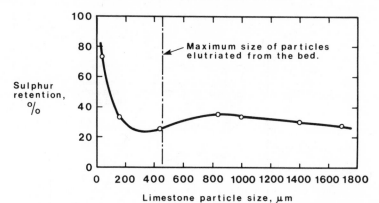

FIG. 2.   Effect of limestone particle size on sulphur retention: data of Ehrlich *et al.*
(Pope, Evans & Robbins, Inc.).

temperature of 815 °C, and at these conditions all particles finer than about 450 μm are elutriated.

2.4 Sorbent type

Sorbents vary widely and unpredictably in their reactivity and in other important characteristics. The greatest variability is exhibited by limestones to the extent that, for an unreactive type of limestone, the Ca/S ratio needed for a given sulphur retention may be two to four times as great as that for a reactive stone. The general rule is for dolomites to be of greater reactivity than most limestones, but at atmospheric pressure the superiority of dolomites is masked by their tendency to decrepitate to a fine powder and suffer elutriation before absorbing much sulphur. Dolomites come into their own when they do not decrepitate; as we shall see, this is in operation under high pressure.

Many of the most reactive limestones have a high content of mineral impurities, and many of the least reactive stones are very pure. Attempts have been made to correlate reactivity with calcium content of the stone, but there are too many exceptions for this to be a workable rule.

2.5 Fuel type

Some variability in sulphur retention has been ascribed to fuel type or sulphur content, usually for the want of another explanation. One possible explanation is that a fuel of low sulphur content requires a smaller mass throughput of sorbent, and this may result in a longer sorbent residence time in the bed and therefore more efficient utilization. Another is that

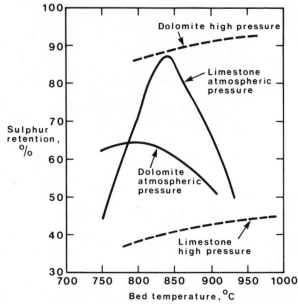

Fig. 3.   Effects of temperature and pressure on sulphur retention. From data for reactive limestones and dolomites sized 0–1600 μm, normalized to Ca/S = 2, $t_s = 0.5$ s.

some fuels have a high free lime content and require less added sorbent. On the whole, however, the effect of fuel type on sulphur retention is small. Even liquid fuels, which might be expected to differ markedly from coals in their sulphur oxides liberation characteristics in fluidized beds, exhibit almost identical dependence of sulphur retention on Ca/S ratio and other variables.

2.6 Bed temperature

The low energy of activation for the calcium oxide sulphation reaction—about 31 kJ/g mol (7·5 kcal/g mol)—would not lead to the expectation of any major effect of combustion temperature. In practice, however, temperature has a dramatic effect on sulphur retention, at least in atmospheric pressure fluidized bed combustion, as shown by the full curves in Fig. 3. The curves have been derived from experimental data for reactive limestones and dolomites sized 0–1600 μm; the data were normalized to a Ca/S ratio of 2 by applying eqn. (1) and to a superficial gas residence time, t_s, of 0·5 s by eqn. (2). The retention curves peak at 800–850 °C, the effect being more pronounced with limestone than with dolomite. It is found that the effect is greater at low Ca/S ratios, i.e. when the fraction of sorbent sulphated, α, is high; in fact laboratory studies show that at values of α below about 0·1 there is no peak.

The rise in sulphur retention efficiency at about 800 °C coincides with the onset and increasing extent of calcination of the sorbent, and is undoubtedly a consequence of the development of porosity upon calcination. The difference in behaviour between dolomite and limestone at lower temperatures is explained by the fact that dolomite half-calcines and becomes porous at low temperatures. These topics are enlarged upon later in this chapter, and explanations for the fall in sulphur retention at high temperatures will also emerge.

2.7 Combustion pressure

Operation at elevated pressures brings about some important changes in behaviour. The most striking of these can be seen by comparing the broken curves in Fig. 3 for high pressure with the full curves for atmospheric pressure; the high pressure curves show no peak in sulphur retention over the temperature range studied (up to about 950 °C).

In Fig. 3 the high pressure data were normalized to the same superficial gas residence time (0·5 s) as the atmospheric data; in fact most pressurized plant would have residence times of 2–3 s and hence better sulphur retention performance than indicated by the broken curves in Fig. 3. However, they serve to show another major effect of pressure. With dolomite there is a general improvement in performance. With limestones on the other hand the onset of calcination is delayed until a higher temperature is reached and their performance, which is generally poorer than at atmospheric pressure, is inferior to that of dolomites on a mol for mol basis even for reactive limestones. However, it has been observed that as a result of limestone having a molecular weight of 100, little more than half that of dolomite (184), the differences in sulphur absorbed per tonne of stone are small. There are indications that precalcined limestone may be as reactive under high pressure as dolomite on a Ca/S mol ratio basis.

3 THE CHEMISTRY OF SULPHUR RETENTION

3.1 Sulphation

Sulphur dioxide can react with sorbent which is present in the bed as uncalcined calcium carbonate:

$$CaCO_3 + SO_2 + \tfrac{1}{2}O_2 = CaSO_4 + CO_2 + 303 \, kJ/g \, mol \qquad (a)$$

However, as we have seen, it is preferable to operate with sorbent in the calcined form, i.e. quick lime (calcium oxide):

$$CaO + SO_2 + \tfrac{1}{2}O_2 = CaSO_4 + 486 \, kJ/g \, mol \tag{b}$$

The sorbent can be fed as quick lime, but since it is generally possible in atmospheric pressure combustors to operate under conditions allowing calcination of the sorbent *in situ*, it is usual to feed sorbent as raw stone.

It is likely that the sulphur dioxide and oxygen do not react simultaneously, as suggested by Reactions (a) and (b) above, but in a sequence of reactions, such as:

$$SO_2 + \tfrac{1}{2}O_2 = SO_3 \tag{c}$$

$$CaO + SO_3 = CaSO_4 \tag{d}$$

Reaction (c) certainly occurs in all combustion systems, giving rise to concentrations of sulphur trioxide that are usually low because, although high temperature favours the rate of reaction, it reduces the equilibrium concentration. In fluidized bed combustion the concentration of sulphur trioxide is usually 20–40 % of that expected at equilibrium; this may be partly a consequence of the intervention of Reaction (d).

3.2 Calcination

Unlike the sulphation reactions, calcination is endothermic:

$$CaCO_3 = CaO + CO_2 - 183 \, kJ/g \, mol \tag{e}$$

The reaction is governed by chemical equilibrium. The equilibrium partial pressure of carbon dioxide, P_e (bar), is related to the absolute temperature, T (K), by the Arrhenius relationship:

$$P_e = 1 \cdot 2 \times 10^7 e^{-E/RT} \tag{6}$$

where E is the activation energy (159 kJ/g mol) and R is the gas constant. Reaction will occur only if the partial pressure of carbon dioxide, which is dependent on total pressure and excess air level, is less than P_e. The dependence of the calcination temperature on these two operating variables is illustrated in Fig. 4, which also shows the equilibrium curve for pure carbon dioxide. The rate of calcination is determined by the extent to which the equilibrium temperature is exceeded.

3.3 Dolomite

Calcination of dolomite is a little more complex. The first step is thermal

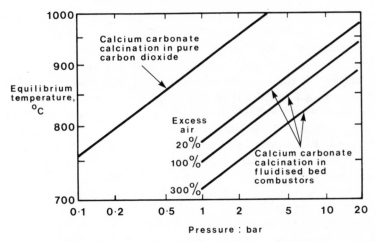

Fig. 4. Dependence of calcination equilibrium temperature on pressure and level of excess air.

decomposition, which occurs at about 620 °C and results in the formation of a mixture of calcium and magnesium carbonates:

$$CaCO_3 . MgCO_3 = (CaCO_3 + MgCO_3) - 32 \, kJ/g \, mol \qquad (f)$$

Magnesium carbonate is unstable under all conceivable fluidized bed combustion conditions, and undergoes rapid calcination:

$$(CaCO_3 + MgCO_3) = (CaCO_3 + MgO) + CO_2 - 100 \, kJ/g \, mol \qquad (g)$$

This is the so-called 'half-calcination' of dolomite. The calcium carbonate in the half-calcined dolomite can then react as in limestone, i.e. direct sulphation by reaction (a) or, if conditions permit, 'full-calcination' (reaction (e)) followed by sulphation of the oxide (reaction (b)).

There is evidence that in some circumstances the magnesia liberated by reaction (g) can react with sulphur dioxide and lime, forming small amounts of a calcium magnesium sulphate, $CaSO_4 . 3MgSO_4$. For most purposes, however, the magnesia can be regarded as inert.

We have already seen that with limestone, calcination is the key to high sorbent reactivity. With dolomite this is true even of the half-calcination (reaction (g)) which produces only calcium carbonate and inert magnesia. The crucial importance of calcination will be discussed later, after considering sorbent reactivity and reaction rates.

4   REACTIVITY OF SORBENTS

The reactivity of sorbents can be investigated by a number of laboratory techniques. The National Coal Board passed gas mixtures containing sulphur dioxide through thin packed beds of the sorbent supported in vertical tubes in a furnace. Possibly the most powerful tool is the Thermo-Gravimetric Analyser (TGA) as used by Argonne National Laboratory; it has also been used by Westinghouse Research and Development Center (Pittsburgh, Pa.) who have performed many hundreds of TGA runs, over a hundred of which have been at pressures up to 20 bar. In the TGA a few milligrammes of solid particles are held in the reacting gas, and heated to reaction temperature, in the pan of a balance that can give a continuous indication of the sample mass.

The results of typical tests of sorbent reaction rates are shown in Fig. 5 for a reactive dolomite, and for a limestone showing poor retention of sulphur in fluidized bed combustors. It can be noted that the reaction rate falls to a negligible value (less than 0·1 % per minute) before the sorbent is fully reacted, particularly with the limestone which had an initial reaction rate little smaller than that of the dolomite.

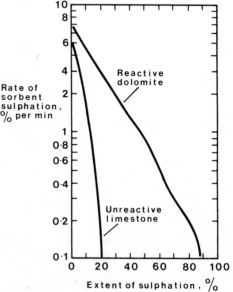

FIG. 5.   Typical sorbent reaction rate curves.

Studies have shown complex effects of reaction conditions on the reaction rate curves. For example, increase in temperature has little influence on the initial reaction rate but it does affect the maximum extent of sulphation attainable. At atmospheric pressure, the maximum sulphation rises with increase in temperature up to about 850 °C and then falls, but at high pressure there is a progressive increase in the maximum sulphation with temperatures up to about 1000 °C. This reflects the variation in sulphur retention in fluidized bed combustors already described and illustrated in Fig. 3.

Increase in pressure has a direct, albeit smaller than anticipated, effect on initial reaction rate, which is approximately doubled by an increase in pressure to 10 bar, but further increases in pressure up to 20 bar have little or no effect. The departure from a first-order effect of pressure is ascribed to control of reaction by the rate of diffusion of reactants to the reaction interface through pores which, as we shall see later, can be severely blocked by reaction product. Since diffusivity is inversely proportional to total pressure the diffusion controlled reaction rate is independent of pressure. This is likely to have an adverse effect on sulphur retention in pressurized combustion unless, as in supercharged boilers and air heaters, the bed volume is increased in relation to total pressure.

Particle size has a major effect both on the initial reaction rate and on the maximum sulphation attained; both are improved with decrease in particle size.

Oxygen and carbon dioxide concentrations have no effect on reaction rate, except in so far as they influence calcination. Uncalcined limestone generally has a somewhat lower initial reaction rate and a much lower maximum sulphation level than calcined limestone, but as the carbon dioxide concentration is reduced there is a progressive improvement in reactivity until conditions favour calcination. Most half-calcined dolomites are highly reactive and fully-calcined dolomites are even more so.

Although, as noted above, total pressure has only a small effect on reaction rate, the initial reaction rate is proportional to sulphur dioxide concentration at any given total pressure, as would be expected for first-order reaction kinetics; for reactive sorbents the first-order dependence on reactant concentration is maintained throughout the sulphation range. For sorbents of low reactivity, however, there is a progressive divergence from first-order kinetics with increase in sulphation. This is ascribed to the effects (which may be adverse or beneficial) of prolonged residence at reaction conditions. It can be concluded from the foregoing that there are no simple characterizing parameters for expressing sorbent reactivity. Some light is shed by considering the physical effects of calcination and sulphation, particularly in relation to porosity characteristics.

5   THE PHYSICAL EFFECTS OF SORBENT REACTIONS

5.1 The development of porosity

The reactivity of a solid reactant in a gas–solid reaction is generally a function of its porosity or of some related characteristic such as specific surface area. Extensive studies of sorbent porosity have been made by the National Coal Board, Argonne National Laboratory, and Westinghouse Research and Development Center.

A simple technique for measuring total porosity is to determine the particle density in mercury and in helium, and to calculate the specific volume (the reciprocal of the density). The specific volume in mercury includes the volume of the internal pores, but that in helium does not, since the helium can penetrate and fill all the pores. The total pore volume is therefore the difference between the mercury and helium specific volumes, and this can be expressed as a percentage of the mercury specific volume.

It was found in work by the National Coal Board that the percentage porosity developed by calcination of the raw sorbent is linearly related to the fraction of calcium present as free oxide, i.e. to the extent of calcination. This is shown by the open points in Fig. 6 for limestones and dolomites covering a range of reactivities. The porosity falls upon subsequent sulphation (closed points in Fig. 6) because the bulkier sulphate fills the pores. The closeness of the points in Fig. 6 for extremes of sorbent reactivity, and the fact that for an unreactive sorbent the reactivity can become negligible at a low sulphation level, i.e. when only a small proportion of the pores have been filled, indicates that percentage porosity alone cannot explain differences in reactivity. The study of pore size distribution proves more rewarding.

5.2 Pore size distribution

The mercury specific volume measured at low pressure includes the volume of internal pores, but if the pressure is increased the mercury can penetrate pores of progressively smaller entrance diameter. Because of the phenomenon of capillary depression, the mercury pressure needed to penetrate and fill a pore is related to the pore entrance diameter provided that the surface is not wetted by the mercury. This technique, known as 'mercury injection porosimetry' employs pressures ranging up to 1000 bar, and

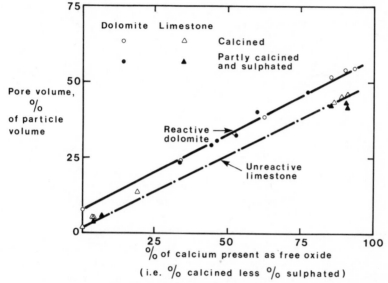

FIG. 6. Variation of sorbent porosity with extent of calcination and sulphation
(data of National Coal Board, 1971).

measures the volume of pores with entrance diameters down to about
0·01 μm. In the course of a determination the pressure is increased in steps,
and the cumulative pore volume is measured and plotted as a function of
decreasing pore entrance diameter.

Examples of cumulative pore volume distribution curves for calcined
sorbents are shown in Fig. 7 for a reactive dolomite, and for an unreactive
limestone. The curves reveal differences between the porosity characteristics of the sorbents. Thus, the reactive dolomite has the greatest abundance
of porosity in the pore entrance diameter range greater than 0·1 μm,
whereas the unreactive limestone has the greatest porosity in the entrance
diameter range below 0·1 μm. This suggests the following explanation for
the difference in reactivity.

The explanation is based on differences in molar volumes of solid
reactant and product. These volumes are, in ml/mol,

Calcium carbonate	37
Calcium oxide	17
Calcium sulphate	52

Calcination halves the molar volume resulting in the generation of

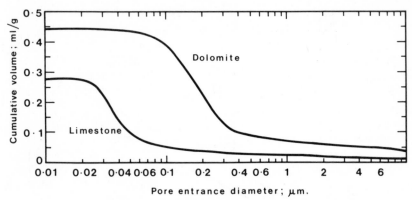

FIG. 7. Typical pore size distribution curves for calcined sorbents.

porosity; sulphation trebles the molar volume, resulting in loss of porosity. The bulky reaction product can readily block the smaller pore entrances, making the surfaces behind those entrances inaccessible to reacting gas which can then react only with the surfaces of larger and more accessible pores. This has a catastrophic effect with a sorbent that has most of its porosity in fine and easily blocked pores.

Pore size distributions of partly sulphated sorbents have shown that whereas unreactive sorbents suffer loss of porosity mainly in the finest pores, the losses in dolomite are more uniformly distributed throughout the pore size range. When the sulphated sorbent particles are broken and the pore size distributions redetermined there is no detectable change with dolomite, but a large increase is seen with unreactive limestone. This provides further confirmation that with such sorbents the reaction product blocks the pore entrances and renders pore surface inaccessible.

Thus, a reactive sorbent is one enjoying a proper balance between large 'transport' pores that give free access to the interior of the particle, and small pores that, since they provide extensive reaction surface, can effect a high degree of sulphation. The 'unreactive' sorbents are those inadequately endowed with large pores. The porosity characteristics developed by calcination depend to some extent on the physical characteristics of the raw sorbent, but they can be profoundly altered by the phenomenon of sintering.

5.3 Sintering

When calcium carbonate undergoes calcination, the calcium and oxygen atoms are left behind in their original positions in the calcium carbonate crystal lattice. This 'pseudo-lattice' is unstable, however, and the atoms migrate to form a more compact stable calcium oxide lattice. At this stage the lime is present as a large number of minute closely-packed crystallites having a high but inaccessible specific surface area. In the next stage the crystallites merge together to form larger crystals of lime. This is the process known as 'sintering' and it occurs at a rate and to an extent that is dependent on the sorbent characteristics. It is accelerated by high temperature and by high concentrations of carbon dioxide. Sintering can be promoted by the presence of certain additives, and this has implications that we shall examine later in this chapter.

Sintering results in a reduction in specific surface area accompanied by the production of larger pores. Thus, the tendency is towards achieving the better balance between large transport pores and fine reactive pores already referred to; but if sintering is prolonged the balance can be tipped too far, the diminution in reaction surface outstrips the increase in accessibility and the reactivity falls. The ultimate condition is that referred to as 'dead burned lime'.

5.4 Size degradation and elutriation

As with all crushed solids, limestone and dolomite contain fines, and this can lead to losses from the fluidized bed by elutriation, and consequently to less efficient utilization of the sorbent. Further fines can be produced by attrition and abrasion during the handling of the stone and in feeding it to the combustor. This problem can be minimized by good design and careful operation of the solids handling and feeding equipment. Once the sorbent has entered the combustor it generally calcines and can fall victim to the processes of decrepitation and of attrition and abrasion in the bed.

Decrepitation may be partly the familiar result of breakage from shock-heating of the particle, but the principal agency for decrepitation is the sudden evolution of water vapour from decomposition of impurities containing combined water, and of carbon dioxide. The effect is most marked at high temperature and low pressure, which favour rapid calcination and sudden generation within the particle of high relative water vapour and carbon dioxide pressures. The most severe decrepitation occurs at atmospheric pressure with dolomites, which evolve twice as much carbon dioxide as limestone; half of the gas is evolved, from the magnesium carbonate, at a high rate and with the generation of a high pressure, because the combustor is so far above the equilibrium temperature for magnesium carbonate calcination. In consequence it is not unusual for all of the dolomite fed to atmospheric pressure combustors to be reduced to a fine powder and elutriated. At high pressures the rates of calcination are lower and the water vapour and carbon dioxide pressures within the particles are also lower with respect to the total pressure, so that decrepitation is greatly reduced.

The possibility of renewed attrition and abrasion of particles occurring within the bed is enhanced by the fact that calcination converts the hard stone particles to a soft, easily breakable material. Degradation rates are dependent on fluidizing velocity and can be particularly high in the neighbourhood of air jets from coal feed or combustion air nozzles.

Sulphation of sorbent brings about two unexpected benefits. The first is that the formation of the sulphate strengthens the particle and helps to reduce the effects of decrepitation and attrition. The second is that the sulphate shell that forms on the surface of a low-reactivity limestone can be broken off by attrition, revealing fresh reaction surface and increasing the utilization of the sorbent.

6 CAUSES AND EFFECTS

6.1 The physical reasons for performance factors
The foregoing review of the chemical and physical behaviour of sorbents allows us to reconsider their sulphur retention performance in fluidized bed combustors, and to attempt explanations for the effects of operating variables. It is not a simple situation; many of the effects can be accounted for by more than one feature of sorbent behaviour, and some sorbent characteristics explain more than one of the performance factors.

6.2 The Ca/S ratio
An increase in Ca/S ratio results in an increase in the sorbent inventory in the bed. Furthermore, since Ca/S ratio (c), fractional sulphur retention (R) and fractional sorbent utilization (α) are related by:

$$\alpha = R/c \tag{7}$$

where $R < 1$ and $c > R$, an increase in c is equivalent to a reduction in α which, as we have already seen in considering sorbent reactivity, brings about an increase in specific reaction rate, $d\alpha/dt$. Hence, in addition to increasing the sorbent inventory, an increase in Ca/S ratio also increases the average reactivity of the sorbent in the bed; sulphur retention benefits on both accounts. The observation that a step change in Ca/S ratio brings about only a slow change in sulphur retention is an indication that reaction involves the whole of the sorbent inventory, and not merely the fresh sorbent entering the bed with the fuel.

6.3 Bed depth and fluidizing velocity
An indication has already been given that although these two variables can be combined as gas residence time, to correlate sulphur retention data, they have other important effects. A reduction in fluidizing velocity results in lower elutriation losses and therefore an increase in the inventory of fine particles which can react more rapidly and more completely. Increase in bed depth or decrease in fluidizing velocity result in increase in sorbent residence time, so that the utilization, α, is increased and hence, from eqn. (7), c is reduced for a given value of R.

6.4 Sorbent particle size
As demonstrated by laboratory studies, reduction in particle size increases the reaction rate of sorbents. Furthermore, since the formation of bulky reaction product hinders further reaction to an extent dependent on the depth of penetration of the reaction zone within the particle, a reduction in particle size increases the particle surface/volume ratio and allows a greater proportion of the sorbent to be utilized. The sulphur retention observed with dolomites at atmospheric pressure appears to be independent of the size grading of the dolomite fed; this is probably because under these conditions, decrepitation reduces the dolomite to a fine powder before sulphation occurs.

6.5 Sorbent type
The great variability seen in the reactivity of different sorbents is matched by a corresponding variability in pore size distribution, the effects of sintering and the propensity for decrepitation and attrition. These factors are to some extent functions of the physical structure of the uncalcined sorbent; for example, although dolomites generally have a well-balanced pore-size distribution that gives good reactivity characteristics, some large-grained dolomites have poor reactivity because most of their porosity is in the form of pores of small entrance diameter. The beneficial effects often conferred by impurities in the sorbent can result from a number of causes including:

(a) The impurity reacts with some of the sorbent giving accelerated sintering—this can sometimes go too far, however, leading to local fusing and loss of reaction surface, or to combination of lime with, for example, silica.

(b)   The impurity provides a 'skeleton' for the sorbent particle which gives it greater resistance to decrepitation and attrition.

(c)   The presence of the impurity may create a degree of local porosity that improves access to the interior of the particle.

6.6 Bed temperature and pressure

Attempts to explain the dramatic effects of these variables on sulphur retention have been made by many workers, including those in the National Coal Board (Leatherhead, UK), Central Electricity Generating Board (Marchwood, UK), Esso Research Centre (Abingdon, UK), Argonne National Laboratory (Argonne, US) and Westinghouse Research and Development Center (Pittsburgh, US).

It is most probable that several, if not all of the causes that have been proposed contribute to the observed effects. We shall consider in detail only those that appear to the author to be the most likely. These ascribe the effects of bed temperature and pressure to:

(a)   Porosity characteristics.

(b)   Sorbent decrepitation.

(c)   Reaction with sulphur trioxide.

6.6.1 Porosity Characteristics

The uncalcined sorbent has very little porosity (Fig. 6), so that at temperatures below the calcination equilibrium temperature, reaction is confined almost entirely to the outer surface of the particle. In a bed operating at just under the calcination equilibrium temperature, the sorbent particles will circulate through zones of low carbon dioxide partial pressure, e.g. near the air distributor, and some calcination will therefore occur at lower bed temperatures than would be expected. The rate of calcination increases with temperature; the activation energy for the rate constant is about 170 kJ/g mol and the reaction rate is further accelerated by increasing remoteness from equilibrium conditions. The increase in calcination rate is matched by an increase in the rate of development of porosity and hence availability of surface for reaction with sulphur oxides. Since dolomite develops porosity at well below the lime calcination temperature as a consequence of half-calcination, it can show high sulphation rates at low temperature.

When the bed temperature is more than about 50 °C above the equilibrium lime calcination temperature, fresh reaction surface is generated in both limestone and dolomite at a high rate, and entrances to the pores undergo blockage by formation of calcium sulphate faster than they can be opened by sintering. This premature blockage of pores becomes increasingly acute with further rises in temperature, and the sulphur retention accordingly falls. Operation under pressure increases the temperature at which calcination first occurs, and gives lower calcination rates but higher sintering rates at high temperature. Sintering is therefore better able to keep pace with the development of porosity, and the high-temperature sulphur retention capabilities of the sorbent are therefore good. Dolomites have an advantage over limestones at high pressure in that they are reactive over the whole FBC temperature range whereas limestones are reactive only in the upper part of the temperature range.

6.6.2 Decrepitation
The extent of this phenomenon is dependent on the rate of heating and on the gas and vapour pressures generated within the particle in relation to the total pressure in the combustor. The consequence of decrepitation is elutriation of the fines produced and hence curtailment of sorbent utilization. These factors help to explain the peak in sulphur retention efficiency at atmospheric pressure, the fact that this peak is flatter for dolomite and at a lower level of sulphur retention than would be expected for a sorbent that is inherently more reactive than limestone, and the disappearance of these effects and the display of the superior reactivity of dolomite at high pressure.

6.6.3 Reaction with Sulphur Trioxide
If the sulphation of sorbent is mainly by direct reaction with sulphur trioxide, the adverse effects of high temperature and low pressure on the equilibrium concentration of sulphur trioxide would be reflected in their influence on sulphur retention. On the other hand, high temperature accelerates the oxidation of sulphur dioxide to trioxide, and there is some experimental justification for the belief that it also increases the tendency for sorbent to react directly with trioxide. This results in a greater likelihood of reaction of the sulphur oxide at the first available reaction sites, i.e. the pore entrances, with the consequential sealing of the pore and inhibition of further reaction that we have already considered.

7   IMPROVING THE PERFORMANCE OF SORBENTS

7.1 The incentive for improvements
The removal of sulphur dioxide in fluidized bed combustion by the methods discussed in this chapter has major technical, economic and

J. E. Stantan

thermal efficiency advantages over other techniques employed in conventional combustion systems. Nevertheless, there are incentives for improving the sulphur retention performance of the sorbents used and so bringing about reductions in:

(a) The costs of fresh sorbent and disposal of spent sorbent—with fluidized bed combustion, the fresh sorbent requirement would generally be higher than with stack gas scrubbing, but although the quantity of spent sorbent would also be greater, its disposal would be easier. Furthermore, compared with stack gas scrubbing, the capital costs of additional equipment needed for desulphurization would be minimal.

(b) Losses in plant thermal efficiency—the calcination of fresh sorbent is endothermic, and spent sorbent removes sensible heat from the system, so that although the sulphation reaction is exothermic, the use of limestone or dolomite, particularly at atmospheric pressure, usually increases the consumption of fuel. However, the losses in thermal efficiency would be smaller than those brought about by stack gas scrubbing which involves cooling and reheating of the gas.

7.2 Plant design and operation

Some of the obvious routes to reduced sorbent consumption—low fluidizing velocity, deep beds, finer particle sizes—bear heavy capital or operating cost penalties, particularly at atmospheric pressure, and are therefore adopted only in pressurized plant. In circumstances where worthwhile improvements in combustion efficiency are obtained by recycling elutriated fines to the bed, it is often found possible to reduce fresh sorbent consumption. This is particularly true under those conditions, e.g. at atmospheric pressure, where most of the sulphur is retained by sorbent that suffers elutriation, but if, as at high pressure, the sulphur is mainly retained by larger particles that do not become elutriated, recycling has little or no benefit. It is doubtful whether the added plant complexity and operating cost would ever justify fines recycle unless the main benefit were improved combustion efficiency. Another operating variable that has fallen from favour in current thinking is the use of finely ground limestone (e.g. finer than 50 μm particle size). This technique improves the reactivity of the sorbent by increasing its external specific surface area (Fig. 2), but the costs of fine grinding, and the difficulties of efficient collection and environmentally acceptable disposal of the spent sorbent are severe.

7.3 Precalcination of sorbent

This has proved beneficial particularly for operation under conditions of high pressure and/or low temperature (see Fig. 4) in which calcination of limestone *in situ* is retarded or prevented. In these circumstances precalcination of limestone may result in a sulphur dioxide absorption performance matching that of dolomite on a Ca/S mol ratio basis, and therefore superior to dolomite on a raw stone to sulphur mass ratio basis.

7.4 The use of additives

It has been shown, in pilot plants by Pope, Evans and Robbins, Inc., and in the laboratory by Argonne National Laboratory, that some additives can improve sorbent reactivity. The most effective additives are sodium carbonate and sulphate, and the chlorides and hydroxides of sodium, calcium and magnesium. Careful control of additive concentration is needed, the optimum quantity being dependent on the additive and on the sorbent.

At concentrations below about 0·5–1·0 mol % the additive forms traces of a liquid phase with lime, accelerating recrystallization and sintering and hence producing a desirable pore size distribution. At higher additive concentrations a second mechanism comes into play; the additive forms significant quantities of a liquid phase in which the lime reacts in solution, and the consequent mobility of reactants and products allows for more extensive sulphation of the sorbent.

The treatment of the sorbent with additive, which has to be sprayed as a solution or slurry on to the crushed sorbent, followed by drying, adds another stage to sorbent preparation. It may be worthwhile if the only cheaply available sorbent has a low reactivity, but a potential disadvantage is that some of the additives are volatile and are high-temperature corrosion agents; depending on the particular application, this would often limit the choice of additive.

7.5 Hydration

Free lime can be hydrated ('slaked') with water:

$$CaO + H_2O = Ca(OH)_2 \qquad \text{(h)}$$

The reaction is reversible: the free lime is recovered by calcination, and it has been shown by the Argonne National Laboratory that this results in a particularly favourable pore size distribution. However, hydrated lime suffers too greatly from attrition unless it is first 'strengthened' by partial sulphation. Repeated hydration and recycling of spent sorbent can give high utilization of even unreactive limestones.

7.6 Regeneration

This is the process of stripping the sulphur from spent sorbent, and since it has the potential for the biggest savings in fresh sorbent requirements and spent material disposal, it has been extensively studied in the US on the laboratory and pilot-plant scales. The spent sorbent is treated at about $1100\,°C$ under mildly reducing conditions, achieved by combustion at close to stoichiometric air/fuel ratio. The reaction occurs in two stages, reduction to sulphide

$$CaSO_4 + 4 \begin{cases} CO \\ H_2 \end{cases} = CaS + 4 \begin{cases} CO_2 \\ H_2O \end{cases} \tag{i}$$

followed by solid–solid reaction to liberate sulphur dioxide:

$$CaS + 3CaSO_4 = 4CaO + 4SO_2 \tag{j}$$

A small progressive loss of reactivity occurs with each regeneration cycle, giving an average useful life in a combustor of five to ten cycles and necessitating a small make-up of fresh sorbent. The deactivation results from incipient melting in the sulphate–sulphide system, but this increases the mechanical strength of the particle—an important factor in a regeneration system which demands high resistance to attrition to ensure prolonged particle integrity. The chemistry of regeneration has ties with that of the gasification of oil in 'Chemically Active Fluidised Beds' of lime developed by Esso Research Centre (Abingdon, UK).

Chemical equilibrium for reaction (j) is favoured by low pressure, and the process should therefore be particularly suitable for operation at atmospheric pressure, which gives higher equilibrium concentrations of sulphur dioxide in the off-gas. Although regeneration has been studied in AFBC plant by Pope, Evans and Robbins, Inc. and Argonne National Laboratory, the most extensive studies were made by Exxon Research and Engineering Co. in their 320 mm diameter pilot-scale PFBC boiler, which was integrated with a 215 mm diameter continuous fluidized bed sorbent regenerator; this so-called 'Miniplant' is illustrated in Fig. 8.

Sulphated sorbent flowed from the combustor to the regenerator which was supplied with air and fuel, and with additional air to the upper part of the regenerator bed to ensure elimination of sulphide. Regenerated sorbent flowed back to the combustor. The two solids transfer ducts in the Miniplant operated under non-fluidized stick-slip gravity flow conditions

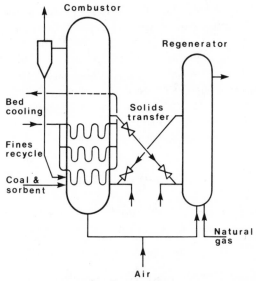

FIG. 8. Exxon 'Miniplant' integrated PFBC and regenerator.

to minimize the need for transport air, and the circulation rate was controlled by a pulsed nitrogen jet at the lower end of one of the transfer ducts. Other solids circulation systems could be used in full-scale plant, however.

Reaction is limited by the effect of high pressure on equilibrium, and by the need to avoid temperatures that could cause 'dead-burning' of the lime or fluxing with coal ash. Further limitations on sulphur dioxide concentrations in the regenerator off-gas would be imposed under conditions of reduced sulphur input to the combustor; in such circumstances it would not always be possible to reduce the flow of regenerator gases in proportion to the sulphur throughput.

There is a choice of processes for recovering the sulphur from the regenerator effluent gases as elemental sulphur or as sulphuric acid. A process that recovers elemental sulphur gives a product that is more versatile commercially than sulphuric acid, or that is safer to dump. Since most of the processes for sulphur or sulphuric acid recovery are variants of flue gas desulphurization processes, some of the advantage of sulphur retention in fluidized beds is lost.

7.7 The penalties
The benefits of processes that reduce consumption of fresh sorbent are

obvious, but they all carry penalties in terms of increased complexity and capital cost of plant, or increased fuel consumption. Nevertheless, these penalties may be an acceptable price to pay in circumstances where the fuel has a high sulphur content, or the locally available sorbent is of poor reactivity, or where disposal of spent sorbent poses particularly difficult practical problems. There is no doubt that there is a choice of technically feasible methods for reducing fresh sorbent consumption in such circumstances, but once-through utilization of untreated sorbent will always be the choice when other restraints do not apply.

ACKNOWLEDGEMENT

This chapter is published with the approval of the National Coal Board but the views expressed are the author's own and not necessarily those of the NCB.

BIBLIOGRAPHY

The following references constitute only a small selection from the literature relating to sulphur retention in fluidized bed combustors. They are arranged according to the organizations from which the papers and reports have emanated, and an indication is given of the relevant subject matter.

Argonne National Laboratory (Argonne, Illinois):
SWIFT, W. M., MONTAGNA, J. C., SMITH, G. W. and SMYK, E. B., Process costs and flowsheets, bed defluidization characteristics, stone reactivity changes and attrition losses for a regenerative fluidized-bed combustion process, Report no. ANL/CEN/FE-78-14 to US Department of Energy, May 1980. (Subject: Regeneration.)
JOHNSON, I. *et al.*, Support studies in fluidized bed combustion: Annual Report, October 1978–September 1979. Report no. ANL/CEN/FE-79-14 to US Department of Energy. (Subject: Hydration of sulphated sorbents.)
JOHNSON, I. *et al.*, Investigation of limestone sulphation enhancement agents and their corrosion rates in FBCs: Annual Report, October 1978–September 1979, Report no. ANL/CEN/FE-79-15 to US Department of Energy. (Subject: Use of additives.)

Cambridge University and Central Electricity Generating Board (Marchwood Engineering Laboratory, Hampshire):
FIELDES, R. B., BURDETT, N. A. and DAVIDSON, J. F., Reaction of sulphur dioxide with limestone particles: The influence of sulphur trioxide, *J. Instn Chem. Engrs.*, **57**, 276–80 (1979). (Subject: Role of sulphur trioxide.)

Esso Research Centre (Abingdon, Bucks):
Moss, G., The mechanisms of sulphur absorption in fluidised beds of lime, *Fluidised Combustion*, Vol. 1, Inst. Fuel Symp. Series No. 1, 1975. (Subject: Role of sulphur trioxide.)

Exxon Research and Engineering Co. (Linden, New Jersey):
Hoke, R. C. et al., Miniplant and bench studies of pressurized fluidized-bed coal combustion: Final report, US Environmental Protection Agency, Report no. EPA—600/7-80-013, January 1980. (Subjects: Effects of operating variables under pressure; regeneration.)

National Coal Board Coal Research Establishment (Stoke Orchard, Glos.) and Coal Utilisation Research Laboratory (Leatherhead, Surrey):
Reduction of atmospheric pollution: Final report on research on reducing emission of sulphur oxides, nitrogen oxides and particulates by using fluidised combustion of coal, NCB Report, Ref. DHB 060971, to US Environmental Protection Agency, September 1971. Main report and eight appendices. (Subjects: Effects of operating conditions at atmospheric and high pressure; sorbent reactivity; sorbent porosity; mathematical modelling.)
Roberts, A. G., Stantan, J. E., Wilkins, D. M., Beacham, B. and Hoy, H. R., Fluidised combustion of coal and oil under pressure, *Fluidised Combustion*, Vol. 1, Inst. Fuel Symp. Series No. 1, 1975. (Subjects: Effects of temperature, pressure and fuel type.)

Pope, Evans and Robbins Inc. (Alexandria, Virginia):
Ehrlich, S., Robison, E. B., Gordon, J. S. and Bishop, J. W., Development of a fluidized-bed boiler, *A.I.Ch.E. Symp. Series*, **68**(126), 231–40 (1972). (Subjects: Effects of operating variables; regeneration.)

Westinghouse Research and Engineering Center (Pittsburgh, Pennsylvania):
Ulerich, N. H., Vaux, W. G., Newby, R. A. and Keairns, D. L., Experimental/engineering support for EPA's FBC program: Final report, Vol. 1, Sulfur oxide control, US Environmental Protection Agency, Report No. EPA-600/7-80-015a, January 1980. (Subjects: Sorbent reactivity; sorbent porosity; decrepitation and attrition; regeneration; mathematical modelling.)

Chapter 6

EMISSIONS OF NITROGEN OXIDES

J. T. SHAW

National Coal Board, Coal Research Establishment, Cheltenham, UK

1 INTRODUCTION

Nitrogen oxides, often written NO_x, are taken to consist of nitric oxide (NO) and nitrogen dioxide (NO_2). Nitrous oxide (N_2O) is not included, and its concentration has seldom been measured. Unlike the others, it is not toxic, but perhaps more attention should be paid to it in the future because of its role in stratospheric chemistry.[1]

The nitrogen dioxide in cooled flue gas seldom amounts to more than 10% of the nitric oxide by volume and is usually much less. However, its measured concentration exceeds what could exist in the hot flue gas, and the balance must therefore be formed as the gases cool, by the oxidation of nitric oxide.

Measurements of nitric oxide were not very reliable at one time, but the advent of the 'chemiluminescence' monitor and of gravimetrically prepared gas mixtures containing known concentrations of nitric oxide in nitrogen, for calibration, have made it easy to obtain an accurate and continuous record of nitric oxide emissions.[2] Almost all the NCB's data quoted here were obtained in this way.

The emissions of nitrogen oxides described in this chapter cover a wide range of concentrations, and there is a good deal of contradictory information. Warning is given that as most of the data emanates from rather small experimental rigs, it may not prove to be a reliable guide to the emission of nitrogen oxides from the large commercial combustors that are in prospect.

At present the largest non-pressurized FBC in Britain is the one at Babcock's Renfrew Works. It can generate 21 000 kg/h of steam. Its NO_x

emissions are in the range 300–400 vpm.[3] A large pressurized FBC[4] was being commissioned by the International Energy Agency at Grimethorpe (Yorkshire) at the time of writing, but it was too early for details of its NO_x emissions to be made known. When the pressure was 0·5 MPa (5 atm), the NO_x concentrations from the pressurized fluidized bed combustors at the NCB's Leatherhead Laboratories depended mainly on the excess air, and were usually in the range 50–250 vpm.[5]

TABLE 1

Calculated equilibrium concentrations of nitric oxide and nitrogen dioxide in flue gas at the temperatures quoted: Combustion of a bituminous coal at atmospheric pressure in 10% excess air

	Temperature (°C)				
	727	927	1 227	1 627	1 827
Approx. vpm nitric oxide	11	68	420	1 900	3 300
Approx. vpm nitrogen dioxide	0·2	0·4	0·7	1·1	1·4

Note: Pressurizing to 5 atm makes very little difference to the calculated concentrations. For example, at 1 227 °C (1 500 K) the concentration of nitric oxide would be altered by less than 0·1 vpm. The concentration of nitrogen dioxide would be roughly doubled at all the temperatures shown, but would still be extremely small. Source: ref. 8.

For comparison, a large modern pulverized-fuel-fired power station boiler would be likely to give off 400 to 600 vpm of nitric oxide at full load.[6] A chain-grate shell boiler, investigated by the author, gave off about 150 vpm.[6] Table 1 shows the concentrations of nitric oxide and nitrogen dioxide that would be expected at various temperatures if the flue gas were in chemical equilibrium.[7]

2   HISTORICAL

In its early days fluidized bed combustion was expected to yield low emissions of NO_x, in accordance with the equilibrium concentrations shown in Table 1 for the appropriate temperature range. However, at about that time Sutton and Starkman, measuring the production of nitric oxide by an ammonia-fuelled automotive engine, had shown that it greatly exceeded the equilibrium values at the temperatures obtained.[9] A little later, Shaw and Thomas showed that the combustion of certain volatile organic nitrogen compounds in a flat $CO-O_2$–argon flame at temperatures similar to those of fluidized combustion could yield nitric oxide concentrations that greatly exceeded the calculated equilibrium values. The nitric oxide must have come from the organic nitrogen compounds.[10] Soon afterwards Jonke proved that in certain conditions the emission of nitric oxide from fluidized bed coal combustion was almost entirely due to the oxidation of the nitrogen that was chemically combined in the coal.[11] The kinetic overshoot that was implied by the super-equilibrium concentration was taken to show that the rate of formation of NO_x from the 'fuel-nitrogen' was much faster than the rate of oxidation of atmospheric nitrogen in similar conditions, but that there was no comparable acceleration of the decomposition or reduction reactions of NO_x.[8]

Many studies of the effects of various conditions on the NO_x concentration have been made and much information, some of it contradictory, has been obtained about the effects of such important variables as the pressure, temperature, oxygen concentration, and addition of sulphur retention stone to the bed. Several mathematical models have been put forward, but none are comprehensive. In what follows, the diversity of the experimental results is briefly described, and this is followed by an outline of the author's views on the underlying reasons for the phenomena.

3   EXPERIMENTAL RESULTS

As part of its development work on fluidized combustion, the National Coal Board has tested a wide variety of solid fuels and sulphur retention stones in fluidized combustors of several designs. It has thus accumulated a good deal of information on the resulting emissions of NO_x and sulphur dioxide. The NCB's data on NO_x are summarized in this section, and they are sufficiently varied to make it unnecessary to cast the net much wider, though information from some other sources has been included.

3.1 At atmospheric pressure
Two results are mentioned at the outset to illustrate the wide range of NO_x concentrations that can be encountered. During the fluidized combustion of a graphitic rock, fewer than 10 vpm of nitric oxide were measured in the flue gas, even when the temperature of the bed was as high as 1200 °C.[12] On the other hand, the combustion at 1000 °C of certain coal fines mixed with a calcareous sulphur acceptor (Ca/S molar ratio of 8/1) gave off no less than 1200 vpm of nitric oxide.[6] The graphitic rock contained less than 0·1 % by weight of chemically combined nitrogen, and the oxidation of atmospheric

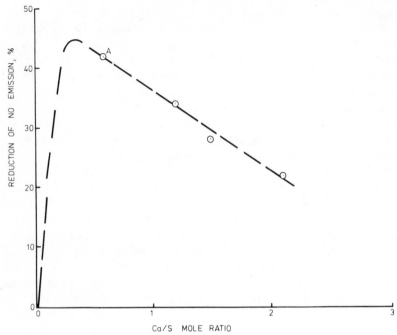

FIG. 1.   Nitric oxide reduction at atmospheric pressure with 325 mesh ($<44\,\mu$m) tymochtee dolomite. (After Jonke[14] by courtesy of Argonne National Laboratory, USA. The portion of the line between point A and the origin has been conjecturally added by the present author as the graph must start at the origin.)

nitrogen was obviously negligible. It is not known why the other test gave such a high result; some oxidation of atmospheric nitrogen may have occurred, although the emission of NO_x did not exceed the equivalent of the 'fuel-nitrogen' which amounted to 0.5% by weight of the mixture as fired.

There has been some disagreement about the effects that sulphur retention stones such as limestone and dolomite have on the NO_x emission, but the NCB has noted several instances in which an unusually high content of free lime in the bed has been associated with an unusually high concentration of NO_x.[6] Also, during tests of the NCB's air-blown fluidized bed gasifiers, the presence of free lime has often caused a large increase in the ammonia content of the product gas.[13] However, there is evidence that moderate additions of sulphur retention stone can sometimes *lower* the NO_x emission from combustors, as shown by Fig. 1 which has been modified from one given by Jonke.[14] It may be significant that in Jonke's work the bed consisted of alumina at the start, whereas in the NCB's work it has been either silica sand or spent sulphur retention stone recovered from previous tests. In Jonke's work the Ca/S ratio was not the only influential factor, as the NO emission in the absence of additive decreased considerably from one experiment to the next, which casts doubt on the value of the result.

The combined-nitrogen content of the fuel is usually influential. In conjunction with the British Petroleum Co. Ltd, the National Coal Board observed a good example of this when firing a combustor variously with propane, with fuel oil and with bituminous coal, having combined nitrogen contents of 0, 0·3 and 1·9 %, by weight, respectively. The NO_x emissions ranged between 15 and 20 vpm for propane, 70 and 160 vpm for the fuel oil, 400 and 500 vpm for the coal.[8] More recently, a combustor was fired with

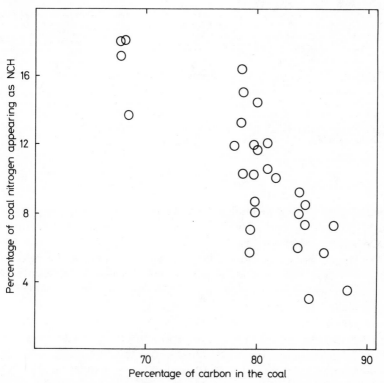

FIG. 2. Percentage of coal nitrogen appearing as hydrogen cyanide on devolatilization for 30 s at 800–850 °C *in vacuo*. (Adapted from Solomon[16] by courtesy of United Technologies Research Center, USA.)

TABLE 2

Brief descriptions of NCB's fluidized bed combustors mentioned in the text

Combustor reference letter	Pressure (MPa; 0·1 MPa = 1 atm)	Plan	Configuration of base	Expanded bed height (m)	Fuel feed
A	0·1	Circular, 0·15 m diameter, above conical base	60° cone. 39 air holes. A 38 mm diameter central hole (normally closed)	Up to 0·91 m above apex of cone	Pneumatic, through a horizontal pipe terminating at side of cylindrical reactor immediately above its junction with the cone
B	0·1	Square, 0·3 m × 0·3 m	Horizontal, flat. Air via 9 capped stand pipes on a 0·076 m pitch	0·7 m controlled by an open-ended stand pipe	Pneumatic, through a central down-facing pipe terminating 0·1 m above the fluidizing stand pipes
C	0·1	Rectangular, 0·56 m × 0·8 m	Horizontal, flat. 70 capped stand pipes for air	Up to 0·3 m above stand pipes	Stainless steel chute discharging within expanded bed, at centre, from above
D	0·1	Circular, 0·15 m diameter	120° cone. 185 air holes. A 25 mm diameter central hole (normally closed)	Up to 0·76 m	Pneumatic, through a horizontal pipe injecting the fuel tangentially, immediately above the base
E	0·1	Circular, 1·52 m diameter	Horizontal, flat. 300 capped stand pipes for air	0·34 m above stand pipes	Coal falls on to the centre of the bed from a 0·1 m diameter pipe which terminates 0·9 m above the expanded bed
F	Up to 0·61 MPa	Circular, 0·3 m diameter	Flat. Bubble cap distributor (20 units)	1·0 m	Pneumatically, horizontally, about 100 mm above the stand pipes, through the 50-mm long arms of a vertical, central T-nozzle

ten different coals sequentially. Their DMMF nitrogen-contents ranged from 1·3 to 1·9% w/w. The NO_x output was strongly linked with the nitrogen-content of the coal.[15]

The mode of combination of the nitrogen may be influential also, but little has been published on how this varies as between one coal and another. Figure 2, after Solomon,[16] shows that there may be important variations.

The effects of excess air, fuel feed rate, fluidizing velocity, bed height and temperature are discussed next, in the light of the NCB's recent work.[6] It is shown that the effects that were found at one test rig were very different from those that were found at another one. The leading characteristics of the rigs are given in Table 2.

In a fluidized bed combustor containing heat exchange tubes, it is possible to vary the mass flows of coal and air, and to vary the bed height,

TABLE 2—*contd.*

Size range of fuel	Bed material	Sulphur retention stone feed	Surplus bed material extracted via	Constructional material in contact with bed	Temperature control by
Up to 1·7 mm	(1) Silica sand (2) Silica sand plus dolomite derivatives	Batchwise through freeboard	38 mm hole at apex of base	Stainless steel	Electrical heaters surrounding the reaction. Removable lagging
Up to 1·7 mm	Partly sulphated, half-calcined dolomite from previous runs	Pneumatic; the feedpipe joins the fuel feed-pipe outside the reactor	51 mm hole in base, also stand pipe	Refractory	Air-cooled field tubes in the bed
Nominally 12 to 25 mm	Silica sand	None	Not applicable	Cast iron water-cooled wall. Refractory wall. Stainless steel	Coal feed rate
Up to 1·7 mm	Various	By premixture with the fuel	25 mm diameter hole at apex of base	Stainless steel	Electrical heaters surrounding the reactor. Water-cooled tubing in the bed
Top size 25 mm 5% below 6 mm	Silica sand with some lumps of shale	None	Hole in base	Water-cooled mild steel	Water-cooled coils, also coal feed rate
Up to 3·2 mm	Mainly Molochite.[a] Some coal ash	None	Via an offtake at side	Refractory	Excess air

[a] Molochite is a mixture containing about 56% of mullite ($3Al_2O_3 . 2SiO_2$) and 44% of amorphous silica glass.

and the bed temperature, all independently. But if the mass flow of incoming fuel-nitrogen is to be held constant (with a given coal) the air to fuel ratio cannot be varied independently of the fluidizing velocity at constant temperature; and if the fluidizing velocity is to be held constant, the air to fuel ratio cannot be varied at constant temperature without altering the mass flow of incoming fuel-nitrogen. For research purposes these interconnections can be removed by arranging to dilute the fluidizing air with varying amounts of an inert gas. Preferably the 'transport' air (used for carrying the coal and the sulphur retention stone into the bed) should be diluted too. As an example of what can be done, the replacement of some of the fluidizing and transport air by the same mass flow of inert gas lowers the air to fuel ratio without affecting either the residence time of the gases in the combustor, or the in-flow of oxidizable nitrogen (i.e. the fuel-nitrogen), as long as the temperature, the bed height and the mass flow of coal are held

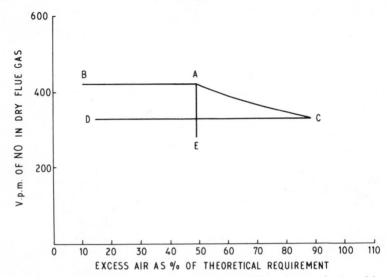

FIG. 3. Generalized representation of experimental results consistent with a constant fractional conversion of fuel-nitrogen and nitric oxide. Combustor 'A', temperature = 935 °C. A, datum; AB, gaseous nitrogen replacing air; AC, addition of air; AE, addition of gaseous nitrogen; CA, increasing coal feed rate; CD, gaseous nitrogen replacing air.

constant. The resulting concentration of nitric oxide can be compared with what was obtained originally and also with what would be obtained if the lower air to fuel ratio was obtained in some other way, i.e. by lowering the air flow without using inert make-up gas, or by increasing the mass flow of coal. Even so, strict comparability is not obtained, because although the addition of inert gas to a given mass flow of air does not affect the air to fuel ratio, it does affect the concentration of oxygen at the gas inlets and throughout the bed. This would be expected to affect the mass of unburnt fuel that would have to be present in order to maintain the overall rate of combustion.

Experiments along these lines have been performed in two of the NCB's experimental fluidized bed combustors but they gave sharply contrasting results, as shown in Figs. 3 and 4. Briefly, these were that in combustor 'A', the fraction of the fuel-nitrogen whose equivalent was emitted as nitric oxide was unaffected by wide variations in the temperature† and in the mass flows of reactants and diluent, whereas in combustor 'B' it was strongly affected. This difference has not yet been convincingly explained but it is

† Data not shown in Fig. 3.

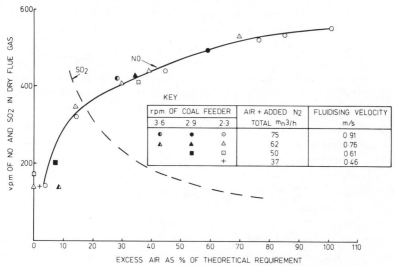

FIG. 4.   Nitric oxide versus excess air at combustor 'B' at 905 °C.

presumably associated with either the differences in design between the combustors (see Table 2) or with the difference in bed material, or both. (The coal was unchanged; it was 'Illinois No. 5'.) In both tests the 'inert gas' was nitrogen, but the results suggest that it may not have been entirely inert in all circumstances, as will be seen, and further work is planned to elucidate this.

In combustor 'A', no sulphur acceptor was used at first; the bed material was silica sand. The nitric oxide concentration was unaffected by temperature changes within the range 800–1000 °C, during a test in which the mass flows of coal and air were kept constant and no diluent 'inert' nitrogen was added. Then the temperature was held within the range 935 ± 15 °C, and the flows of fuel, air and additional gaseous nitrogen were varied widely. Some of these changes caused the concentration of nitric oxide to change but others did not. The results showed that if the gaseous nitrogen was indeed inert then the fraction F of the fuel-nitrogen that was given off as nitric oxide was approximately constant, throughout all the changes of temperature and mass flow. The data for 935 °C are generalized in Fig. 3 on the basis that F had the value of 0·15.

Some further experiments on nitric oxide concentration were performed with combustor A. The effect of varying the sand bed height by $\pm 50 \%$ was found to be nil. But when about 40 % of the sand was replaced batchwise by fresh cold dolomite the nitric oxide concentration was immediately more than doubled and, although the temperature was restored to normal and all the dolomite had been calcined within a few minutes, the nitric oxide concentration came down only very gradually and was still 15% above normal 7 h later. The peak emission was not more than could theoretically have been provided by the 'fuel-nitrogen', but the possibility that the freshly calcined dolomite was helping to fix gaseous nitrogen has not been ruled out. Here it may be recalled that the high nitric oxide concentration of 1200 vpm mentioned in an earlier paragraph was obtained with a fuel containing an unusually high proportion of calcareous sulphur retention stone, though the particle size was smaller than normal and there is a possibility that this was influential. These results show that the nature of the bed material should always be carefully described, when measurements of NO_x emission are being reported.

In combustor 'B', a sulphur acceptor (dolomite) was added continuously. The molar ratio of the feed rates of the calcium in the dolomite and the sulphur in the coal was 2 to 1. Figure 4 shows the results that were obtained when the bed temperature was 905 ± 10 °C. The nitric oxide concentration depended strongly on the air to fuel ratio, but to a good approximation it depended only on this, irrespective of large changes in the coal feed rate, in the flow of air, and in the quantity of added 'inert' nitrogen. To the same approximation it follows that, when the air to fuel ratio was fixed, the *mass flow* of nitric oxide was directly proportional to the fluidizing velocity, no matter what the air to 'inert' nitrogen ratio might have been. One implication of this is that at a given temperature, and in the absence of added nitrogen gas, the fraction F of the incoming fuel-nitrogen whose equivalent was emitted as nitric oxide was governed by the air to fuel ratio alone, irrespective of the coal feed rate and the fluidizing velocity; whereas when 'inert' gaseous nitrogen was added, but the air and fuel feed rates were kept constant, F increased in proportion to the fluidizing velocity. This raises the question whether the gaseous nitrogen was really inert in the conditions of the test. Data obtained at temperatures close to 810 °C and to 1020 °C supported the conclusions drawn from the work at 935 °C. However, the effect of temperature was complex, as the plots of nitric oxide concentration against air to fuel ratio for the three temperatures crossed each other. When the temperature was varied progressively between 850 and 1025 °C whilst the mass flows of air and coal were held steady, the nitric oxide emission was least at about 910 °C, averaging 300 vpm. At 850 °C and again at 1025 °C it averaged 450 vpm. These values are the averages from several tests in which the temperature was varied both upwards and downwards. Gaseous nitrogen was not added. One possible but unconfirmed explanation of the results is this: the concentration of NO_x could have been lowered by reduction reactions with unburnt fuel, and also, since it was above equilibrium, by catalyzed decomposition. It seems to be possible that as the temperature was raised from 850 °C the catalytic activity increased, but that from 910 °C upwards this was counteracted by the lowering of the concentration of unburnt fuel that would have resulted from faster combustion, and by the fact that the equilibrium concentration of nitric oxide would have been rising sharply. Another possibility is that the composition of the sulphur acceptor was influential; presumably it would have been affected by the variations in temperature and in the concentration of unburnt fuel.

Interesting results were obtained in combustor 'C', when Ollerton washed small coal was burnt in a shallow bed about 0·15 m deep and 0·56 × 0·80 m in plan.[17,18] The bed material was mainly silica sand, and there was no sulphur acceptor. The mass flow of air was kept constant but the coal feed rate was altered in steps. This caused large changes in the temperature of the bed and in the oxygen content of the flue gas, nevertheless the mass flow of nitric oxide was directly proportional to the coal feed rate. The fraction F of the incoming fuel-nitrogen that was emitted as nitric oxide was close to 0·25 throughout. A consequence was that the concentration of nitric oxide increased with rising temperature and with decreasing excess air, but the real governing variable was, as stated, the mass flow of coal. The latter was true of curve CA on Fig. 3 (combustor 'A') where, however, the temperature was constant. These results may be compared with those found at the chain-grate boiler[6] which has a shallow fuel bed though it is not fluidized. Here the percentage of excess air was varied at more than one coal feed rate and it was found that the nitric oxide concentration was directly proportional to that of carbon dioxide, i.e. a constant proportion of the fuel-nitrogen was emitted as nitric oxide, if it can be assumed that no atmospheric nitrogen was oxidized. (No sulphur retention stone was present.)

Combustion tests of chars (i.e. devolatilized coals) in the first and second combustors ('A' and 'B') have revealed nitric oxide concentrations in the range 400–640 vpm. Tests of a Texan lignite in combustor 'A' and of a Turkish lignite and of some Canadian lignitous fuels in yet a fourth combustor 'D', showed nitric oxide concentrations of about 300 vpm for the lignites and of less than 200 vpm for the others. (The temperature in the fourth combustor is not allowed to exceed about 800 °C.[18])

Tests of Illinois No. 6 coal in combustor 'D' showed that the fraction F of incoming fuel-nitrogen that was emitted as nitric oxide increased with air to fuel ratio and with the temperature in the range 730–820 °C.[18] A consequence was that when the excess of air was progressively increased, at constant temperature, from zero, the nitric oxide concentration rose at first but then fell, as the dilution overtook the increased value of F. The highest value of F was 28 % in this work. From other observations it would seem that F seldom exceeds 30 % except when much limestone or dolomite is being added.

Various workers have sought evidence on how far the NO_x emissions originate with the nitrogen-bearing volatile matter of the coal, and how far they come from the nitrogen that remains in the char. The action of volatile matter and of char to reduce or decompose nitric oxide has also been studied. References 16 and 19–48 are relevant. The NCB's combustion tests show that char can make plenty of NO_x, ranging in a selected example from 400 vpm at 840 °C to 640 vpm at 1000 °C, with no intermediate minimum.[18] At 950 °C the value of F was about 30 % in this example.

Ammonia addition has been recommended as a means of destroying NO_x in flue gas.[30,49–54] The special conditions that are required were not present when the NCB added ammonia to a fluidized bed burning a bituminous coal. The outcome was that the NO_x concentration was increased in direct proportion to the ammonia addition. About 10 % of the ammonia was emitted as nitric oxide; the bed temperature was 950 °C.[13]

Tests of various coals in combustor E which is a commercial, shallow-bed FBC were being carried out at the time of writing.[15] The bed consisted mainly of silica sand with some lumps of coal ash. There was no limestone or dolomite. The concentration of nitric oxide in the flue gas averaged about 350 vpm when the combustor was burning its normal coal at a bed temperature of 900 °C, with 5 % v/v of oxygen in the flue gas. This coal contained 1·3 % w/w of nitrogen (DMMF basis). With other coals containing up to 1·9 % w/w of nitrogen (DMMF basis), burned in similar conditions, higher emissions were seen, ranging up to 750 vpm. The average value of F was 26 %. (Compare combustor C which gave 25 %.) The coals were sized in the range 10–25 mm.

3.2 Two-stage combustion

Low nitric oxide emissions have been secured by supplying insufficient air for complete combustion to the bed, with extra air to complete the combustion coming in above the bed.[32,33,39,41,55,56] Application of this method would probably lead to problems of sulphur retention and of corrosion, and the ash might be offensive if it contained calcium sulphide.

A method of two-stage combustion at atmospheric pressure, in which the

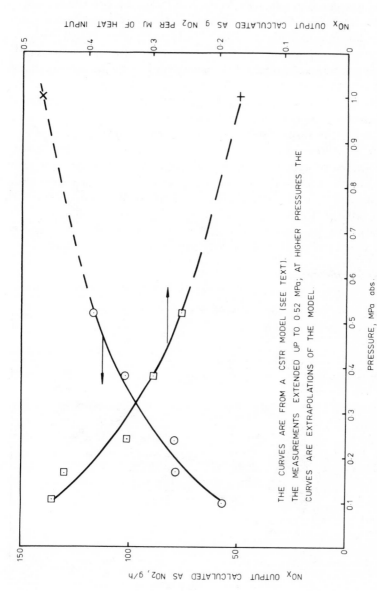

FIG. 5. Effect of pressure on NO$_x$ emissions from one of NCB's combustors (*F* in text).

gases pass through two fluidized beds in series, has been described.[56a] The NO$_x$ emission was optimized at about half that of the comparable single-stage combustor. The carbon monoxide emission was also lowered, and the combustion efficiency reached 98 %. The authors envisaged that a de-sulphurizing sorbent could be used in the second bed, and as this runs in oxidizing conditions it would overcome the disadvantages due to calcium sulphide mentioned above.

3.3 Starting-up at atmospheric pressure
In most types of furnace starting from cold entails that products of low-temperature carbonization are given off at first. In the fluidized bed this can be minimized by appropriate design and operation, but there may be an accumulation of unburnt fuel in the bed at first. When this begins to react, as the temperature rises, it is apt to give off appreciable quantities of what appears to be hydrogen cyanide, as well as carbon monoxide and hydrocarbons. With further rise in temperature the cyanide emission comes down and is replaced by nitric oxide—presumably by combustion.[6]

3.4 At elevated pressure
A small number of experimental, pressurized fluidized bed combustors

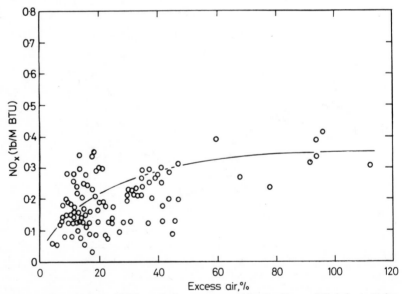

FIG. 6.   Correlation of NO$_x$ emissions measured in the Exxon Miniplant. (After Nutkis[61] by courtesy of Cambridge University Press.)

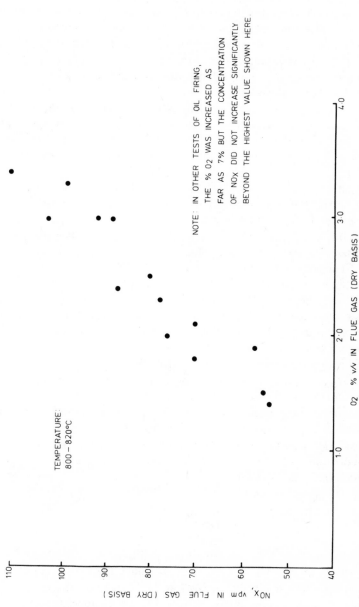

NOTE: IN OTHER TESTS OF OIL FIRING,
THE % O2 WAS INCREASED AS
FAR AS 7% BUT THE CONCENTRATION
OF NOx DID NOT INCREASE SIGNIFICANTLY
BEYOND THE HIGHEST VALUE SHOWN HERE

TEMPERATURE:
800 – 820°C

O_2 % v/v IN FLUE GAS (DRY BASIS)

NO_x, v.p.m IN FLUE GAS (DRY BASIS)

FIG. 7. Relationship between NO_x and oxygen during oil firing at 0·5 MPa during a period of low excess-air operation.

exists.[4,56-61] All but one are fairly small. The pressures are up to 12 bar. The small combustors apparently all give lower nitric oxide concentrations than are usual at atmospheric pressure, and with one of these combustors it was shown that raising the pressure from 1 to 5 bar did markedly lower the nitric oxide concentration.[12] The nitric oxide concentration was roughly proportional to $1/p^{1/2}$ at given temperature and excess air (p = pressure). As shown in Fig. 5, the mass flow of nitric oxide was roughly doubled, but as the heat output was roughly quadrupled the mass flow of nitric oxide per unit of heat output was roughly halved. The conditions were: temperature, 900 °C; fluidizing velocity, 1·2 m/s; expanded bed height, 1·0 m. There were no heat exchange tubes or other fitments in the bed apart from the coal feeder.

The effect of excess air has been investigated by NCB[5,58] and by Exxon.[56,60,61] Both concluded that increasing the excess of air from zero increased the concentration of nitric oxide, though the graphs were not always convincing (Fig. 6). Data obtained by the NCB in conjunction with the British Petroleum Company Ltd, for fuel oil-firing, are shown in Fig. 7.[5] The temperature was varied during the NCB's work on both coal firing and oil firing, but no significant effect was seen in either case.[5,58] The pressure was about 0·5 MPa (5 atm absolute). The oil contained 0·22 % of nitrogen by weight whereas the coal contained 1·4 %. Nevertheless, the NO_x concentrations were similar as long as the concentration of oxygen in the flue gas was kept below 4 %. Beyond this the NO_x concentration from the coal continued to increase, but that from the oil remained steady. The conversion to nitric oxide of the nitrogen combined in the oil was much higher than that of the 'coal nitrogen', but there were signs that it would reach an upper limit at a value of about 50 %. The bed materials in the two cases were different: for the oil the bed was a mixture of 'Molochite' ($\equiv 56$ % mullite ($3Al_2O_3.2SiO_2$) + 44 % of an amorphous silica glass) and partly spent dolomite, whereas the bed for the coal experiments contained partly spent dolomite and coal ash.[5]

4   DISCUSSION

Several mathematical models have been published[32,40,41,46,48,62,63] but the diversity of the phenomena has prevented the construction of a comprehensive one so far. This discussion does no better. Nevertheless, it is possible to suggest some of the reasons for what happens, and this is attempted below.

Nitrogen dioxide is ignored in what follows, as appreciable concentrations of it have not been observed.

The concentration of nitric oxide that is emitted is the result of the competition between the processes of its formation and its destruction. These processes will now be described.

4.1 Formation of nitric oxide

In principle nitric oxide could be formed in five ways, but it is clear that at least two of these are almost completely ineffective in fluidized combustion. The five ways are described below.

(1) From the air, by Zel'dovich's mechanism:[64]

$$O + N_2 \rightleftharpoons NO + N$$
$$N + O_2 \rightleftharpoons NO + O$$

As a reversible mechanism, this would yield the equilibrium concentration of nitric oxide. The fact that the concentrations are usually far above equilibrium shows that it can be neglected. At the other extreme this is shown also by the experiment in which graphitic rock, burned at 1000 °C, yielded only 10 vpm of nitric oxide, a value that was far below equilibrium.

(2) From the air, by mechanisms involving atoms and radicals released during combustion (these may include oxygen and nitrogen). References 24 and 25 are examples of the voluminous literature on this subject.

The very low yield of nitric oxide (not more than 20 vpm) obtained from the combustion of propane in a fluidized bed[8] rules these mechanisms out as a significant source. The graphitic rock experiment[12] shows the same, for a different fuel.

(3) and (4) From air, and/or from the 'fuel-nitrogen', via a solid intermediate such as calcium cyanamide (CaNCN). This suggestion is unproved but is made because the addition of large excesses of calcium-based sulphur retention stone seems to enhance the nitric oxide emission very considerably.[6] However, it is possible that such additions really exert their effect either by displacing better catalysts of N + N recombination, or of nitric oxide decomposition, or by lowering the concentrations of nitric oxide reductants in some way.

(5) From 'fuel-nitrogen' in ways not involving a solid intermediate. In some instances, but possibly not in all, this source accounts for virtually all the nitric oxide.[11,39] Combustion tests of volatile nitrogenous compounds, and of devolatilized char, suggest that both the volatile and involatile nitrogenous constituents of the coal can contribute nitric oxide to the flue gas.[6,10,21,23,34,35,37]

4.2 Destruction of nitric oxide

There are four ways of accomplishing this, not counting Zel'dovich's mechanism which is shown above to be ineffective.

(1) By reactions with volatile matter such as hydrocarbons. (Orange plumes containing nitrogen dioxide can be entirely 'de-NOxed' by burning hydrocarbons in them, though for economy's sake the reduction is not usually carried beyond nitric oxide which is colourless.[65])

(2) By reactions with char.[31,39,41,46,48]

(3) By reactions such as:[21,23,66-69]

$$NO + NH_2 = N_2 + H_2O$$

(4) By heterogeneous catalysis (provided the nitric oxide concentration is above equilibrium). Such catalysis is evidently not very effective in existing fluidized bed combustors, but its occurrence cannot be ruled out at present.

In order to make nitrogen, each of these processes must incorporate some step in which two molecules or radicals, each containing a nitrogen atom, react, or in which one species reacts with a nitrogen atom in the surface of the char.

4.3 Net result

Because coal contains oxygen as well as nitrogen, it is likely that the volatile matter that soon surrounds an incoming particle of coal will contain nitric oxide as well as other volatile nitrogen compounds or radicals:

$$NX + OY = NO + XY \tag{1}$$

The mutual destruction reaction,

$$NO + NX = N_2 + OX \tag{2}$$

may then take place throughout the envelope of volatile matter. The thicker the envelope, the more this reaction (and others) will degrade the volatile 'fuel-nitrogen' to nitrogen.

At the outer surface of the envelope, molecular oxygen is present:

$$NX + O_2 = NO + OX \tag{3}$$

This brings in the possibility of further mutual destruction via reaction (2).

Reactions (2) and (3) could lead to a factor F of 33 % if they were the only ones, and if nitric oxide formed at the outer surface of the envelope by reaction (3) diffused equally inwards and outwards, since the stoichiometry would then be:

$$3NX + 2O_2 = NO + N_2 + 3OX \tag{4}$$

The char particle that is left after the volatile matter has been evolved contains involatile nitrogen. The mechanisms that cause some of this to leave as nitric oxide and some as nitrogen are not known, but the net result may be analogous to reaction (4).

The nitric oxide that gets away from the envelopes of volatiles and from the particles of char does not necessarily all reach the exit from the bed. Some of it may come into contact with surfaces that can decompose it catalytically, as its concentration is above equilibrium. Also, a weaker source can act as a sink for a stronger source, if it comes so near to it that the concentration gradients of nitric oxide around the weaker source are reversed and the nitrogen atoms or nitrogeneous compounds and radicals in the weaker source can then act to destroy nitric oxide coming from the stronger one, i.e. to lower the mass flow of nitric oxide rather than to enhance it. This hints at the importance of coal concentration in relation to combustion rate.

Much may depend on whereabouts in the bed and freeboard the volatiles and char are mainly found. In most fluidized beds, the burn-out time of the char is long in comparison with the rate of mixing of the solids. Consequently the dispersion of the char should be nearly uniform. However, this is probably not true of the volatile matter. Although devolatilization times as long as 16 s for 1 mm diameter particles in a bed at 875 °C have recently been noted by the NCB, using a shallow bed (9 cm expanded bed height) of 'Molochite' (this contains about 56% mullite $(3Al_2O_3 . 2SiO_2)$ and 44% of an amorphous silica glass), it was also seen that there was a strong tendency for these (and larger) coal particles to be buoyed up near the top of the bed until devolatilization was complete, as judged by the disappearance of the halo.[70] On the other hand, in deep beds in which particle movement is obstructed by tube banks, the devolatilization may occur mainly near to the coal feed points. When the fuel is coarsely sized, e.g. has diameters 10–25 mm, the heating and devolatilization time is probably long enough for the volatiles to be more widely dispersed, unless the particles tend to float, and back-mixing of the gases is poor.

5 MODELLING THE EFFECT OF PRESSURE

The effect of pressure on the mean NO concentrations from combustor F operated by the NCB is given in Table 3 (see also Fig. 5). Also tabulated are

TABLE 3

Combustor F—Effect of pressure on nitric oxide concentration: Comparison between an experimental result and two correlations based on it

	Pressure (MPa)					
	0·10	0·17	0·24	0·38	0·52	1·01[a]
Actual mean nitric oxide conc. in vpm (wet basis)	366	300	214	175	146	Not det'd
CSTR Model[b] (eqn. (a)) vpm	367	280	233	178	146	91
Empirical correlation (eqn. (b)) vpm	348	268	225	178	153	109

[a] 10 atm. absolute.
[b] CSTR, Continuously Stirred Tank Reactor.
Notes on Table 3.
Equation (a) is:

$$c = \frac{10^6 \times \text{g. atoms of N per kg of coal as fired}}{P\left(\dfrac{104\,000}{\text{Exit } O_2 \text{ conc., } \% \text{ v/v}} + \dfrac{63\,000}{r}\right)}$$

where c is the concentration of NO (vpm), P is the pressure (MPa), r is the coal feed rate (kg/h) and $r = 29P + 1\cdot64$ in the conditions of the test. The 1·64 represents heat losses from the casing, which were constant because the temperature was fixed. There were 0·83 gram atoms of nitrogen per kilogram of coal as fired.

Equation (b) is:

$$c = 110/\sqrt{P}$$

(This is not the best empirical correlation, but it is a simple one.)

the predictions from two correlations based on the experiment. The conditions were:

Pressure	up to 0·52 MPa (75 psia)
Temperature	900 °C
Fluidizing velocity	1·2 m/s
Expanded bed height	1·0 m
Diameter of bed	0·30 m
Bed material	Mainly molochite
Coal	Illinois No. 6
Coal feed rate	up to 17 kg/h
Size distribution of the coal	Invariant
Oxygen content of the flue gas	11·9 to 14·5 % v/v (calculated)
Fitments in the bed	None apart from coal feed pipe

The empirical correlation can be improved, at the cost of assuming that the concentration of nitric oxide does not tend to zero as the pressure tends to infinity.

The CSTR model is based on the ideas about the formation and destruction of nitric oxide that have been described. It is discussed below. The same ideas have been used to build up a model embodying plug flow of gases but well mixed solids. This gives nearly the same results as regards the effect of pressure but it is not discussed further.

It should be realized that the assumption of a CSTR means that the physics of the bed are assumed to be unaffected by the changes of pressure. The reduction in the emission of nitric oxide is brought about by purely chemical means, in the model, with no help from better mixing between the sources and sinks of nitric oxide as the pressure is increased. (This is true of the plug flow model also.†) The other assumptions are these:

(1) The partial pressure of nitric oxide leaving the bed is the result of competition between processes of formation and of destruction;
(2) The rate of liberation of nitric oxide from the given fuel is proportional to the fuel feed rate only;
(3) The rate of degradation to nitrogen of the liberated nitric oxide is proportional to the product of the partial pressure of nitric oxide and the active mass of fuel in the bed;
(4) The combustion of the fuel is diffusion-controlled;
(5) Reactions in the freeboard are negligible (alternatively, the freeboard is part of the CSTR).

The following conditions applied to the experiment being modelled:

(6) The bed was of fixed dimensions;
(7) The fluidizing velocity was fixed; ⎫ These conditions
(8) The temperature was fixed; ⎬ validate assumptions
(9) The size distribution of the fuel was fixed; ⎭ (2) and (3) above.
(10) The fuel feed rate could be expressed by the equation:

$$r = 29P + 1.64 \qquad (r \text{ in kg/h}; \ P \text{ in MPa})$$

(11) There were 0·83 gram atoms of nitrogen per kilogram of the fuel as fired.

† It is obvious that if there was a variation in the quality of the mixing, it would affect the NO_x emission. Some years ago the present author believed that the effect of pressure on NO_x emission was thus explained. However, independent evidence of the required improvement in mixing has not been forthcoming. Therefore, the author sought an alternative, chemical explanation, as above.

5.1 Nomenclature

Fuel feed rate (kg/h)	r
Partial pressure of nitric oxide at the exit (and so throughout the CSTR) (MPa)	[NO]
Partial pressure of oxygen at the exit (and so throughout the CSTR) (MPa)	$[O_2]$
Mass of unburnt fuel in the CSTR (kg)	M
Overall pressure (MPa)	P
Therefore, mole fraction of oxygen at the exit	$[O_2]/P$
Concentration of nitric oxide (vpm)	c
Various constants of proportionality	k_1, k_2, etc.

5.2 Working

Equating mass flows of nitric oxide yields that:

$$k_1 r = k_2[NO]M + k_3[NO] \tag{5}$$

where $k_1 r$ represents the nitric oxide that is liberated from the fuel and $k_2[NO]M$ is the nitric oxide that is degraded to nitrogen. The constant k_2 represents a weighted average of the rate constants of all the reactions that tend to degrade nitric oxide to nitrogen. No distinction is made between the effects of volatile matter and of char, and none is needed in this model because in the conditions of the test the ratio of unburned volatiles to unburned char must have remained approximately constant. (The air to fuel ratio did not vary much.)

$k_3[NO]$ is what is emitted as nitric oxide. From eqn. (5),

$$[NO] = k_1 r/(k_2 M + k_3) \tag{6}$$

To find the value of M, note that in the steady state,

$$\text{Rate of combustion} = \text{Fuel feed rate} \tag{7}$$

i.e.

$$k_4 M[O_2]/P = r \tag{8}$$

Since $r = 29P + 1.64$, and $[O_2]/P$ does not vary much (calculations used here but not given) it is clear that M increases with P albeit not in direct proportion. Hence (in this model), the degradation reactions between nitric oxide and M are favoured by increasing the pressure.

Equation (8) makes use of the assumption that the rate of combustion is under diffusion control. This simplifies the working but the model is not sensitive to it, as an appreciable element of kinetic control can be added without significantly altering the predicted form of the curve of nitric oxide concentration versus overall pressure. The kinetic resistance becomes progressively less important as the pressure is increased, but Pillai has shown that it is not usually negligible at atmospheric pressure.[70] The diffusion rate is taken as proportional to $[O_2]/P$.[71] It will be remembered that the particle sizing of the fuel was invariant.

By definition,

$$c = 10^6([NO]/P) \tag{9}$$

From eqns. (6), (8) and (9), it follows that:

$$c = \frac{10^6 k_1}{\left(\dfrac{k_2 P}{k_4 [O_2]} + \dfrac{k_3}{r} \right) P} \tag{10}$$

Fitting this to the experimental data assuming that if k_1 is taken as 0.83, and the mole fraction of oxygen $[O_2]/P$ is expressed as a percentage by volume, gives $k_2/k_4 = 104\,000$ and $k_3 = 63\,000$.

The model contains no input of data for variations in oxygen concentration independently of pressure, but it can be used to predict the effect. For example, if P is put equal to 0.1 MPa (i.e. 1 atm) and r is put equal to 1, 2, 3, etc., kg/h up to 10, the result is as shown in Table 4.

The peak in the curve of nitric oxide against per cent oxygen, and the near proportionality of the fraction F to the per cent oxygen as shown by

TABLE 4

Predictions of CSTR model at 0·1 MPa and various coal feed rates; fluidizing velocity and temperature kept constant

Coal feed rate (kg/h)	Per cent oxygen in wet gas	vpm nitric oxide in wet gas	Fraction F, of nitrogen in coal emitted as nitric oxide	Column 4 ÷ Column 2
1	19·0	121	0·480	0·025
2	16·9	219	0·438	0·026
3	14·9	297	0·398	0·027
4	13·0	349	0·353	0·027
5	11·0	376	0·306	0·028
6	9·1	378	0·257	0·028
7	7·1	352	0·207	0·029
8	5·3	300	0·155	0·029
9	3·4	219	0·101	0·030
10	1·5	111	0·046	0·031

the final column, are both typical of an unpressurized combustor that was tested by the author[18] (combustor D) but it has not yet been possible to test these predictions in the pressurized combustor 'F' in Table 2 from which the model was derived.

Such a model must predict that F (this F is the percentage of fuel-nitrogen that is emitted as nitric oxide) depends on the air to fuel ratio. An invariant F can come by putting $M = $ constant, or $k_2 = 0$, or by making k_2 vary inversely as M, in eqn. (5) of the working, but these changes must lead to a departure from the CSTR model.

5.3 Plug flow

An attempt is made in this section to show how if F is constant in a given isothermal shallow bed, it nevertheless would be expected to vary with conditions if the bed is deepened (if the coal is well mixed throughout the deep bed).

Consider an array of nitric oxide sources spaced out in a horizontal plane, i.e. a very shallow bed, in which they are burning isothermally. They are all burning in air that initially contains 21 % by volume of oxygen, but no nitric oxide, and on average their value of F is F_1. Now let the average spacing of the array be widened by reducing the coal feed rate. The conditions surrounding each nitric oxide source are exactly as they were, because each source was already burning in air, not in partly vitiated air. F, therefore, still equals F_1. The concentration of nitric oxide leaving the bed has been lowered in proportion to the coal feed rate.

Now with the original coal feed rate, let the bed be made much deeper. The array is now three dimensional. The sources at the very bottom exhibit $F = F_1$. *Ex hypothesi*, the plug flow of gases does not preclude lateral mixing in this example. Consequently, the sources at higher levels in the bed all encounter less oxygen but more nitric oxide than those at lower ones. Both these changes must tend to lower their value of F. If now the coal feed rate is reduced, it lessens the change in oxygen and nitric oxide concentration with height in the bed, and so the average value of F for the bed is increased.

The effect of temperature cannot be so simply explained. In one of the NCB's experiments F remained constant over the bed temperature range 800–1000 °C.[6] The mass flows of coal and air were kept constant. The temperature range of the bulk of the bed was so wide that it is not reasonable to believe that the constancy of F could indicate a constant temperature of the burning particles. It is necessary to suppose that their temperature did vary but that there were opposing effects of this, which balanced each other. In the experiment quoted, a rise in the temperature of combustion would entail the following.

(1) In the fluidized bed as a whole, and at a given fuel feed rate, the ratio of the mass flow of all forms of 'fuel-nitrogen' to the mass of unburnt fuel contained in the bed would increase, to an extent that depends on the increased rates of combustion of the individual 'packets' of volatiles and particles of char. If the restricted model of a devolatilized burning char particle, given by Field *et al.*,[72] is adopted, and the mechanism factor (which depends on the ratio of CO/CO_2 formed at the char surface) is constant, then the rate of liberation of carbon from the surface of a particle of given diameter burning under diffusional control varies as $pT^{0.75}$, where p is the partial pressure of oxygen in the free stream and T is the mean temperature (K) in the boundary layer.[73] This must also apply to the involatile 'fuel-nitrogen'. The rate of combustion of a given 'packet' of evolved volatile matter will also, with similar assumptions, vary as $pT^{0.75}$. If the diffusion coefficients of nitric oxide and its precursors through the other gases resemble those of oxygen, carbon dioxide and water through nitrogen, they vary approximately as $(T/T_0)^{1.75}$.[73] Then the powers of T cancel from the diffusion equation,† and it follows that in the event of the nitric oxide and its precursors not undergoing any mutual destruction after liberation, or undergoing it in some fixed proportion, their concentration gradients around a given particle or 'packet' of volatiles will be unaffected by changes of temperature, and so will F. Presumably, this would apply to a very shallow bed. However, given a well-mixed fluidized bed, the mass of unburnt fuel that must be present in order to maintain the overall rate of combustion equal to a given fuel feed rate varies as $1/pT^{0.75}$ if the size distribution of the fuel does not vary. For a temperature change from 850 to 950 °C this means a decrease in the fuel inventory of 6·6 %. If the ratio of volatiles to char in the bed remains unchanged, the distribution of sizes of

† In a mixture of two gases A and B, the diffusive flux of component A across a surface in the fluid is given by :[74]

$$G_A = \frac{M_A D_{AB}}{R^1 T_g} \frac{dP_A}{dx}$$

where G_A is the mass flux of A per unit area (g/cm²s), M_A is the molecular weight of gas A, D_{AB} is the binary diffusion coefficient of gases A and B (cm²/s), R^1 is the universal gas constant (82·06 atm cm³/mole K), T_g is the gas temperature (K), x is the distance in direction of concentration gradient (cm) and P_A is the partial pressure of gas A (atm).

This flux is measured not with respect to a stationary frame of reference but with respect to a surface across which there is no net molar flux of the two gases.

the char particles and 'packets' of volatiles is not likely to vary appreciably. Accordingly the ratio of the mass flow of the 'fuel-nitrogen' to the mass of unburnt fuel in the bed varies as $pT^{-0.75}$. The mutual reduction reactions of nitric oxide and its precursors (including the involatile nitrogen in the surface of the char) could be affected by this.

The restricted model of combustion, used above, refers to small particles around which the gases are stagnant. This will not be true of the 10–25 mm size lumps of coal that are being fired in some installations.

Amongst other consequences of a rise in temperature are the ones given below.

(2) Volatile matter would be evolved more quickly, and the proportion of volatile matter obtained from the fuel would increase (probably only a little).

(3) The specific rates of the gas-phase reactions in which nitric oxide and its precursors are degraded to nitrogen would probably increase. (Cowley and Roberts[47] have denied that nitric oxide is effectively destroyed by gas-phase reactions.)

(4) The same is true of the NX–char nitrogen degradation reaction.

(5) The equilibrium concentration of nitric oxide would increase.

(6) The catalytic activity of the various solid surfaces in the bed and freeboard might alter.

(7) The chemical nature of the bed may alter; it would do so if sulphur retention stone were present.

The complexity of this list perhaps makes it more surprising that constant F can occur, than that F is sometimes affected by the temperature.

When interpreting temperature effects it is as well to bear other conditions in mind. It is usual to run a fluidized bed at some fixed fluidizing velocity. Then if the temperature is raised, the mass flow of fluidizing air must be lowered. Then, if the coal feed rate is not altered, the air to fuel ratio is lowered. Alternatively if the air to fuel ratio is preserved, the in-flow of 'fuel-nitrogen' is lowered. In some of the NCB's work the interdependence of these factors was removed by the device of adding nitrogen to the fluidizing air but, as already mentioned, even this led to a lack of comparability in some respects.

5.4 Particle size
The smaller the char particles the faster the diffusion of gases to them and from them, unless there is a high resistance in series. If this led to faster combustion it should result in an increased value of F, since the inventory of unburnt fuel in the bed would be reduced. For a devolatilizing particle

the smaller it is, the thinner the layer of volatiles might be expected to be. This should also lead to increased F.

Unequivocal evidence of an effect of fuel particle size on nitric oxide emission has not been obtained by the NCB, but other experimenters have reported that smaller fuel particle size leads to enhanced nitric oxide emission.[75] Reference 46 reports calculations tending to show that for a given fluidizing velocity, reduction in bed solid particle size coupled with increased coal feed size would result in a higher carbon load in the bed, hence favouring the reduction of NO_x, but would not cause a loss of combustion efficiency. However, as mentioned previously, the NCB has evidence that the firing of 25-mm diameter, i.e. large, lumps of coal does not necessarily lead to low nitric oxide concentrations in the flue gas.

The low nitric oxide emissions from the chain-grate shell boilers investigated by the NCB deserve further mention. At these installations F is less than 0·1, and seems to be a constant. The beds are shallow but not fluidized. The fuel is washed smalls, i.e. it has a fairly low ash content and is sized up to about 6 mm diameter. To secure the most efficient combustion the fuel is kept slightly moist. The concentration of fuel on the bed is high for at least part of its length and there is a long flame of volatiles above the bed, burning in the secondary air.

Given the previous discussion it is not surprising that F is low. Indeed, the conditions approach those of two-stage combustion.

6 EFFECT OF NO_x CONTROL MEASURES ON SULPHUR DIOXIDE EMISSIONS, AND VICE VERSA

Not much is known about this. In an experiment by the NCB using combustor 'B', the sulphur dioxide emission curve was the inverse of the nitric oxide emission curve, as shown in Fig. 4, i.e. the lower the excess air and nitric oxide, the higher the sulphur dioxide. This presumably reflected: (1) the fact that the sulphur retention reactions need oxygen, and (2) a tendency to regenerate lime from calcium sulphate, with release of sulphur dioxide, when low excess air caused the fuel content of the bed to increase. The calcium/sulphur molar ratio in the feed was 2/1. As mentioned earlier, there is conflicting evidence of the effect on NO_x emission of moderate increases or decreases in this ratio, but a high calcium/sulphur ratio of 8/1, which reduced the sulphur dioxide emission virtually to zero, was associated with an unusually high NO_x emission[6] although this result may have been influenced by the small particle size of the coal feedstock.

Increasing the pressure reduces NO_x emissions and makes it easier to control the sulphur dioxide.

There is an optimum temperature for sulphur retention with any particular stone, but the same will usually not be true of the NO_x emission.

Preliminary tests of two-stage combustion in the presence of sulphur retention stone showed low emissions of both nitric oxide and sulphur dioxide, the sulphur being retained as calcium sulphide[55] but this work does not appear to have been pursued. As the calcium sulphide could hardly be dumped it would have to be oxidized in a separate vessel, causing complications. A process involving successive fluidized beds has been suggested.[56a]

7 CONCLUSIONS AND RECOMMENDATIONS

The NO_x emission from coal combustion in fluidized beds has been shown to be very varied. It is not yet possible to predict what it will be in a particular case, and each case must be investigated experimentally. Nevertheless, the following statements may be made:

(1) The NO_x is nearly all nitric oxide as it leaves the combustor.

(2) The low temperature of fluidized bed combustion does not guarantee low NO_x emissions; consequently, it is incorrect to attribute the low emissions, when they do occur, to the low temperature.

(3) In some cases it was proved that virtually all the NO_x came from the 'fuel-nitrogen', i.e. the nitrogen that was chemically combined in the fuel, rather than that of the combustion air.

(4) There is 'fuel-nitrogen' in both volatiles and char, and both contribute to the NO_x.

(5) The configuration of the combustor, possibly the nature of the bed material and possibly also the materials of construction of the plant and the particle size and type of the coal, influence the effects that variations of the temperature and of the excess air have upon the fraction of the incoming 'fuel-nitrogen' (or its equivalent) that is emitted as NO_x. The uncertainty of this statement derives from the wide variety of the results.

(6) Increasing the pressure reduces the NO_x emissions.

(7) Measures to control sulphur dioxide emission may affect NO_x emission, and vice versa.

(8)   In partial explanation of the results, it is suggested that the combination of both nitrogeneous and carbonaceous matter of the coal in both the volatiles and the char acts as both source and sink of nitric oxide. Mathematical models based on this idea can explain the effect of pressure without having to postulate better mixing as the pressure is increased.

(9)   There are many possible influences on NO_x emission apart from pressure, temperature and excess air. Some are listed below, and it is recommended that they receive further study.

   (i)   Configuration of the combustor;
   (ii)   Bed material—its composition and particle size;
   (iii) Materials of construction;
   (iv)   Particle size of the coal;
   (v)   Coal type (including the proportion and chemical nature of the combined nitrogen);
   (vi)   Fundamentals of combustion.

ACKNOWLEDGEMENT

The author thanks the National Coal Board for permission to publish this work. Any views expressed are those of the author and not necessarily those of the National Coal Board.

REFERENCES

1. SMITH, I., Nitrogen oxides from coal combustion: Environmental effects, Report no. ICTIS/TR10, IEA Coal Research, London, 1980. (A review.)
2. ALLEN, J. D., BILLINGSLEY, J. and SHAW, J. T., Evaluation of the measurement of oxides of nitrogen in combustion products by the chemiluminescence method, *J. Inst. Fuel*, **47**, 275 (December 1974).
3. BEACHAM, B. and MARSHALL, A. R., Experiences and results of fluidised bed combustion plant at Renfrew, *J. Inst. Energy*, 59 (June 1979).
4. BROADBENT, D. H. and WRIGHT, S. J., A technical description of plant design and project progress report, *Fifth International Conference on Fluidised Bed Combustion*, Washington, 1977.
5. SHAW, J. T., Reduction of air pollution by the application of fluidised-bed combustion under pressure, with special reference to emissions of NO_x, *Second International Conference, Control of Gaseous Sulphur and Nitrogen Compound Emissions*, University of Salford, 1976.
6. Author's measurements.

7. SHAW, J. T., Comment on recent papers. Oxides of nitrogen: their occurrence and measurement in flue gas from large coal-fired boilers. *BCURA Mon. Bull.*, **34**, 252 (October 1970).
8. SHAW, J. T., Progress Review No. 64: A commentary on the formation, incidence, measurement and control of nitrogen oxides in flue gas. *J. Inst. Fuel*, 170 (April 1973).
9. SUTTON, E. and STARKMAN, E. S., Oxides of nitrogen in engine exhaust with ammonia fuel, Thermal Systems DIV. R TR 7, TS 66 4, Contract DA 04 200 AMC 791 X 27P JUN.66, California University, Berkeley, 1966.
10. SHAW, J. T. and THOMAS, A. C., Oxides of nitrogen in relation to the combustion of coal, *Seventh International Conference on Coal Science*, Prague, 1968.
11. JONKE, A. A., Reduction of atmospheric pollution by the application of fluidised bed combustion, Monthly Progress Report No. 8 to the National Air Pollution Control Administration, USA, ANL/ES/CEN-F08, Argonne National Laboratory, 1969.
12. ROBERTS, A. G., STANTAN, J. E. and PILLAI, K. K., NCB private verbal communication, 1980.
13. DUXBURY, J., NCB private communication, 1980.
14. JONKE, A. A., Reduction of atmospheric pollution by the application of fluidised bed combustion, Monthly Progress Report No. 19 to the National Air Pollution Control Administration, USA, ANL/ES/CEN-F019, Argonne National Laboratory, 1970.
15. MENDOZA, M. P., NCB private communications, 1980, 1981.
16. SOLOMON, P. R., Investigation of the devolatilisation of coal under combustion conditions, Fourth Quarterly Report for July 1–September 30, 1979, US/DOE/FE-3167-T4, United Technologies Research Center, East Hartford, Connecticut.
17. MENDOZA, M. P.; SHAW, J. T., NCB private communications, 1974.
18. SHAW, J. T., Nitric oxide emissions during the fluidised combustion of various solid fuels, Paper P, *I. Chem. E. Symposium Series No. 57, The Control of Sulphur and Other Gaseous Emissions*, Salford, 1979.
19. Various authors, in (eds.) COUGHLIN, R. W., SAROFIM, A. F. and WEINSTEIN, N. J., *Air Pollution and Its Control. Section 2, Nitrogen Oxides*, A.I.Ch.E. Symp. Series 126, **68** (1972).
20. DUXBURY, J. and PRATT, N. H., A shock tube investigation of the fuel nitrogen-nitric oxide problem, *Combustion Institute (European) Symposium*, Sheffield, Academic Press, London, 1973, p. 433.
21. DE SOETE, G. G., An overall mechanism of nitric oxide formation from ammonia and amines added to premixed hydrocarbon flames, *Combustion Institute (European) Symposium*, Sheffield, Academic Press, London, 1973, p. 439.
22. MYERSON, A. L., The reduction of nitric oxide in simulated combustion effluents by hydrocarbon–oxygen mixtures, *Fifteenth Symposium (International) on Combustion*, Tokyo, The Combustion Institute, 1974, p. 1085.
23. DE SOETE, G. G., Overall reaction rates of NO and N_2 formation from fuel nitrogen, *Fifteenth Symposium (International) on Combustion*, Tokyo, The Combustion Institute, 1974, p. 1093.

24. HAYNES, B. S., IVERACH, D. and KIROV, N. Y., The behaviour of nitrogen species in fuel-rich flames, *Fifteenth Symposium (International) on Combustion*, Tokyo, The Combustion Institute, 1974, p. 1103.

25. HAYHURST, A. N. and McLEAN, H. A. G., Mechanisms for producing NO from nitrogen in flames, *Nature*, **251**, 303 (1974).

26. MULVIHILL, J. N. and PHILLIPS, L. F., Breakdown of cyanogen in fuel-rich $H_2/N_2/O_2$ flames, *Fifteenth Symposium (International) on Combustion*, Tokyo, The Combustion Institute, 1974, p. 1113.

27. PEREIRA, F. J., BEÉR, J. M., GIBBS, B. and HEDLEY, A. B., NO_x emissions from fluidised bed coal combustors, *Fifteenth Symposium (International) on Combustion*, Tokyo, The Combustion Institute, 1974, p. 1149.

28. AXWORTHY, A. E., SCHNEIDER, G. R., SHUMAN, M. D. and DAYAN, V. H., Chemistry of fuel nitrogen conversion to nitrogen oxides in combustion, US EPA report EPA-600/2-76-039, February 1976.

29. CARETTO, L. S., Modelling the gas-phase kinetics of fuel-nitrogen reactions, Final report on research project 240, Electric Power Research Institute, Palo Alto, Calif., 1976.

30. MUZIO, L. J. and ARAND, J. K., Homogeneous gas-phase decomposition of oxides of nitrogen, Final report on research project 461-1, Electric Power Research Institute, Palo Alto, Calif., 1976.

31. BEÉR, J. M., SAROFIM, A. F., CHAN, L. K. and SPROUSE, A. M., No reduction by char in fluidised combustion, *Fifth International Conference on Fluidised Combustion*, Washington, DC, December 1977, p. 577.

32. HORIO, M., MORI, S. and MUCHI, I., A model study for the development of low-NO_x fluidised bed coal combustion, *Fifth International Conference on Fluidised Combustion*, Washington, DC, December 1977, p. 605.

33. SAKAMOTO, K., Control of nitric oxide and carbon monoxide emissions in fluidised bed combustion, *Fifth International Conference on Fluidised Combustion*, Washington, DC, December 1977, p. 594.

34. MENDOZA, M. P. and SAGE, P. W., The production of NO_x during the combustion of solid fuels at moderate temperatures, NCB private communication, 1977.

35. BLAIR, D. W., WENDT, J. O. L. and BARTOK, W., Evolution of nitrogen and other species during controlled pyrolysis of coal, *Sixteenth Symposium (International) on Combustion*, The Combustion Institute, 1977, p. 475.

36. PERSHING, D. W. and WENDT, J. O. L., Pulverised coal combustion: The influence of flame temperature and coal composition on thermal and fuel NO_x, *Sixteenth Symposium (International) on Combustion*, The Combustion Institute, 1977, p. 389.

37. POHL, J. H. and SAROFIM, A. F., Devolatilisation and oxidation of coal nitrogen, *Sixteenth Symposium (International) on Combustion*, The Combustion Institute, 1977, p. 491.

38. AXWORTHY, A. E., DAYAN, V. H. and BLAIR MARTIN, G., Reactions of fuel-nitrogen compounds under conditions of inert pyrolysis, *Fuel*, **57**, 29 (January 1978).

39. FURUSAWA, T., HONDA, T., TAKANO, J. and KUNII, D., Nitric oxide reduction in an experimental fluidised-bed coal combustor, *Fluidisation*, Cambridge University Press, 1978, p. 314.

40. PEREIRA, F. J. and BEÉR, J. M., A mathematical model of NO formation and destruction in fluidised combustion of coal, *Fluidisation*, Cambridge University Press, 1978, p. 401.

41. SAROFIM, A. F. and BEÉR, J. M., Modelling of fluidised bed combustion, Colloquium on coal combustion, *Seventeenth Symposium (International) on Combustion*, Leeds, England, August 1978, pp. 189–204.

42. SOLOMON, P. R. and COLKET, M. B., Evolution of fuel nitrogen in coal devolatilisation, *Fuel*, **57**, 749 (December 1978).

43. VOGT, R. A. and LAURENDEAU, N. M., Preliminary measurements of fuel nitrogen formation in a pulverised coal transport reactor, *J.A.P.C.A.*, **28**(1), 60 (January 1978).

44. PERSHING, D. W. and WENDT, J. O. L., Relative contributions of volatile nitrogen and char nitrogen to NO_x emissions from pulverised coal flames, *Ind. Eng. Chem. Process Des. Dev.*, **18**(1), 60 (1979).

45. MORRISON, G. F., Nitrogen oxides from coal combustion—abatement and control, Report No. ICTIS/TR11, IEA Coal Research, London, 1980. (A review.)

46. BEÉR, J. M., SAROFIM, A. F., SHARMA, P. K., CHAUNG, T. Z. and SANDHU, S. S., Fluidised coal combustion: The effect of sorbent and coal feed particle size upon the combustion efficiency and NO_x emission, in *Fluidisation, Proceedings of the 1980 International Fluidisation Conference*, sponsored by the Engineering Foundation, and held at Henniker, August 3–8 1980, Grace, J. R. and Matsen, J. M. (eds.), Plenum Press, New York, 1980.

47. COWLEY, L. T. and ROBERTS, P. T., The mechanism of NO_x generation during the fluidised combustion of coal, *Fluidised Combustion Conference*, Energy Research Institute, University of Cape Town, 28–30 January 1981.

48. BEÉR, J. M., SAROFIM, A. F. and LEE, Y. Y., NO formation and reduction in fluidised bed combustion of coal, *J. Inst. Energy*, **54**(418), 38 (March 1981).

49. ANON. Thermal process knocks NO_x, *Chem. Engng.*, 85 (June 19, 1978).

50. MATSUDA, S., TAKEUCHI, M., HISHINUMA, T., NAKAJIMA, F., NARITA, T., WATANABE, Y. and IMANARI, M., Selective reduction of nitrogen oxides in combustion flue gases, *J.A.P.C.A.*, **28**(4), 350 (April 1978).

51. ENGA, B. E., Catalyst systems for the control of nitrogen oxide emissions, I. Chem. E. Symposium Series No. 57, The control of sulphur and other gaseous emissions, Salford, 1979, Paper S.

52. KIOVSKY, J. R., KORADIA, P. B. and LIM, C. T., A new zeolite catalyst for selective catalytic reduction of NO_x with NH_3, I. Chem. E. Symposium Series No. 57, The control of sulphur and other gaseous emissions, Salford, 1979, Paper Q.

53. LYON, R. K., VARGA, G. M., JR., RUTERBORIES, B. H., TOMSHO, M., HARDY, J., FREUND, H., MUZIO, L. J. and ROBINSON, J. M., The Exxon Thermal DE NO_x Process, I. Chem. E. Symposium Series No. 57, The control of sulphur and other gaseous emissions, Salford, 1979, Paper T.

54. SLACK, A. V., Status of NO_x removal from combustion gases, I. Chem. E. Symposium Series No. 57, The control of sulphur and other gaseous emissions, Salford, 1979, Paper O.

55. JONKE, A. A., Reduction of atmospheric pollution by the application of fluidised bed combustion, Monthly Progress Report No. 26 to the National Air

Pollution Control Administration, USA, ANL/ES/CEN-F026, Argonne National Laboratory, December 1970.

56. HOKE, R. C., BERTRAND, H. R., NUTKIS, M. S., RUTH, L. A., GREGORY, M. W., MAGEE, E. M., LOUGHNANE, M. D., MADON, R. J., GARABRANI, A. R. and ERNST, M., Miniplant studies of pressurised fluidised coal combustion, 3rd Annual Report, EPA/600/7-78/069, April 1978. Purchasing code PB-284 534/5WE.

56a. HIRAMA, T., TOMITA, M., ADACHI, T. and HORIO, M., An experimental study for low-NO$_x$ fluidised-bed coal combustor development. 2. Performance of two-stage fluidised bed combustion, *Envir. Sci. Technol.*, **14**(8), 960 (1980).

57. HOY, H. R. and ROBERTS, A. G., Fluidised combustion of coal at high pressures, *Paper to 64th Annual Meeting, American Institute of Chemical Engineers*, San Francisco, Nov–Dec, 1971. Also printed in *Air Pollution and Its Control*, Coughlin, R. W., Sarofim, A. F. and Weinstein, N. J. (eds.) A.I.Ch.E. Symposium Series 126, **68** (1972).

58. ROBERTS, A. G., STANTAN, J. E., WILKINS, D. M., BEACHAM, B. and HOY, H. R., Fluidised combustion of coal and oil under pressure, *Institute of Fuel Symposium Series No. 1: Fluidised Combustion*, London, 1975, Paper D4.

59. VOGEL, G. I., SWIFT, W. M., MONTAGNA, J. C., LENC, J. F. and JONKE, A. A., Application of pressurised fluidised bed combustion to reduction of atmospheric pollution, *Ibid*, 1975, Paper D3.

60. HOKE, R. C. and BERTRAND, R. R., Pressurised fluidised bed combustion of coal, *Ibid*, 1975, Paper D5.

61. NUTKIS, M. S., Pressurised fluidised bed coal combustion, *Fluidisation*, Cambridge University Press, 1978, p. 252.

62. GIBBS, B. M., PEREIRA, F. J. and BEÉR, J. M., Coal combustion and NO formation in an experimental fluidised bed, *Institute of Fuel Symposium Series No. 1: Fluidised Combustion*, 1975, Paper D6.

63. RAJAN, R. R. and WEN, C. Y., A comprehensive model for fluidised bed coal combustors, *A.I.Ch.E. Journal*, **26**(4), 642 (1980).

64. ZEL'DOVICH, YA. B., The oxidation of nitrogen in combustion and explosions, *Acta Phys.-Chim URSS*, **21**(4), 577 (1946).

65. KIRK-OTHMER, *Encyclopaedia of Chemical Technology*, 1967 edn., Vol. 13, p. 806.

66. BAMFORD, C. H., The reaction between nitric oxide and some nitrogeneous free radicals, *Trans. Faraday Soc.*, **35**, 568 (1939).

67. ADAMS, G. K., PARKER, W. G. and WOLFHARD, H. G., Radical reactions of nitric oxide in flames, *Disc. Faraday Soc.*, **14**, 97 (1953).

68. GRAY, P. and LEE, J. C., Recent studies of the oxidation and decomposition flames of hydrazine, *Seventh Symposium (International) on Combustion*, Butterworths, London, 1958, p. 61.

69. FENIMORE, C. P. and JONES, G. W., Oxidation of ammonia in flames, *J. Phys. Chem.*, **65**, 298 (1961).

70. PILLAI, K. K., NCB private communication, 1981.

71. FIELD, M. A., GILL, D. W., MORGAN, B. B. and HAWKSLEY, P. G. W., Combustion of pulverised coal, The British Coal Utilisation Research Association, Leatherhead, 1967, p. 347.

72. *Ibid*, p. 189 ff.
73. *Ibid*, p. 348.
74. *Ibid*, p. 346.
75. GIBBS, B. M. and HEDLEY, A. B., Combustion of large coal particles in a fluidised bed, *Fluidisation*, Cambridge University Press, 1978, p. 235.

Chapter 7

THE OXYGEN DONOR GASIFICATION PROCESS

G. MOSS

Esso Research Centre, Esso Petroleum Co. Ltd, Abingdon, UK

1 INTRODUCTION

In Chapter 5 an account was given of the ways in which chemically active fluidized beds may be used to retain and concentrate the sulphur contained by the fuels burned in them. Sulphur may be retained under oxidizing conditions as calcium sulphate and under reducing conditions as calcium sulphide. In both cases regeneration, the process of stripping sulphur from the sorbent, may be accomplished via the formation of thermally unstable calcium sulphite, which decomposes at temperatures above 1000 °C to produce a concentrated stream of sulphur dioxide. Since calcium sulphite is a transitional compound between calcium sulphate and calcium sulphide it follows that in both cases the regenerating reactions compete with alternative reactions which do not release sulphur dioxide. Equations (1)–(4) illustrate pairs of competing reactions. In the cases which have been considered the operating conditions were chosen to minimize the formation of the alternative stable compound. It is quite possible however to use operating conditions within two reactors which exchange bed material (as in Fig. 1) so that, on the one hand calcium sulphate is reduced to calcium sulphide with a minimum release of sulphur dioxide, whilst on the other hand calcium sulphide is oxidized to calcium sulphate also without appreciable decomposition. It is, in fact, possible to arrange matters so that there is a net gain of sulphur from the fuel oxidized over the cycle of oxidation and reduction.

When, as in Fig. 1, the oxidation and reduction reactions are carried out

261

G. Moss

in adjacent reactors which exchange bed material the overall effect is to transfer oxygen in the solid state, i.e. as sulphate, from the oxidizing reactor to the reducing reactor. In the case in which both reactors are fluidized with air the resultant oxygen enrichment within the gasifier can be used to produce a gas having a calorific value higher than that of a low Btu gas (120–140 Btu/SCF) (4·5–5·3 MJ/m^3) but lower than that of a synthesis gas

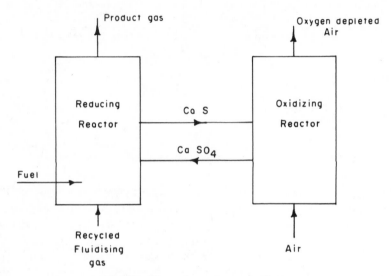

Fig. 1. The oxygen donor gasifier.

(300 Btu/SCF) (11·2 MJ/m^3). It has, however, been demonstrated that if the air fluidizing the gasifier is replaced by recycled gas or steam or a mixture of the two, then synthesis gas may be produced with the use of a relatively small amount of steam. It is, in principle, possible to produce a gas comparable in quality with that made by any steam/oxygen process but to do so more efficiently and more cheaply. In this chapter an account is given of the exploratory experimental work which has established the feasibility of the oxygen donor gasification process.

$$2CaS + 3O_2 \rightarrow 2CaO + 2SO_2 \tag{1}$$

$$2CaSO_4 + C \rightarrow 2CaO + 2SO_2 + CO_2 \tag{2}$$

$$CaS + 2O_2 \rightarrow CaSO_4 \tag{3}$$

$$CaSO_4 + 4CO \rightarrow CaS + 4CO_2 \tag{4}$$

2 THE OXIDATION OF CARBON BY CALCIUM SULPHATE

Since both carbon and calcium sulphate are solids it is necessary to effect the transfer of oxygen via gas phase reactions. The reactions which are readily available are:

$$C + CO_2 \rightarrow 2CO \qquad (5)$$

$$4CO + CaSO_4 \rightarrow 4CO_2 + CaS \qquad (6)$$

$$H_2O + C \rightarrow CO + H_2 \qquad (7)$$

$$4H_2 + CaSO_4 \rightarrow 4H_2O + CaS \qquad (8)$$

Reaction (5) is known to be a slow reaction since the carbon monoxide/ carbon dioxide ratios obtained in the product gases of gasifiers operating at temperatures in the region of 900 °C are always well below the thermo-dynamic equilibrium values. The steam/carbon reaction on the other hand is rather faster and when reaction (7) is combined with reaction (8) so that hydrogen is removed as it is formed the effect should be to speed up the gasification rate even more.

If calcium sulphate is mixed with carbon and is then heated in an inert atmosphere to temperatures above 900 °C the calcium sulphate starts to decompose at a rapid rate because the sulphite produced by thermal decomposition reacts with the carbon and is reduced to sulphur dioxide with the production of carbon dioxide and carbon monoxide, but largely carbon dioxide. In a reducing atmosphere however the calcium sulphate tends to reduce directly to calcium sulphide and no decomposition to calcium oxide occurs. In the first case reaction (2) predominates whereas in the second case reaction (8) predominates. In order to check the feasibility of this approach the apparatus shown in Fig. 2 was used. A tubular reactor packed with a mixture of charcoal and calcium sulphate with a stoichio-metric excess of carbon was heated in a tubular furnace initially in an inert atmosphere and, after the temperature in the working section of the reactor had reached 850 °C, in a stream of hydrogen. The gas was metered into the system via a gas meter and the tailgas downstream of a water condenser was likewise metered. Hydrogen was used for two reasons: first, because it is easier to meter than steam, and, secondly, because of the need to establish reducing conditions in order to suppress the decomposition of calcium sulphate. These reasons apart, steam, or a mixture of steam and carbon monoxide, would have served equally well. At temperatures below 850 °C the equilibrium partial pressure of sulphur dioxide over lime is quite low and therefore the rate of decomposition of $CaSO_3$ in an inert atmosphere is

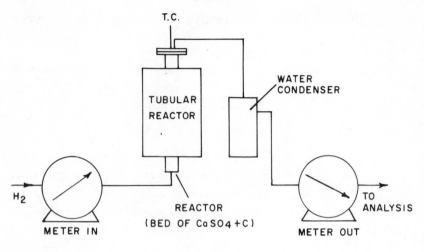

FIG. 2. Oxygen donor gasification bench apparatus.

small. It turned out, however, that reactions (4) and (5) were in evidence at very low temperatures. The carbon dioxide and carbon monoxide contents of the tailgas varied with temperature, as shown in Table 1.

It is an interesting point that whilst the carbon dioxide contents shown in Table 1 are all far too high to be accounted for by the formation of sulphur dioxide they do not rise between 550 and 750 °C, whereas the concentration of carbon monoxide rises quite rapidly with temperature over the whole range from 450 to 750 °C. It seems likely that reactions (4) and (5) accounted for the relatively high concentrations of carbon oxides in the tailgas but, since there was no oxygen present other than that in the calcium sulphate, it would seem equally likely that some degree of decomposition releasing sulphur dioxide triggered the gasification reactions. The equilibrium partial pressure of sulphur dioxide with calcium oxide increases rapidly as the temperature rises but the formation of sulphur oxides is suppressed by the presence of carbon monoxide and this effect might possibly account for the fact that the total carbon content of

TABLE 1

Oxidation of carbon by calcium sulphate in nitrogen

Temperature ($^\circ$C)	Carbon dioxide in tailgas (%)	Carbon monoxide in tailgas (%)
450	9·0	0·3
550	14·0	0·5
650	14·0	1·4
750	13·0	2·5

the gas remained more or less constant between 550 and 750 $^\circ$C. At 850 $^\circ$C the carbon monoxide content of the gas was greater than 4 % by volume and the carbon dioxide content was greater than 20 % by volume, the equilibrium partial pressure of sulphur dioxide at this temperature in the absence of oxygen being in the region of 0·125 % by volume. At this point hydrogen was admitted to the system in place of nitrogen and there was a dramatic change both in the composition of the tailgas and in its volume. The progress of the experiment after the admission of hydrogen is charted in Fig. 3 which shows the volumetric ratio of the output gas to the input gas against expired time, together with the proportion of calcium sulphate remaining unreacted and the reactor temperature. Also shown are the compositions of four bag samples of gas.

The most interesting effect was that in the early stages of the experiment the volume of the gas emerging from the apparatus was about four times that of the hydrogen entering it. Very little hydrogen was present in the gas leaving the reactor at this point, about 1·53 % by volume as hydrogen and the equivalent of 3·62 % by volume as methane. When the volumetric ratio is taken into consideration this unoxidized hydrogen accounts for 20·6 % of the input gas, the remainder being lost as water. If it is assumed that the oxygen was transferred from the calcium sulphate to the carbon by reactions (6), (7) and (8) then the effective volume of water forming carbon

G. Moss

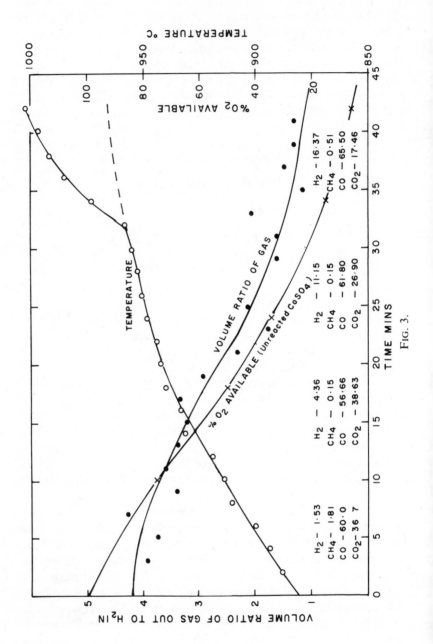

Fig. 3.

monoxide was equivalent to 3·9 times the volume of hydrogen entering the reactor and when the final formation of water is taken into account it follows that each molecule of hydrogen was oxidized on average 4·7 times during its passage through the reactor. Since the nominal residence time of the incoming gas was of the order of 1 s, the reactions were very fast. A high ratio of output gas to input gas is a very desirable feature of the system since in a practical process the input gas has to be heated and pumped and the efficiency of the process will therefore be related to the output to input ratio. The heat which energized the gasification reactions in the bench experiment was of course supplied by the furnace and it is interesting to see that as the oxygen was depleted and the gasification rates became diffusion limited there was a tendency for the rate of temperature rise to increase.

3   THE PILOT-SCALE GASIFICATION EXPERIMENT

In the case of a continuous process it is necessary to be able to add carbon and sulphate continuously to the gasifier in order to ensure steady state working conditions. The only practicable methods to achieve this involve the use of two fluidized bed reactors or a combination of a fluidized bed and a transfer line reactor. A simplified flow diagram for a fluidized bed reactor scheme is shown in Fig. 4.

If the process is regarded overall as a black box which receives air and delivers a stream of oxygen depleted air and another of partially oxidized fuel then it is clear that heat is generated in the process as a whole. However, the gasification step involving as it does the reduction of calcium sulphate to calcium sulphide is endothermic whilst the oxidation of calcium sulphide to calcium sulphate is highly exothermic. It is, consequently, necessary to transfer a considerable amount of heat from the oxidizer to the gasifier as well as oxygen and calcium sulphate and this heat is conveniently transferred as sensible heat in the bed material. The rate at which bed material must be transferred will depend upon the temperature difference between the two reactors. However, the maximum temperature at which the oxidizer may be operated may be limited by the tendency for calcium sulphide to be oxidized to calcium oxide + sulphur dioxide though the presence of excess air tends to suppress this reaction. As shown in Fig. 3 the gas composition improves as the temperature of the gasifier bed rises and good carbon monoxide/carbon dioxide ratios require operating temperatures in excess of 950 °C.

It is possible to fluidize the gasifier with either steam or else a mixture of

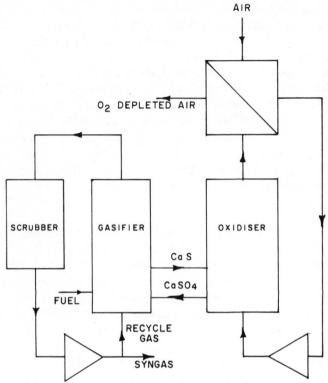

FIG. 4.   Fluidized bed oxygen donor gasifier.

steam and recycle gas. Whilst steam might be expected to give the larger ratio of output to input gas, the use of steam alone does pose some operating problems and a mixture of steam and recycle gas results in a more controllable apparatus which is much easier to start-up and turn-down.

The pilot-scale Chemically Active Fluidized Bed (CAFB) gasifier at the Esso Research Centre at Abingdon comprised two fluidized bed reactors which exchanged bed material, albeit at a comparatively slow rate. It was, therefore, possible to convert it to an oxygen donor gasifier by utilizing the regeneration reactor as the gasifier and the low Btu gasifier reactor as the oxidizer. The relative sizes of the two reactors were not correct, the oxidizing reactor being far too large in cross-section. This was not, however, regarded as a serious drawback, since the oxidizer could be fired in order to supply the heat required to bring the excess air up to its operating temperature. The large gas flow through the oxidizer did however impose a limitation on the working temperature of the reactor since the loss of sulphur from the bed material for a given sulphur dioxide concentration in the oxygen depleted tailgas was greatly increased. This in turn, taking into account the restricted bed transfer rate, imposed limitations on the working temperature of the gasifier.

The objectives of the continuous gasification experiment were as follows:

(1) To demonstrate the feasibility of gasifying residual fuel oil *without* introducing either oxygen, air or appreciable quantities of steam into the gasifier.

(2) To establish whether or not the gas so produced would be of acceptable quality.

(3) To establish whether or not the rate of gasification would be acceptable.

(4) To determine the degree of desulphurization which might be achieved.

(5) To discover whether carbon transfer from the gasifier to the oxidizer via surface deposits of the sulphated stone would present a serious problem.

The conversion of the gasifier to the oxygen donor mode of operation involved the installation of a gas recycle circuit incorporating a wet scrubber. It was anticipated that if the temperature of the recycle gas was kept at a reasonable value, say 70 °C, then the saturated vapour pressure would ensure an adequate supply of water to the gasifier.

In the first instance it was assumed that the gas made would be similar in composition to a normal CAFB gas, but without the nitrogen. This gave the assumed composition for a gas made at a temperature in the region of 900 °C (Table 2).

TABLE 2
*Composition of nitrogen-free CAFB gas
made at 900°C*

Component	Per cent by volume
H_2	22·05
CH_4	16·41
CO	25·56
CO_2	24·36
C_2H_4	11·63
C_2H_6	0·31

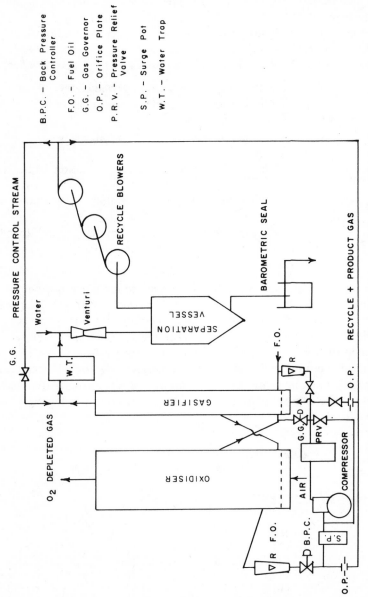

B.P.C. – Back Pressure
Controller

F.O. – Fuel Oil

G.G. – Gas Governor

O.P. – Orifice Plate

P.R.V. – Pressure Relief
Valve

S.P. – Surge Pot

W.T. – Water Trap

Fig. 5. Configuration of oxygen donor gasification pilot plant.

Associated with this gas would be approximately 27·5 % by weight of the oil as condensable tars which could be recovered and recycled in a commercial unit.

On the basis of this assumed gas composition, on the assumption of a 0·5 to 1·0 recycle ratio, and a gas recycle rate of 5 SCFM (0·14 m³/min), it turned out that with a ΔT between the reactors of 170 °C it should have been possible to treat 14·3 lb (6·5 kg) of fuel per hour. The relatively low gas ratio was assumed because only the hydrogen and the water components of the recycle gas were expected to take an active part in promoting oxygen transfer.

A simplified flow diagram for the pilot plant is shown in Fig. 5. The main control problem was the maintenance of a pressure balance between the two reactors over a wide range of gas flows through the gasifier. This pressure balance was necessary in order to allow the bed transfer system to function properly because the transfer system is valveless and uses pulses of gas to induce the flow of bed material. This problem was dealt with by sending some of the recycle gas round the scrubber circuit and using a gas governor valve to automatically adjust this flow so that the pressure upstream of the venturi scrubber was held constant. It was also necessary to maintain a constant gas delivery pressure and this was done by means of a back pressure control valve. For convenience the gas was vented into the oxidizer which served to incinerate it. Gas chromatography was used to analyze the product gas and gas flows and temperatures were monitored by a data logger. The fuel input was measured accurately by weight and a nitrogen bleed at a fixed rate enabled the materials balance to be closed by reference to the gas analysis. This was an important factor in relation to carbon balance.

In practice it was not possible to reduce the recycle gas to 5 SCFM (0·14 m³/min). About 7·5 SCFM (0·21 m³/min) was admitted to the gasifier plenum but this flow was augmented with another 3·5 SCFM (0·698 m³/min) which was used to inject the fuel and to transfer the bed material. In addition to this a further 2·0 SCFM (0·056 m³/min) of nitrogen was injected. During the experiment the oxidizer temperature was held at about 1000 °C and the bed transfer rate was maintained at its maximum value. The oil injection rate was varied in four steps from 6·6 lb (3 kg)/h to 12·5 lb (5·7 kg)/h and gas compositions and carbon balances were obtained for each condition. At the conclusion of the test programme the recycle circuit was stripped down and the carbon and tar recovered was weighed. The gas compositions which were obtained are shown in Table 3, together with other relevant data.

TABLE 3
Continuous gasification: Test results

	Run number			
	1	2	3	4
Oil input (lb/h)	6·6	7·6	9·7	12·4
Product gas composition on a dry nitrogen free basis				
H_2	7·7	11·11	13·7	16·03
CH_4	5·59	8·44	11·32	14·84
CO_2	68·36	56·02	46·60	35·60
CO	16·67	21·84	25·56	28·60
C_2H_4	1·53	2·38	3·09	4·50
C_2H_6	0·13	0·21	0·26	0·44
Gasifier temperature (°C)	926	917	909	894
Gas recycle ratio	4·7	3·3	2·9	2·5
Recycle/Product				
Per cent carbon in gas	81	94	87	73
Per cent hydrogen in gas	25	45	54	60
Per cent sulphur in gas	36	33	18	14

It can be seen that in all of the tests the recycle ratio was very high, varying from 4·7 at the lowest fuel rate to 2·5 at the highest fuel rate. As a consequence of this, the product gas was always subjected to the oxygen donor environment several times and a comparison of the gas analyses in Table 3 with each other and with the typical nitrogen free CAFB gas analysis shown in Table 2 indicates that these repeated exposures resulted in the cracking and subsequent oxidation of the ethene component of the gas. In line with this it will be seen that the carbon monoxide/carbon dioxide ratio of the gas fell with repeated exposure despite the fact that the gasification temperature rose with the recycle ratio.

The high recycle ratios which were used resulted from the small scale of the experiment and the relatively high proportion of recycle gas which was required to inject the oil and to transfer the bed material. In a large-scale unit recycle ratios well below 1·0 should be easily attainable and under these conditions the carbon monoxide/carbon dioxide ratios should be more in line with those obtained in the bench tests.

The proportion of fuel carbon present in the gas was, in general, rather higher than was expected, for whilst the gas was recycled the tars were not, being removed by the scrubber and lost from the process. Normally tars would be expected to account for about 30 % of the carbon in the fuel but in tests 1, 2 and 3 the amount of carbon present in the gas exceeded 70 % by a handsome margin. When the carbon recovered on strip-down was taken into account the carbon balance was virtually closed. This indicates that very little carbon was transferred from the gasifier to the oxidizer on the surface of the bed material, an essential condition for the efficient operation of the process.

So far as the sulphur content of the gas was concerned, this could only have been measured accurately by burning a slipstream of the gas and measuring the resultant carbon dioxide/sulphur dioxide ratio. As a slipstream burner had not been installed, only the hydrogen sulphide content of the gas could be measured and the results listed in Table 3 suggest that a substantial degree of desulphurization may have occurred. Under similar operating conditions CAFB gas is normally desulphurized to the extent of about 80 %.

The maximum fuel rate of 12·4 lb (5·6 kg)/h was equivalent to about 40 lb/h ft² of bed, or about 700 000 Btu/ft² h (2158 kW/m² h). In fact under these conditions only 70 % of the fuel was gasified since the tars were not recycled; on the other hand the recycle ratio was unrealistically high and given an adequate bed transfer rate it should be possible to at least double this output. The quality of the gas produced will depend on the nature of the fuel, the gasifier temperature and the recycle ratio. In cases where the end product is simply a desulphurized gas which is to be directly fired in a furnace or a gas turbine then the objective will simply be to produce a clean gas with as high a calorific value as possible and consequently a high hydrocarbon content will be advantageous. In cases where synthesis gas or hydrogen are the desired end products then it will be desirable to minimize the hydrocarbon content of the gas and this may best be done by choosing a fuel with a relatively low hydrogen content.

4 FUTURE DEVELOPMENTS

The work described in this chapter has demonstrated the feasibility of the oxygen donor gasification process, both with liquid and solid fuels. Whilst there may be some limited applications for the gasification of liquid fuels, the main requirement is likely to be for the production of synthesis gas from coal or lignite. The next logical step therefore is to demonstrate the gasification of coal or lignite on a continuous basis. This was not attempted in the first instance because the design of the existing pilot-scale gasifier is not suitable for this purpose.

It has been shown that when oil is gasified the transfer of carbon from the gasifier to the oxidizer on the surface of the stone is not a serious problem. This is presumably because the cracking oil lays down a thin film of carbon on the surface of the sulphated lime and under normal operating conditions this thin film is rapidly oxidized. When coal is gasified a different situation arises because the carbon being oxidized is not in such intimate contact with the stone. Under normal fluidizing conditions the char will tend to become dispersed throughout the fluidized bed and unless remedial measures are taken the char will be transferred with the bed material passing from the gasifier to the oxidizer.

Another practical problem which must be dealt with when coal is gasified is the removal of the coal ash from the reactors system, since the dilution of the bed material with a substantial proportion of coal ash can be expected to result in undesirable consequences.

Previous experience with the CAFB pilot plant has shown that when lignite is the fuel, the tendency is for the ash to be blown out of the gasifier bed. It follows that ash sizing, by suitable grinding of the coal feed, is a possible solution to the ash problem. The problem of char retention is somewhat related to the problem of ash removal because when a mixture of two bed materials differing in density but having the same particle size is fluidized, then at low fluidizing velocities it is possible to induce a degree of stratification within the bed so that the lighter material tends to float to the top. Stratification can be enhanced by making the low density component smaller in particle size than the high density component. Low ash coal has a specific gravity in the region of 1·4 whereas lime has a specific gravity in the region of 1·8. It is therefore possible in principle to float coal on a fluidized bed of lime. It has however been demonstrated that when the fluidizing medium contains hydrogen or steam, then at suitably high temperatures carbon is very rapidly oxidized by sulphated lime and will produce a large volume of additional gas. This gas production should serve to sharpen the boundary between the lime-rich and char-rich layers of the bed.

Given the possibility of stratification it becomes a simple matter to restrict the transfer of carbon from the gasifier to the oxidizer by ensuring that bed transfer occurs in such a way that material is withdrawn from the bottom of the gasifier. Carbon transfer may be further reduced by the use of a vertical partition within the gasifier bed which will prevent the surface layer of char from spreading to the zone above the bed material exit point. The gas fluidizing this zone of the bed may also contain a large amount of steam so that any carbon which enters the zone is rapidly oxidized whilst the amount of gas recycled in this region is minimized.

A proposed configuration for a coal fuelled oxygen donor gasifier is

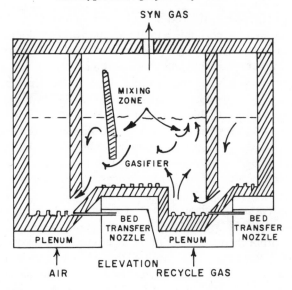

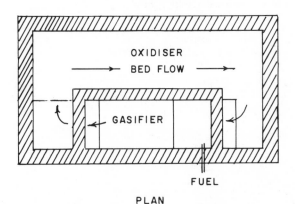

FIG. 6. Oxygen donor gasifier.

shown in Fig. 6. As can be seen, a rectangular configuration is used, the gasifier bed being bounded on three sides by the oxidizer bed. This arrangement reduces heat losses from the gasifier and also facilitates the transfer of bed material in a manner which enables a longitudinal carbon concentration gradient to be established in the gasifier. The bed transfer system is designed to minimize the transfer of gas with the bed material and is also designed for compactness and simplicity. The driving force for bed transfer is gravitational, the bed material being allowed to slump into an angled duct which directs it into a fluidized sump in the adjoining bed. The re-fluidized bed material then wells out of the sump and flows to the return port. With the arrangement shown the bed transfer ducts can extend over the full width of the gasifier bed and very high bed transfer rates should be attainable.

It may be remarked that the overall size of the gasifier should be comparable with that of a simple CAFB gasifier giving the same output. The two devices differ in operation largely to the extent that in the CAFB gasifier the nitrogen in the gasifying air is mixed with the product gas whereas in the case of the oxygen donor gasifier it emerges as a separate stream.

The oxygen donor gasifier will need to accommodate a somewhat larger air flow than the CAFB gasifier because of the need to maintain oxidizing conditions within the oxidizer in order to suppress the release of sulphur dioxide. Also, since the temperature of the oxidizer must be higher than that of the gasifier so that heat may be transferred as well as oxygen, it is necessary to recoup some of the sensible heat in the oxygen depleted air in order to achieve overall thermal balance.

Chapter 8

THREE BOLD EXPLOITERS OF COAL GASIFICATION: WINKLER, GODEL AND PORTA

A. M. SQUIRES

Department of Chemical Engineering, Virginia Polytechnic Institute and State University, Blacksburg, USA

1 INTRODUCTION

It is a privilege to be able to write about Douglas Elliott. It is difficult to do this without becoming autobiographical. Indeed, my own progress in the past two decades would have been very different if I had not met Douglas during a visit to Marchwood in 1963. I am certain, therefore, that Elliott would approve my decision to write here about three bold inventors who have given us radical departures in arts that relate to the gasification of coal.

Before addressing my three subjects, I would first comment: all three innovations that I will describe provide examples of the slowness and difficulty with which radical innovation travels across national boundaries and broad oceans. As an example, Winkler's fluidized bed gasification process operated commercially in 1926, yet W. K. Lewis and Edwin Gilliland did not have it on their minds when they invented fluid cracking in late 1938.[1] Similar remarks could be made about the first commercial Ignifluid boiler operated in 1955, and Porta's 'Gas Producer Combustion System' (GPCS) which operated in steam railway locomotives nearly 25 years ago in Argentina, and more recently in South Africa. Our professional societies do a beautiful job of keeping us abreast of developments in arts already widely practised within our own nations, but seem to lack mechanisms for heralding radical innovation abroad.

I would offer another comment: all three innovations that I will describe were placed into service quickly. So much can often be told about the robustness of an idea—its simple 'rightness'—by following its fortunes in the development process. Developments that get bogged down in complexity often do not make it to the market place. The fortunes of Winkler's, Godel's and Porta's ideas put into the shade many developments on which various national governments have spent scores or even hundreds of millions. It is not true, of course, that *all* government-managed development takes forever: John Highley's shallow fluidized beds with floating, burning coal particles, an elaboration of Elliott's earlier work on combustion in shallow beds, is a notable exception. Other examples could be given: radar, gaseous diffusion, synthetic rubber in World War II; the nuclear submarine in the 1950s. Yet how many counter-examples make us mistrustful of governmental management of technological change!

2   WINKLER'S GASIFICATION SYSTEM

On December 16, 1921, Fritz Winkler of Germany introduced gaseous products of combustion into the bottom of a crucible containing coke particles.[2] He saw the particles lift under the drag force exerted by the gas, and the mass of particles took on a liquid-like appearance. At a gas flow where the pull of gravity downward upon each particle was just cancelled by the upward drag exerted by the gas, the particles became free of one another. As Winkler increased the gas flow further, much of the gas rose through the 'fluidized' mass of particles in the form of bubbles, and the mass took on the character of a vigorously boiling liquid.

I greatly doubt that Winkler was the first person to observe this phenomenon. Almost certainly fuels engineers had seen it long before. Airflows to grate furnaces (for burning coal) can sometimes be too great for the size of particles of coal provided, or air can become badly distributed to the grate, so that coals are buoyed by a locally excessive rate of air flow. Mayers[3] described this phenomenon in terms making it evident that fluidization on a grate furnace was not newly observed at the time he wrote, i.e. 1941.

Winkler's distinction is that he was the first to recognize that the phenomenon could be put to practical use. He invented the 'fluidized bed'. In 1922, he filed patent applications for two embodiments of this new chemical engineering unit operation: one for producing activated carbon by action of steam on coke, and a second for gasifying coal fines with steam and air.

Winkler built a large atmospheric model, many feet in diameter; and as well as making measurements he observed the appearance of the upper surface of his fluidized coke. This apparently simple procedure reflects one of the most important lessons to be learnt when studying fluidized beds, namely that, the eyeball is the first, and often, in the end, the best teacher.

Commercial operation of Winkler's gasifier began in 1926 in a unit having 12 square metres of horizontal bed area. Four additional units of double the area were placed in service by 1929, at what is now the great chemical complex at Leunawerke in East Germany. The five air-blown units provided 200 to 230 000 Nm^3/h of power gas ('low-Btu gas') to gas engines that compressed ammonia synthesis gas and hydrogen for tar hydrogenation. The power generated by these gas engines must have been of the order of 100 MW, and in only seven years from invention![2,4]

Figure 1 reproduces the drawing from Winkler's first gasification patent,[5,6] showing his initial concept for coal gasification. It was to be a cyclic operation, like a cyclic generator of water gas from large particles of coke in a fixed bed, and was already old in the gasification art. A bed of lignite coke was to be blown first with air, to raise its temperature, and later with steam, producing water gas. The profile of the fluidized bed in Fig. 1 probably arose from Winkler's practical studies in his large atmospheric model, since the widening of bed diameter just above the grid plate for introducing fluidizing gas is a good idea when coarse particles, such as those Winkler used, are to be fluidized. A bed in a vessel with straight walls

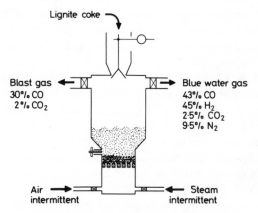

Fig. 1.   Winkler's initial gasification concept.

is apt to develop dead 'shoulders' resting on the grid plate near the sides. This is a difficulty that development groups working on Geldart's Group B and D particles seem constantly to be rediscovering, right up to the 1970s, and often it is because they have omitted building a cold model in which they can see what is happening!

Winkler built a large pilot unit for water gas production at Oppau in 1925, but results were disappointing. Production of power gas—'blast gas' in Fig. 1—was five times greater by volume than production of water gas, and the water gas contained 1 % methane, making it less than ideal for use in ammonia synthesis. The pilot unit provided design data used for the first commercial unit in 1926 making power gas.[4]

The first commercial 'Winklers' were fitted with a travelling grate to remove 'slag' that sank to the bottom of the fluidized bed. These units were idle during the hard times of 1930 and, when reactivated, were blown with oxygen and steam to yield gas for ammonia synthesis. For removal of the slag, each unit was 'fitted with a stirrer in place of a travelling grate'.[4] Leuna continued to work on the Winkler design after World War II, and Fig. 2 shows a configuration arrived at by the mid-1950s.[7] No doubt the conical bottom with a screw to remove ash from a relatively narrow opening was an improvement on earlier arrangements. The Leuna Winklers shut down in 1971.[8]

The Leuna Winklers when blown with steam and oxygen made a synthesis gas containing only 0·4% methane, leading to a much lower purge loss from a loop for ammonia synthesis than if gas with 1 % methane needs to be used. This might be due to the introduction of secondary oxygen by the 18 nozzles (seen in Fig. 2) for this purpose. This oxygen and the higher temperature that it causes in the space above the fluidized bed of Fig. 2 may well be a factor in reducing methane content of the make-gas, but the stated purpose of the secondary oxygen is to improve carbon efficiency by burning up carbon fines.[4]

Revision of the Leuna units for blowing with steam and oxygen brought about a reduction in temperature of both fluidized bed and overhead space; i.e. in air-blown operation in 1929, the bed had been at 950 °C and the overhead space at 1000 °C. Reduction of these temperatures to 800 and 900 °C, respectively, got rid of problems associated with formation of 'bird's nests' near the gas outlet at the higher temperature of the earlier operation. These were accumulations of loosely sintered ash matter, which fell off into the fluidized bed from time to time.

The reduction in temperature that was possible in the steam–oxygen practice was a result of: steam's higher reactivity toward carbon relative to

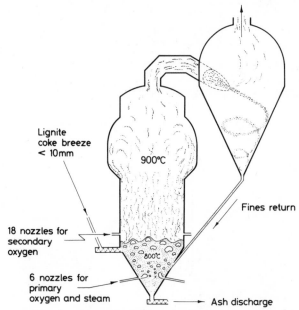

Lignite
coke breeze
< 10mm

900°C

Fines return

18 nozzles for
secondary
oxygen

800°C

6 nozzles for
primary
oxygen and steam   → Ash discharge

FIG. 2.   Late-model Winkler gasifier at Leuna.

carbon dioxide, and also of the steam's far higher partial pressure relative to that of carbon dioxide, in the presence of nitrogen (from the air used earlier to fluidize the bed).

A different path of development arose in West Germany from Pintsch Bamag's (GmbH) desire to provide a gasifier blown with steam and oxygen that could treat lignites of higher rank and lower reactivity than the brown coal of Leuna.[9] Professor Kurt Hedden states that no Winkler ever operated commercially on a coal of higher rank than lignite. The Bamag designers found it necessary to operate the fluidized bed at about 1000 °C and to raise the temperature of the overhead space to about 1100 °C. Tall reactors were provided, as seen in Fig. 3, and boiler surface was placed within the reactor near the top to reduce the exit temperature to about 900 °C, to avoid growth of troublesome ash deposits in the gas-outlet system. Two Winklers of this type have been operated at the nitrogen fertilizer factory at Kütahya, Turkey. They were tall vessels relative to diameter. It is possible that cooler gas next to the boiler surface dumped by convection into the hotter zone below, perhaps reaching all the way to the fluidized bed. The design probably can not be scaled up in size very easily. Albert Godel had discovered that ash matter does not stick to cold metal so

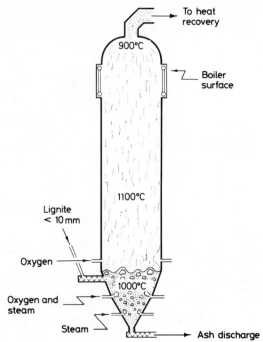

FIG. 3.   Pintsch Bamag Winkler gasifier.

the designers should have perhaps simply provided water-cooled surfaces
at the gas exit in order to eliminate the problems of deposits. There could be
something magic about 900 °C. Kaye[10] reported fusing and bonding of fly
ash particles to a ring playing air upon the surface of the fuel bed of a small
coal furnace when the metal of the ring was at 1050 °C under a reducing
atmosphere. The effect was much less marked at 950 °C and disappeared at
850 °C.

A major difficulty and disadvantage of the Winkler is the great amount of
carbon fines that it produces. During air-blown operation at Leuna in 1929,
a typical exit dust loading was 60 g/Nm^3.[11] The dust carried three-quarters
of the coal's ash, and contained from 33 to 55 % carbon. The 'bottom ash'
contained 33 to 44 % carbon. The unit utilized only 83 % of the carbon fed
to it; power gas contained 60 % of the fuel's heat value as potential energy,
and 20 % as sensible heat. The dust was fine. The upper curve of Fig. 4 is a
particle size distribution for 'raw gas dust' from a Winkler unit, probably a
Pintsch Bamag design.[8] Leuna devoted much effort to ways of utilizing the
dust, and succeeded in burning it in pulverized-fuel combustion. The units

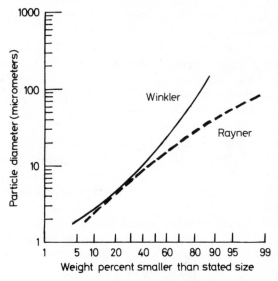

FIG. 4.   Distributions of particle sizes from two fluidized bed gasification systems operating below about 1000 °C.

at Kütahya sent the dust to pulverized-fuel combustion together with additions of pulverized lignite.

Superficial fluidizing-gas velocities at Leuna were high. During air-blown operation in 1929, the velocity at the grate was 6·7 m/s; that leaving the bed, about 3 m/s; and above the bed, after introduction of 25 % secondary air, about 4 m/s.

After World War II a number of groups, noting the success of fluid catalytic cracking, undertook development of fluidized bed gasification systems at a relatively low fluidizing-gas velocity, typically around 0·4 m/s. An example of these developments was ICI's 'moving burden process'. The interesting point is that even at a low fluidizing-gas velocity there was still a fines problem. Rayner[12] published a careful study of losses of carbon from ICI's process, and implicated particle decrepitation: 'Each particle fed to the plant gives rise to several thousand very small particles [after a critical fraction of the particle has been consumed].' The gasification reactions occur throughout the volume of the particle, and voids within the particle grow larger as the reactions proceed. Rayner believed that walls between voids become so thin after about 80 % of the carbon has been removed that the particle no longer has strength to withstand even the slightest impact; and it falls into tiny pieces. Rayner published data on the distribution of particle size in two carbon dusts recovered from two gas streams leaving the ICI process, and the lower curve in Fig. 4 shows Rayner's data as if the two dusts had been mixed. At smaller sizes the similarity to the distribution of sizes in Winkler dust is remarkable. The lesser amount of larger particles in Rayner's dust can be understood when one remembers his much lower fluidizing-gas velocity.

Twenty-five Winkler gas generators were commissioned before the end of World War II at nine locations. Between 1953 and 1960, Pintsch Bamag built ten additional generators at seven locations. None have been purchased since 1960. In the late 1960s, the Government of India conducted a careful study of coal gasification systems suitable for installation for ammonia synthesis, and decided Koppers-Totzek to be the preferred system.

The Winkler system probably can not now be revived. The carbon dust problem is too serious a difficulty. Rheinische Braukohlenwerke AG is currently studying a 'high-temperature Winkler' process, and has built a pilot unit to gasify 1300 kg/h of dry lignite at temperatures up to 1100°C and pressures up to 10 bar. It was disappointing to hear a report on this effort (Adlhoch and Theis[13]) in which no new suggestions were offered for ways to deal with the dust problem.

3   THE IGNIFLUID COMBUSTION SYSTEM

It does not detract from Albert Godel's idea to wonder why no one before him thought to fluidize a bed of burning coals above the grate of a classic travelling-grate furnace. It has been mentioned earlier that fuels technologists had recognized that fluidization could occur in this situation, and often did occur;[3] but they wished to avoid the phenomenon. Godel had his idea in about 1950, and it led him quickly to his Ignifluid boiler. He worked with one young assistant, Marcel Clovis Vaille, just graduated from the Ecole Polytechnique. Their effort and expenditure were modest in comparison even with pre-1967 British and American expenditures on fluidized bed combustion at temperatures below about 900°C. Societé Francaise des Constructions Babcock & Wilcox recognized Godel's and Vaille's accomplishment to be commercially viable, and this company placed the first Ignifluid boiler into service in 1955. It was rated at about 2·3 MW thermal, and burned refuse from the mining and preparation of anthracite at La Mur near Grenoble, France.

French Babcock & Wilcox placed five additional Ignifluid boilers into service by the end of 1958, and one Ignifluid combustion system was installed to dry coal. The early Ignifluid boilers included four units at about 4·4 MW thermal and one unit at about 12·6 MW. The early drier was rated at 5·9 MW thermal. Several of these units were still operating in August 1972.

French Babcock sold 17 additional Ignifluid units during the 1960s to customers in France, Morocco, Vietnam, and Scotland—the latter to the Lurgi gasification plant at Westfield, where three Ignifluid boilers burn coal rejected from the preparation of a clean coal for Lurgi gasification. Two of the units sold in the 1960s were driers rated at about 4·7 and 8·8 MW thermal. The remainder were boilers ranging in size from about 11·7 to 73·2 MW thermal, the rating of the two largest units, installed at an electric power plant in Casablanca, Morocco.

Albert Godel and French Babcock were slow to promote their new boiler for use in large sizes. They originally thought the Ignifluid system to be useful only in small boilers and for special fuels of low reactivity or high ash content. Godel believed he lost many years through failure to realize that his system might go into large utility boilers. A mature technique, such as pulverized-fuel firing, tends to become surrounded by an aura of inevitability that inhibits invention and protects it from new, competing ideas.

In addition, Godel's invention came at a poor moment to introduce a new system for burning coal. American Babcock & Wilcox looked at the Ignifluid just when Babcock was beginning to mount a new programme in nuclear energy; American Babcock saw no commercial opportunity in the United States for a new, small coal-fired boiler! Some of the Ignifluid boilers installed in France, indeed, shut down after only a few years of running, because cheap gas arrived from Lacq and Gröningen.

Fives-Cail Babcock now markets the Ignifluid boiler. Table 1 lists ten units commissioned or sold since 1970.[14,15] Dukla constructed the units sold in Czechoslovakia under license from Fives-Cail Babcock. The first Ignifluid for Niger started up early in 1981, and was operating routinely in June on a coal having a mineral matter content fluctuating in the region of 50 weight %. The highest average mineral content for a 24-h operating period has been 57%.[15]

In the Ignifluid, an air-blown fluidized bed of coarse particles of coke generates a low-Btu fuel gas, which is burned in a secondary combustion in free space above the fluidized bed.[16-19] Figure 5 is a schematic cross-section of the lower part of an Ignifluid boiler, showing the fluidized bed for gasifying coal and nozzles to provide secondary air to burn low-Btu fuel gas

TABLE 1
Ignifluid boilers commissioned or sold since 1970 (Source: Cosar[14,15])

Location	Type of coal	Year of start-up	Approximate rating	
			MW thermal	Millions of Btu per hour
Czechoslovakia	Lignite	1974	35	120
Czechoslovakia	Lignite	1974	58	200
Czechoslovakia	Lignite	1974	48	165
Czechoslovakia	Lignite	1977	10	35
Czechoslovakia	Lignite	1977	10	35
Vietnam	Anthracite	1978	100	350
Niger	Bituminous	1981	52	175
Niger	Bituminous	1982	52	175
Paris, France	Bituminous	1981	123	420
Paris, France	Bituminous	1983	123	420

from the gasification. The boiler itself is a conventional water-tube design. Coke particles in the fluidized bed arise from crushed coal supplied to the bed at a single point. Volatile matter from the coal is burned in the secondary combustion if it is not first converted to fuel gas in the bed. Coal need merely be crushed—in general, to a top size of about 20 mm although with coals of low reactivity, such as anthracite, the top size should be about 6 mm.

Superficial velocities of fluidizing gas in the Ignifluid are an astonishing

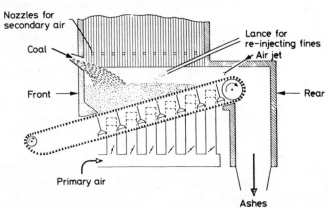

FIG. 5.   Schematic cross-section of lower part of Ignifluid boiler, showing fluidized bed for gasifying coal with air.

10 to 15 m/s. I do not know of another example of a fluidized bed of coarse particles (Geldart B or D) operating at such a high multiple of the minimum fluidization velocity. The air flow to the fluidized bed is about 65 % of the total stoichiometric air needed for complete combustion. At this air flow, the temperature of the fluidized bed falls out at something between about 1200 and 1400 °C. There is no control on bed temperature, nor is any needed! The split between primary and secondary air is set when an Ignifluid boiler is commissioned, and total air flow is controlled to follow coal feed rate so that the total air amounts to about 130 % of stoichiometric. In one Ignifluid, burning Lorraine bituminous coal at Solvay-Usine, Dombasle, near Nancy, it was found to be necessary to introduce a small amount of steam to reduce the bed temperature, in order to prevent slag accumulation at one point in the bed; but once set, the steam flow was not varied in normal operation.

The fluidized bed, when viewed from the 'front end', where the bed has a depth of about 1 m, has a typical rolling, boiling appearance. Much geysering and splashing is to be seen as bubbles burst from the surface but ejection of particles into the freeboard is an unsteady, intermittent phenomenon. When viewed from the rear, where the bed is shallow and indeed vanishes as the grate emerges from the bed, ejection of solid from the bed occurs in a steady stream of particles hurled upward by rising gas. The appearance at the rear is somewhat like a spouted bed. Solids are ejected high into the freeboard, and a large proportion falls back onto banks of coal alongside the fluidized bed (see Fig. 6).

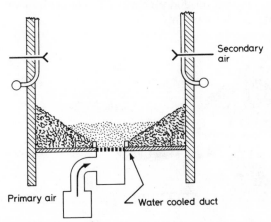

FIG. 6.   Cross-section of lower part of Ignifluid boiler at right angle to direction of travel of the grate, showing how fluidized bed is retained between two banks of coal.

At the temperature of the Ignifluid's fluidized bed the ash matter of all coals is sticky, and a less bold engineer than Godel might have expected that an attempt to operate at such an extreme temperature would produce a catastrophically huge clinker. On the contrary, Godel discovered that small ash agglomerates appear throughout the bed and remain fluidized and grow in size without risk.[20,21] Apparently the high fluidizing-gas velocity produces an effect much like the continuous use of a poker. As a coal particle burns or gasifies, ash is released. Ash sticks to ash and not to coal or coke, and agglomerates of ash grow without incorporating much carbon.

Air is introduced into the bed from compartments beneath an escalating grate. Figure 6 shows how the bed is contained at each side by a talus slope of coal itself, held back by a small water-cooled duct. Ash does not stick to *cold* metal. It will stick to almost anything else that is hot, except to carbon. In Ignifluids built at Casablanca, the French designers provided an arrangement for the water-cooled 'dam' whose purpose is to hold back the talus slope. This poor arrangement is shown in Fig. 7, where a shelf-like agglomerate has grown in the horizontal direction into the fluidized bed from a metal shroud that got too hot because some of the metal was too far away from the cooling effect of water in the tubes. It is important to note, however, that the shelf-like agglomerates did not interfere with the unit's operation.

The fluidized bed at the edges of the grate, next to the water-cooled duct, is relatively stagnant, and an ash agglomerate that by chance comes to rest near one edge of the grate tends to remain. It is possible that the circulation of material thrown from the fluidized bed onto the talus bank plays a role in

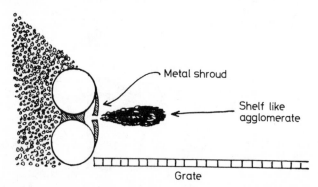

FIG. 7.   Undesirable arrangement of 'dam' holding back talus slope of static coal alongside fluidized bed of Ignifluid boiler at Casablanca, Morocco.

delivering agglomerates to the edge of the grate. Especially near the rear, shallow end of the bed, there is a steady avalanche of solid down the banks. Agglomerates that come to rest near the edge of the grate are sticky, and they capture other agglomerates as they happen along. It probably takes only one hit for an agglomerate to be immobilized upon the grate by earlier arrivals. A pad of clinker develops, and it continues to capture still more of the fluidized agglomerates as they move about in the bed and accidentally strike the pad. The effect is much like fly paper capturing flies. The grate is operated at a speed so that a pad of clinkers develops that is about 50 mm thick, and this is dumped into an ash pit. Godel observed the grate near the deep end through a pipe of 100 mm diameter, cooled with flowing nitrogen. He saw only an occasional unfluidized clinker being carried along. It is probable that the pad of clinkers develops only toward the shallow end of the bed.

At the high fluidizing-gas velocities used in the Ignifluid combustion system, there is a large traffic of coke particles upward through the boiler into mechanical dust-collecting devices, and thence into a lance for reinjection at high velocity into the deep end of the fluidized bed. In the Ignifluid at Solvay-Dombasle, the operators measured the rate of circulation of carbon dust to be about one-half of the bituminous coal feed rate; the operators stated that carbon utilization efficiency depended upon careful aim of the lance toward the deep end of the bed. Even an exceedingly small particle of carbon injected about 0·5 to 1 m below the surface of the deep end of the bed would remain in the bed for several minutes, and such a particle probably cannot survive this long in the presence of oxygen from air at the high temperatures of the bed. Carbon utilizations in the Ignifluid routinely run beyond 99 %, except for coals of unusually high ash content. It should also be noted that the 'fly carbon' trafficking through the boiler, dust collectors, and reinjection lance is mostly carbon. It is not sufficiently sticky to foul boiler surfaces. Experienced boiler engineers observing an Ignifluid in operation have remarked that the surfaces of water-tubes look like those in a gas-fired boiler!

Operators of the Solvay-Dombasle boiler provided a sample of fly carbon to Graff et al.,[19] who measured its distribution in particle size to obtain the upper curve of Fig. 8. The sample came from the hopper of a set of high-velocity multicyclones. There was an earlier stage of gravity knockout, of a coarser dust, that could not be sampled. From the appearance of the stack when Graff et al. visited the unit, the cyclones were judged to have been working at relatively high collection efficiency. If this

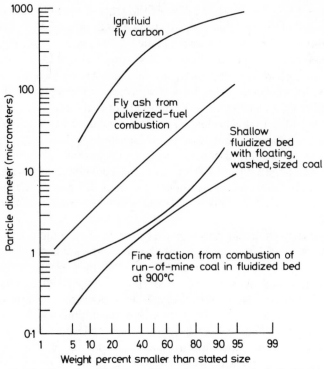

FIG. 8.   Distribution of particle sizes of dusts from four methods for burning coal.

judgement is correct, it would appear that the dust represented by the upper curve of Fig. 8 contained very little material smaller than about 2 μm.

Table 2 lists design and operating data for two of the larger Ignifluid boilers commissioned before 1970: Solvay-Dombasle and Casablanca. Up to that date, the two units at Casablanca were the only Ignifluids in which primary air was preheated. At a bed temperature of 1200 °C, the inlet superficial air velocities, based upon grate areas, are about 23 m/s and 15 m/s for Solvay-Dombasle and Casablanca, respectively. The velocity of gas leaving the upper surface of the fluidized bed is, of course, lower, and an exact calculation of this velocity is difficult because of uncertainties in the precise geometry of the upper surface of the bed. Nevertheless, the fluidizing gas in the Ignifluid may be characterized as being in the neighbourhood of ten times the minimum fluidization velocity for the coke particles that comprise the bed.

Longitudinal mixing in the bed is insufficient to maintain temperature

TABLE 2

Some design and operating characteristics of Ignifluid boilers at Solvay-Dombasle and Casablanca

	Solvay-Dombasle	Casablanca
Maximum rated capacity—steam vaporization, metric tons/h	55	115
Steam superheat, °C	300	538
Total flow of air (primary and secondary), kg/h	86 000	175 000
Average inlet temperature of primary air, °C	20	185
Average inlet temperature of secondary air, °C	20	265
Length of combustion chamber, m	4·77	5·54
Width of combustion chamber, m	3·54	5·08
Length of grate between axes, m	7·5	8·3
Length of grate above primary air compartments, m	4·5	5·23
Width of grate, m	0·55	1·4
Inclination of the grate, degrees	10	10

uniformity. Each Ignifluid displays a temperature gradient with a higher temperature toward the rear, shallow end of the bed. This higher temperature can be understood to result from the smaller carbon inventory available to sustain the endothermic reaction of carbon dioxide with carbon to form carbon monoxide. Gas samples secured directly above the rear end of the bed would give higher readings of carbon dioxide than samples secured directly above the deep end.

It is highly probable that temperatures are relatively uniform from top to bottom of the fluidized bed at any given location.

Graff *et al.*[19] came away from their visits to Ignifluid boilers with an impression that the design is extremely flexible with respect to latitude in approach by its operators. The Casablanca unit was operated with a relatively shallow bed of relatively coarse particles. Since the bed provided, therefore, relatively less carbon surface to sustain the slow, endothermic reaction of carbon dioxide with carbon, temperatures of the Casablanca unit generally ran about 100 °C hotter during the visit there than temperatures in Ignifluids at LaTaupe and Solvay-Dombasle, which were visited later. No pyrometer was available at Casablanca, but M. Vaille of Fives-Cail Babcock estimated a temperature of about 1200 °C at the front, deep end and about 1400 °C at the rear, shallow end.

The LaTaupe Ignifluid was judged to be at about 1200 °C at the front end, and at about 1300 °C at the rear. Using a pyrometer, two observers

TABLE 3

Percentage by weight of ash agglomerates in Ignifluid bed samples

Location in the bed	Casablanca	LaTaupe		Solvay-Dombasle	
		First sample	One hour later	First sample	One hour later
Front	19·8	13·2	17·3	13·6	13·2
Middle	18·4	15·2	13·9	8·9	8·1
Rear	4·6	23·2	18·9	7·0	4·0

judged the Solvay-Dombasle unit to be between 1180 and 1200 °C at the front end; both observers found 1320 °C at the rear.

Graff et al.[19] secured bed samples from each of the three Ignifluid boilers they visited. A steel cup, about 100 mm in diameter and 200 mm deep, was welded to a long steel rod. A sample was taken by inserting the cup into the fluidized bed, withdrawing the cup, placing a brick promptly upon the cup to shut out air, and placing the cup in water for cooling. Samples were taken at the front, middle and rear of each bed. The rear sample was taken by holding the cup in a horizontal position resting on the pad of clinkers moving on the grate, just ahead of the point at which clinkers emerge from the fluidized bed. The depth of the bed at the rear sampling location was estimated at 100 mm.

By hand, Graff et al.[19] separated ash agglomerates from coke particles in each sample. Table 3 lists the percentage of ash agglomerates found in the various samples. To get an idea of variability with time, two samples were taken about 1 h apart from both the LaTaupe and Solvay-Dombasle units. As Table 3 shows, there can be a significant difference in percentage of ash agglomerates in samples secured at different times. In two of the units, the percentage of ash agglomerates declines toward the rear. This is consistent with the view expressed earlier that the transfer of ash agglomerates from the bed to the pad of clinkers largely takes place near the rear end of the grate's travel. Why the LaTaupe samples do not reflect this tendency is unknown.

Figure 9 gives curves showing distribution in size of ash agglomerates in samples from the Casablanca Ignifluid at the front, middle and rear of the bed. Figure 10 repeats the middle curve for the Casablanca unit, and illustrates how much larger the agglomerates were at Casablanca than they were at LaTaupe and Dombasle. There was little variation in size of ash agglomerates at the latter two units from front to middle to rear, and the curves in Fig. 10 are typical of these units.

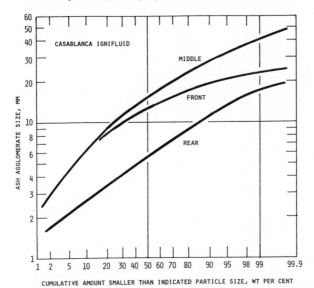

FIG. 9. Distribution in size of ash agglomerates collected from fluidized bed of Ignifluid boiler at Casablanca.

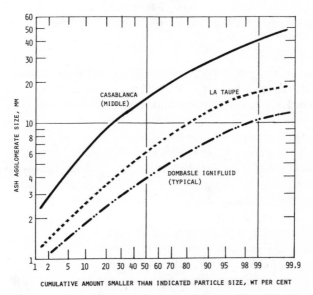

FIG. 10. Distribution in size of ash agglomerates collected from fluidized beds of Ignifluid boilers at Casablanca, LaTaupe and Dombasle.

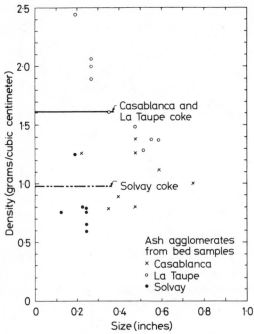

Fɪɢ. 11.   Densities of coke particles and ash agglomerates collected from fluidized beds of Ignifluid boilers at Casablanca, Solvay-Dombasle and LaTaupe.

In light of the generally higher temperatures at Casablanca, it is no surprise that agglomerates from Casablanca have a different character than those from LaTaupe and Solvay-Dombasle. Not only are the Casablanca agglomerates larger, but also they appear to represent a joining together of smaller agglomerates into a composite. The robustness of the ash-agglomeration phenomenon exploited in the Ignifluid combustion system is perhaps best displayed by reflecting that the tendency for agglomerates to merge in the Casablanca unit, which M. Vaille regarded to reflect relatively poor operating practices, did not overcome the ability of fluidizing air to keep the agglomerates freely fluidized in the mass of carbon until they finally joined the pad of clinkers on the grate.

LaTaupe agglomerates were typically roughly spherical and compact. Agglomerates from Solvay-Dombasle were small, friable, and porous, and tended to be irregular in shape. Figure 11 gives data on densities of ash agglomerates from the three units, illustrating the generally high densities at LaTaupe and low densities at Solvay-Dombasle. Figure 11 also shows densities of coke particles from the three units. The Solvay-Dombasle coke, produced from Lorraine bituminous coal, was considerably lighter than cokes produced from anthracites in the other units. Yet, the fluidized agglomerates at Solvay-Dombasle tended to be lighter still, denying the possibility that gravity settling can be the mechanism for transfer of agglomerates from the fluidized bed to the pad of clinkers on the travelling grate. Casablanca agglomerates were also lighter than Casablanca coke. Densities of clinkers from the three units were the same, about $1.7\,\text{g/cm}^3$, and were higher than densities of any of the agglomerates except a few of the most dense ones from LaTaupe. The agglomerates apparently densify after they arrive upon the grate.

Table 4 gives data on carbon content of coke, ash agglomerate, and clinker samples from the three units and Fig. 12 gives size distributions. The carbon content of ash agglomerates was generally low, and was comparable to carbon levels of ash clinkers. The data indicate that the Ignifluid's high utilization of carbon does not depend significantly upon burning out carbon from ash agglomerates after they reach the grate. Additional carbon analyses are found in Harvey *et al.*[22]

The air rate to each of the three Ignifluid boilers visited was about 130 % of stoichiometric. Design intention is to supply about one-half of this air to the fluidized bed, but the primary air could acceptably fall in the range from 45 to 55 %. A primary air rate of 40 % of the 130 % total in respect to stoichiometric air—i.e. a primary air rate that is $130 \times 0.4 = 52\%$ of stoichiometric—would be acceptable if the boiler were to operate only at a design steam flow, with no capability for turn-down. Turn-down capability is acceptable at the 45 % primary air rate. The valves of several wind chests that supply primary air to several regions of the fluidized bed are set once and for all during start-up. Turn-down capability is good: at the time of the Graff *et al.* visit, LaTaupe routinely operated to furnish 18 tons/h of steam during the night and 33 tons/h during the day.

TABLE 4

Average carbon content (weight %) of coke and ash samples from Ignifluid boilers

Sample	Casablanca	LaTaupe	Solvay-Dombasle
Coke	90·2	88·7	88·7
Ash agglomerates from front of bed	4·47	7·50	2·7
Ash agglomerates from middle of bed	5·84	6·40	3·8
Ash agglomerates from rear of bed	7·55	8·43	2·10
Discharged ash clinkers	5·1	4·1	1·8

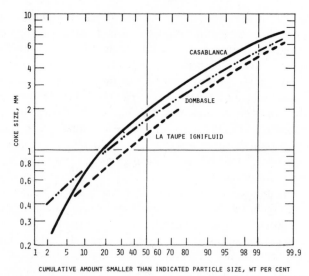

COKE SIZE, MM

CUMULATIVE AMOUNT SMALLER THAN INDICATED PARTICLE SIZE, WT PER CENT

Fig. 12.   Distribution in size of coke particles collected from fluidized beds of
Ignifluid boilers at Casablanca, Dombasle and LaTaupe.

Coal feed rate is set to maintain desired depth of bed, which is judged by
the pressure of an air bleed to the deep part of the bed. This air bleed was
inoperative at Casablanca, however, and here the operators apparently
governed coal feed by eyeball judgement of bed depth.

Primary air rate controls the steam pressure, and secondary air is set to
maintain a desired carbon dioxide analysis in stack gas.

Start-up is effected by establishing an air flow and firing above the bed
with oil or gas to heat it by radiation from above. Graff et al.[19] observed a
start-up at Casablanca that followed a crash shutdown. After the
shutdown, the operators did not stop the grate motion. As a consequence
the bed inventory was entirely dumped into the ash pit, and start-up
commenced from an empty bed. The time which elapsed from coal feed to
establishment of a smoothly rolling bed was approximately 30 min. It was
said that the smaller LaTaupe bed could be started in 20 min.

The crash shutdown at Casablanca occurred as a result of difficulties in
feeding coal. The unit had not been operated on coal for approximately six
months, anthracite production in Morocco not being sufficient to maintain
a supply for the unit all year round. Coal in the bunkers had become non-
uniform in moisture content, and maintaining an even flow of the coal
proved impossible. Smooth operation was re-established after the bunkers
had been completely emptied and refilled with fresh coal.

Graff *et al.*[19] saw an intentional crash shutdown at Fives-Cail Babcock's Ignifluid pilot unit at LaCourneuve, near Paris. The operators at LaCourneuve stopped the motion of the grate when the primary air was turned off. The bed slumped and presented a smooth, level upper surface. About 10 min later flow of primary air was started, and within seconds the fluidized bed was re-established and coal feed was commenced. This was an impressive display of the Ignifluid's capability to be banked overnight and re-started. It is believed that a large Ignifluid could be re-started almost in an instant after remaining banked for at least 24 h.

In test work at LaCourneuve, bed temperature has sometimes been held down when treating special materials by recirculation of flue gas. In one series of tests on a carbonaceous material containing zinc, lead and nickel and displaying an 'ash' softening temperature of about 900 °C, the LaCourneuve bed was successfully operated at a temperature of 750 °C.

If a material of low ash fusion temperature is run at too high a temperature, slagging occurs on the grate that leads to fouling and blockage.

At Solvay-Dombasle, the talus banks were composed of crushed brick bearing a shallow layer of coal. In early operation with banks completely comprising of coal, agglomerates developed on the banks that grew outward in stalactite-like growths. It is not clear whether or not this was because bituminous coal was used to form the banks, or if there was undue leakage of air into the banks causing coal to burn and slag within the banks. In any case, leakage of air through the banks must be scrupulously prevented, and this has occasionally caused problems.

It was during the shutdown of Casablanca, when the operators allowed the grate to carry bed solids away to the ash pit, that the shelf-like agglomerates of Fig. 7 were seen. There were three such 'shelves', about 75 to 300 mm in length and 150 mm in width, near the front, deep end of the bed.

A rare type of coal ash gives trouble. When coal and ash are present in alternating thin layers like a wafer sandwich, the ash that evolves is light and tends to float on the bed. It will not sinter and agglomerate.

It remains to discuss the curves of Fig. 8 giving distributions of dust from four methods of burning coal.

The surprising feature of the Ignifluid combustion system was the coarseness of the fly carbon, illustrated by the top curve of Fig. 8. Comparison with the curves of Fig. 4 shows how very much coarser the fly carbon from Godel's fluidized bed gasifying coke at 1200 °C is in comparison with the fine carbon dust from earlier fluidized beds gasifying coke below about 1000 °C. Research is needed to elucidate reasons for the difference, but one may speculate that a particle of carbon finer than a certain size simply does not survive more than a single trip through the boiler and reinjection lance.

The curve for fly ash from pulverized-fuel (PF) combustion in Fig. 8 is based upon a sample provided by Consolidated Edison Co. of New York, from a boiler burning an Eastern bituminous coal and fitted with an electrostatic precipitator working at well beyond 99 % collection efficiency. (The percentage of extremely fine sizes shown in the curve, e.g. sizes smaller than 1 μm, can be understated by less than 1 %.) The curve is close to the 'most probable distribution' given for PF fly ash by Fennelly *et al.*[23] and also by Bubenick *et al.*[24] The curve is based upon sieving the ash down to 325 mesh, and sedimenting the minus-325 fraction in a water suspension (with sodium pyrophosphate dispersant) in an Andreasen pipette.[25]

The curve for a shallow fluidized bed with floating, washed, sized coal was given by Gill,[26] who also gave data for dust from a chain-grate stoker burning such coal. The latter data reflect particle sizes roughly like those in Fig. 8 for fly ash from PF combustion.

The lowermost curve in Fig. 8 is for a fine fraction of dust from a small Stone-Platt Fluidfire fluidized bed hot water boiler (AFBC), rated at about 0·3 MW thermal, that my colleagues and I have operated at Virginia Polytechnic Institute[27] burning a Virginia caking coal of high swelling index. The curve is based upon data from a Sierra impactor working on dust sampled at about 800 °C in the exit line, ahead of any equipment for dust collection. The loading of dust in an isokinetic sample was 8·9 g/Nm³. A cyclone pre-separator removed 85 % of the dust, and so the curve in Fig. 8 represents a 'loading' of 1·3 g/Nm³. Screen analysis showed 20 % of the dust from the cyclone pre-separator to be smaller than 325 mesh—i.e. 1·5 g/Nm³—and so the total dust smaller than 325 mesh was probably about 1·5 + 1·3 = 2·8 g/Nm³.

The comparisons of Fig. 8 provide only a rough guide to the problem of collecting dust from the four methods of combustion. A better guide would be a set of curves of dust loading as a function of particle size, but such curves cannot be generated from the data at hand. Figure 8 does suggest that dust collection will be progressively more difficult as one moves from an Ignifluid combustion system to stoker firing to PF combustion and finally to AFBC combustion. This impression is consistent with the physical behaviour of mineral matter in coal in the several methods of combustion.

In both Ignifluid and stoker firing, carbon is exposed for many minutes to a temperature at which intrinsic mineral matter 'sweats' from the interior of the carbon and accumulates in surface agglomerates relatively free of carbon. Yerushalmi *et al.*[20] and Kolodney *et al.*[21] published scanning electron microscope photographs of carbon particles in Ignifluid samples bearing surface accumulations of ash matter still attached to the carbon particles via roots penetrating into the particles. At the moment the samples were collected, ash matter in the roots was moving by capillarity toward the surface accumulations. The surface tension of silicate glasses is high, about 300 dynes/cm, and similar levels of surface tension in semifluid mineral matter was the driving force generating the surface accumulations seen in Ignifluid samples. One accumulation was found, still attached to a carbon particle, that was more than 500 μm in size. Obviously, such an accumulation eventually becomes free of carbon, as the carbon is burned or gasified. Intrinsic mineral matter is probably released from carbon particles in both Ignifluid and stoker firing in the form of agglomerates having a wide range of particle size. Further growth in size of the smaller agglomerates probably occurs in any fuel bed in which relative motion of particles takes place on account of any phenomenon that 'stirs' the bed. The Ignifluid fuel bed, of course, is vigorously stirred by fluidization. Agglomerates also undoubtedly stick to any particles of extrinsic mineral matter with which the agglomerates come into contact. All of this leads to discharge of much of the mineral matter present in a coal which is burned in a stoker in the form of ash agglomerates or clinkers. Essentially, all mineral matter in a coal burned in the Ignifluid combustion system leaves the system in clinkers.

In PF combustion, some agglomeration of ash matter occurs,[28,29] but not nearly as much as in a stoker or especially an Ignifluid. A relatively large particle of coal burning in a PF boiler probably produces only one fly ash particle, which represents an agglomeration of the intrinsic ash matter originally present in the particle of coal.

In fluidized bed combustion at a temperature below about 900 °C, there is no agglomeration of mineral matter whatsoever. A relatively large particle of coal may produce a myriad of tiny particles, each arising from one moiety of the finely disseminated intrinsic mineral matter in the original coal particle. In addition, of course, the far larger particles of extrinsic mineral matter also appear essentially unaltered in size, either in dust leaving the fluidized bed, or in matter withdrawn from the bed to maintain the solids inventory constant.

4   LIVIO DANTE PORTA'S GAS PRODUCER COMBUSTION SYSTEM

It may be best to describe Porta's new system by telling how it began.[30] In 1956, Argentina built a narrow-gauge railway at the far southern edge of the nation to haul a low-rank sub-bituminous coal (almost lignite) from Rio Turbio in the mountains next to Chile to port facilities at Rio Gallegos on the South Atlantic. Porta says,

> 'I was chief and manager and master of the Rio Turbio Ry and circumstances allowed me to do a fine work in spite of appalling shortage of resources in the middle of a frozen Patagonia. The urgency was a result from the fact that coal burning locomotives were operating a service far below standards, i.e. shutting off steam for cleaning by hand heavily clinkered fires.... I made a courageous step piercing the firebox of the first locomotive to give secondary air, and also introduced exhaust steam in the ashpan together with some other non-structural improvements. There was an initial startling success. Minor improvements of practical nature were introduced, but since, the system kept practically unchanged.'

What Porta had done was to cut the flow of air to the coal bed to about 30 % of the total! Seventy per cent entered a combustion zone above the firebed. Steam from the exhaust of the engine's cylinders was mixed with air to the firebed, with three results (Click[31]):

'1.   the reactions taking place through the firebed result in a gas producer situation so that carbon monoxide and hydrogen appear above the grate;
2.   the firebed temperature remains far lower than would otherwise be the case so that fusion of ash to a clinker can be prevented; and,
3.   the disturbance of the firebed is greatly reduced so that unburnt fuel carry-over to the chimney is very low. With this comes higher boiler efficiency and reduced pollution of the surrounding countryside.'

Click wrote of his visit to Patagonia to see Porta's locomotives with their 'Gas Producer Combustion System' (GPCS):

> 'The author is happy to say that although he went to Argentina full of doubts about GPCS, he left convinced that it fully met the claims made for it; and, correctly applied and operated, gave high efficiency, pollution-free combustion.
> Of [great] interest in the dark was the amount of sparking from the

chimney top. The locomotive was clean to ride on but just how little was leaving the chimney top partly burnt was surprising after dark. Even on the last assault up to the Rio Turbio Station... when the locomotive was working very hard in 50–55% cut-off with the regulator wide open at about 18 mph (where the author would have expected a trail of fire from the chimney) there were only a few (almost countable!) pea-sized sparks with an occasional one the size of an olive that lay, after bouncing, a few seconds, on the frozen ground before becoming invisible.

It did not go unnoticed either, that that locomotive after many hours of steaming blew off at the safety valve the instant the regulator was shut in the Rio Turbio yard!

Though he did not see GPCS used on any fuel save the Rio Turbio coal (where the high volatile content of that fuel seemed to make its use particularly advantageous) the author is confident that the system lends itself to use with other coals as well as with more unusual fuels on which Ing. Porta has proved it to work admirably. These fuels have included imported Polish coal, wood, sawmill rejects (again firewood), and charcoal fines mixed with fuel oil; all of which have been tried in Argentina. Other fuels Ing. Porta would like to try include peat (turf), sawdust mixed with fuel oil, and the residue from sunflower seed pressing!'

Porta's GPCS has been installed on 20 engines of the Rio Turbio Railway, and in 1958, on an engine from Argentina's main line railways. South Africa still employs steam locomotives, and Wardale[32,33] has converted two engines to GPCS there.

Current GPCS designs provide a refractory roof above the firebed, resembling, just a little, a beehive with a hole in the top. This retains heat in the space above the bed, and produces an intense combustion there. Porta believes that radiation from the flame of the secondary combustion onto the firebed provides a significant part of the endothermic heat needed for the gasification reactions, whereby steam and carbon dioxide react with carbon to form hydrogen and carbon monoxide. Secondary air is introduced tangentially to produce something of a cyclone effect within the refractory 'beehive', to cast large particles of coke back onto the firebed. Several nozzles inject high-pressure, superheated steam into the secondary combustion zone to help mixing of secondary air and fuel gases rising from the firebed.

The GPCS locomotives in South Africa apparently never form clinkers in the firebed, but ash material simply accumulates toward the bottom of the bed where it can be removed by motion of grate bars. Click[31] reports ash from locomotives of the Rio Turbio Railway to be of a spongy appearance and open for passage of air.

GPCS emissions of nitrogen oxides should be low.

Giammar and Coutant[34] published data on the addition of limestone to a firebed in a small experimental rig simulating conditions in a stoker-fired furnace. They found that limestone achieved significant reductions in sulphur dioxide emissions if the firebed was operated with a sub-stoichiometric flow of air. This is encouraging for Porta's GPCS, which might afford an opportunity for development of a small furnace emitting both nitrogen and sulphur oxides at low levels.

It is of interest to note that combustion scientists who studied classic firebed behaviour may have been inching their way as long as 40 years ago towards Porta's ideas. Mayers[3] outlined his thoughts for a new underfeed stoker design to afford higher burning rate of coal per unit firebed area. He made two suggestions evocative of both the Ignifluid and GPCS developments:

(1)   To operate the stoker at air rates 'close to full teeter'—i.e. close to a velocity that would fluidize the fuel bed.

(2)   To operate with only from 40 to 75 % of the air for combustion supplied as primary air to the fuel bed, and with the remainder to be injected over the fire to support secondary combustion.

The Rio Turbio Railway engines feed a free-burning coal onto a static grate, partially with the help of a stoker but with some intervention by hand. The South African locomotives with GPCS apparently also have stationary grates.

There is work to be done to determine how GPCS ought to be developed further for a wide range of types of coals. A travelling grate may be best for free-burning coals, but for the highly-caking bituminous coals of the Eastern United States, we may need to develop a version using an underfeed stoker, in fulfillment of Mayers' prophecy. GPCS sounds so promising that a study of it in depth is well justified.

REFERENCES

1. GILLILAND, E. R., Personal communication, September, 1970.
2. FEILER, P., *Die Wirbelschicht*, Badischen Anilin- & Soda-Fabrik AG, 67 Ludwigshafen/Rhein, West Germany, 1972.

3. MAYERS, M. A., Flow processes in underfeed stokers, *Trans. A.S.M.E.*, **63**, 479–89 (August 1941).
4. GRIMM, H. G., The gasification of fine-grained coal in the Winkler Gas Producer, *Proceedings of the Third International Conference on Bituminous Coal*, Carnegie Institute of Technology, Pittsburgh, Pennsylvania, November 1931, Vol. I, pp. 874–81.
5. German Patent 437970, December, 1926.
6. US Patent 1582718, April, 1926.
7. VON PORTATIUS, B., *Freiberg. Forschungsh. Reihe A.*, **69**, 5 (1957).
8. BANCHIK, I. N., Personal communication, August, 1973.
9. FLESCH, W. and VELLING, G., *Erdoel Kohle Erdgas Petrochem.*, **15**, 710 (1962).
10. KAYE, W. G., Domestic heating appliance development, *Chemistry and Industry*, No. 29, 946 (July 1970).
11. MINISTRY OF FUEL AND POWER (GREAT BRITAIN), *Report on the Petroleum and Synthetic Oil Industry of Germany*, HMSO, London, 1947, p. 21.
12. RAYNER, J. W. R., *J. Inst. Fuel*, **25**, 50 (1952).
13. ADLHOCH, W. and THEIS, K. A., The Rheinbraun high-temperature Winkler Process, in *Conference Proceedings: Synthetic Fuels—Status and Directions*, May 1981, Electric Power Research Institute, Palo Alto, Calif., WS-79-238, Vol. 2, Section 29.
14. COSAR, P., Personal communication, June, 1980.
15. COSAR, P., Personal communication, September, 1981.
16. GODEL, A. A., Process for the gasification of granulated fluidized bed of carbonaceous material, over moving, sloping, horizontal, continuous grate, US Patent 2866696 (December 1958).
17. GODEL, A. A., *The Most Advanced Method for Burning Coal: 'Ignifluid' Fluidized Bed Combustion Process*, private publication, Paris, France, 1965.
18. GODEL, A. A. and COSAR, P., The scale-up of a fluidized-bed combustion system to utility boilers, *A.I.Ch.E. Symp. Series*, **67**(116), 210–18 (1971).
19. GRAFF, R. A., YERUSHALMI, J. and SQUIRES, A. M., Studies toward improved techniques for gasifying coal, Final Technical Report from Grant GI-34286A-1 from National Science Foundation to City College of City University of New York, July 1976.
20. YERUSHALMI, J., KOLODNEY, M., GRAFF, R. A., SQUIRES, A. M. and HARVEY, R. D., Agglomeration of ash in fluidized beds gasifying coal: The Godel phenomenon, *Science*, **187**, 646–8 (1975).
21. KOLODNEY, M., YERUSHALMI, J., SQUIRES, A. M. and HARVEY, R. D., The behaviour of mineral matter in a fluidized bed gasifying coal—The Ignifluid process, *Trans. Brit. Ceram. Soc.*, **75**, 85–91 (1976).
22. HARVEY, R. D., MASTERS, J. M. and YERUSHALMI, J., Behavior of coal ash in gasification beds of Ignifluid boilers, Illinois Minerals Note No. 61, Illinois State Geological Survey, Urbana, September 1975.
23. FENNELLY, P. F., HALL, R. R., YOUNG, C. W., ROBINSON, J. M., KINDYA, R. J. and HUNT, G., Update on emission measurements from fluidized-bed combustion facilities, see ref. 24, pp. 1236–44.
24. BUBENICK, D. V., LEE, D. C., HALL, R. R. and FENNELLY, P. F., Control of particulate emissions from fluidized-bed combustion: Fabric filters or electrostatic precipitators, *Proceedings of the Sixth International Conference on*

Fluidized Bed Combustion, April 9–11, 1980, Atlanta, Georgia, issued by Office of Coal Utilization, US Department of Energy, Washington, DC, Vol. III, pp. 1245–59.

25. LEE, K. C., Filtration of redispersed power-station fly ash by a panel bed filter with puffback, Ph.D. Dissertation, The City University of New York, 1975.

26. GILL, D. W., The potential of fluidised-bed combustion for emission control, *The Chemical Engineer*, 278–80 (June 1981).

27. METCALFE, C. I., FEGLEY, K. E., HALNON, T. D. and SQUIRES, A. M., The operation of a small industrial coal fired fluidized bed hot water heater, see ref. 24, Vol. II, pp. 136–44.

28. FLAGAN, R. C. and FRIEDLANDER, S. K., Particle formation in pulverized coal combustion—A review, Paper presented at meeting of American Institute of Chemical Engineers, Atlantic City, New Jersey, August 29–September 1, 1976.

29. FLAGAN, R. C., Ash particle formation in pulverized coal combustion, Presented at Spring Meeting Western States Section, The Combustion Institute, University of Washington, Seattle, April, 1977.

30. PORTA, L. D., Personal communication, July 1981.

31. CLICK, J. G., Report on the transportation of coal by the Rio Turbio Railway, to the United Nations Industrial Development Organization, Vienna, Austria, December 1977.

32. WARDALE, D., The steam locomotive: Motive power for the future?, Paper presented to South African Railway Engineering Society, August 1978.

33. WARDALE, D., Personal communication, July 1981.

34. GIAMMAR, R. D. and COUTANT, R. W., Experimental studies on the feasibility of in-furnace control of SO_2 and NO_x emissions from industrial stokers, report from Battelle Columbus Laboratories, Columbus, Ohio, December 1975.

Chapter 9

COMBUSTION OF GASES IN FLUIDIZED BEDS

J. BROUGHTON

GEC Gas Turbines Ltd, Whetstone, Leicester, UK

NOMENCLATURE

a	Ratio
C_e	Exit concentration
C_0	Inlet concentration
$\bar{C}_p$	Mean specific heat at constant pressure
$\bar{d}_p$	Particle harmonic mean diameter
D_B, D_0	Bubble diameter, initial bubble diameter
$\mathscr{D}$	Diffusion coefficient
f_u	Fraction of fuel unburned
g	Acceleration (gravitational)
h_D	Heat transfer coefficient (bed to distributor)
K	Bubble growth constant (eqn. (6))
K_L	Distributor plate, laminar flow resistance
K_T	Distributor plate, turbulent flow resistance
L	Height above distributor plate (bed depth)
L_{mf}	Height of bed at minimum fluidization
n	Frequency of pressure fluctuations
P	Pressure
P_0	Pressure in windbox
Q	Gas exchange flow rate
S	Windbox height
St	Stanton number ($h/\rho U C_p$)
T_B	Bed temperature
T_0	Gas inlet temperature

T_S Surface temperature of distributor
U Gas velocity (superficial)
U_B Bubble velocity
U_i Velocity of single isolated bubble
U_{mf} Minimum fluidization velocity
U_0 Gas velocity (superficial) through distributor
V_B Bubble volume
X Bubble-phase to dense-phase exchange factor
Z Function of gas velocity $(U - U_{mf})/U$
γ Gas specific heat ratio
ρ_B Bulk density of fluidized bed
ρ_0 Density of gas

1   INTRODUCTION

Gaseous fuels are reacted in fluidized beds in a number of processes, some of which will be discussed in more detail later. Some typical applications are:

(i)   Gas firing of boilers and furnaces.
(ii)   Gas burning for start-up or modulation of fossil fuel fired power plant.
(iii) Incinerating malodorous vapours, such as those found in enamelling ovens and paint driers.
(iv)   Heating fluidized particles for heat treatments of solids, as in metallurgical furnaces, driers or solids processors, such as briquetting in the Auscoke process.

Thus, it is important to understand the reaction processes occurring in order to predict performance and exit gas compositions other than by experience. Other important theoretical approaches concern the prediction of heat transfer to the distributor to preclude gas ignition, and the formation of oxide of nitrogen in combustion. The final points for consideration are general combustion safety.

2   FUEL INTRODUCTION TO BEDS

Gas fuels can be introduced in many ways, the most important groupings being:

(a)   Separate fuel and air injection, and
(b)   premixed fuel injection.

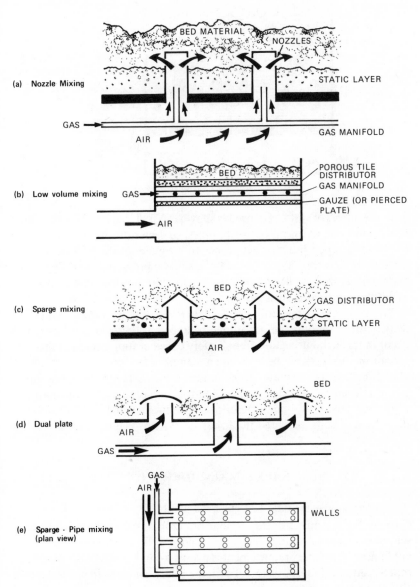

FIG. 1. Some gas introduction methods.

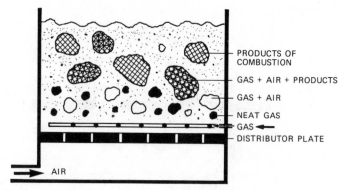

FIG. 2.   Bubble mixing (schematic).

Some examples of systems in use are shown in Fig. 1. Separate fuel injection relies upon fuel diffusing out of bubbles, oxygen diffusing into fuel-filled bubbles and upon bubble coalescence to bring about the production of a flammable mixture (Fig. 2). This can be achieved reasonably easily in deep beds, but in shallow beds many feed points are needed since lateral gas mixing in fluidized beds is not good.

Premixed gas combustion, in which the gases are mixed below the distributor plate, has the advantage of allowing uniform combustion to occur in the bed, but an additional safety consideration occurs in that fuel gas/air mixtures below the bed pose one more safety problem than mixing above the distributor plate. Some important aspects of safe operation of gas-fired systems are considered in Section 3, while the modelling of aspects of the combustion system is described in Sections 4 and 5. More detailed discussion of some gas-fired fluidized bed systems is given in Chapter 10.

3   SAFETY CONSIDERATIONS

3.1 Safety in general fluid bed gas-fired plant

Gaseous fuel/air mixtures can explode or detonate under some conditions; these conditions must obviously be precluded as far as possible in gas-fired appliances. This safety requirement is met by a combination of good design practice and the incorporation of suitable safety controls. In general, good design practice consists of:

(a)   minimizing the volumes of unignited gas/air mixture,
(b)   having adequate flame proving systems,

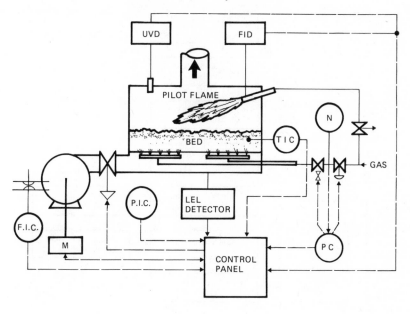

FIG. 3.   A typical control system.

(c)   precluding all gas leakage,
(d)   purging the system before and after gas has been fired,
(e)   ensuring that any explosions can be vented by fitting explosion doors, and that ignition pressure rise can be accepted,
(f)   igniting the minimum practical volumes and flow rates of fuel.

The safety controls must:

(a)   Prevent any unintentional gas introduction to the burner, and in larger units must prove that there is no leakage; for example, by proving that there is no leakage from two closed solenoid valves before gas can be introduced.

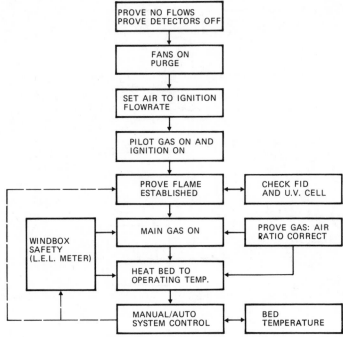

FIG. 4.   Typical logic sequence for light-up.

(b)   The gas must be proved to be ignited within about 5 s of opening the isolating valves.

(c)   If a pilot burner is used, this must be proved both to have lit and to be of an acceptable length before gas is introduced into the main burner. It should also be continually proven while any gas is flowing into the main burner, unless the in-bed ignition is acceptably proven.

(d)   Proving that the fuel/air ratios are correct. It is to be recommended that any gas burner conforms to all reasonable safety requirements above the minimum code of practice requirements. A typical system design for a premixed system is shown in Fig. 3. A typical logic sequence for light-up is shown in Fig. 4.

3.2 Overbed space safety

The general rules above should be applied, in particular it is essential that an adequate pilot flame should be proved and that the furnace volume is adequately purged. It is not easy to prove that combustion in a fluidized bed is occurring. Space does not permit this subject to be dealt with here; however, suffice it to say that safety may be ensured by controlling the fuel/air ratio closely and proving that the pilot flame is maintained properly.

3.3 Below bed safety

With non-premixed systems, safety in operation is straightforward. When the gas and air mix below the distributor plate there are some extra potential hazards due to the existence of a gas–air mixture below the bed; this is a potential|ignition source.

Design requirements for large gas-fired units can require substantial windbox volumes, due to the large areas of the bed required and the depth of windbox needed to allow good air distribution. To ensure safety in premixed systems, the mixing should ideally occur in the plate as shown, for example, in Fig. 1(a).

When mixing occurs below the plate, a flame detection device should be fitted below the plate and the volume should be packed to preclude explosion. Ignition of the fuel/air mixture below the bed can occur for several reasons depending upon the design of the plant and the mixing method; an appreciation of these allows preventative design of the safety system:

(i) *With Porous Tile or Packed Bed Distributor.* Combustion can be initiated below the plate due to thermal ignition if the plate can overheat, so that this must be considered at the design stage. Flashback can also occur due to a mechanical failure of the plate. The volume of unignited gas and air mixture must always be minimized, for example, as in Figs. 1(b) and 1(c).

(ii) *With Nozzle Mixing.* The methods in Fig. 1 allow very small mixing volumes to be used. The main flashback mechanisms are overheating or failure of the plate, and hot or burning solids flowback. Provided nozzle volumes are small the possibility of ignition inside a nozzle is extremely low and it is not dangerous (although it may damage the nozzle). The hazard to be avoided is mixing of fuel and air before the nozzle. This cannot be guaranteed absolutely since nozzles can block and the flows in windboxes are complex; however, correct pressure settings can preclude mixing of substantial quantities. A further check to be commended is the use of a fuel/air explosion limit detector for the windbox gases. This should be coupled first to an alarm warning of any build-up of gas and, secondly, a gas lock-out system to operate at a higher concentration. As is usual with safety systems, a regular checking procedure is needed for this equipment.

3.4 Closure on safety

The general rules given, and a thorough understanding of codes of practice, are fundamental to safe operation. However, in many respects, safe operation and good practice are complementary, e.g. the overheating of a distributor may not be hazardous but could damage the plate. Safety standards can be improved by a good understanding of the basic principles of the design of bed, distributor and ignition/control system.[1]

4   THEORETICAL CONSIDERATIONS

4.1 Premixed combustion

The problem to be considered here is that of calculating the amount of fuel-gas passing through the fluidized bed unburnt. The method used is based on the two-phase theory of fluidization; it is assumed that all of the fuel that enters the dense phase is reacted immediately, while that in bubbles does not react below the bed surface. These assumptions, while they are not strictly true, are reasonable for many practical purposes (they do not cover the very high temperature regime where bubbles ignite explosively in the bed). The aspects of the two-phase theory used are mainly those covered in the chapter on coal burning.

The fraction of fuel passing through a bed unburnt when the reaction rate is infinite is given by (see Chapter 3 and ref. 2):

$$f_u = \frac{C_e}{C_0} = \frac{(U - U_{mf})}{U} e^{-X} \tag{1}$$

where X is the bubble-phase to dense-phase exchange factor,

$$X = \int_0^L \frac{Q \, dL}{U_B V_B} \tag{2}$$

From the two-phase theory, we can write:

$$\frac{Q}{V_B} = 4 \cdot 5 \frac{U_{mf}}{D_B} + 5 \cdot 85 \frac{\mathscr{D}^{1/2} g^{1/4}}{D_B^{5/4}} \tag{3}$$

The relationship between D_B, U_B and L cannot be completely generalized; in particular there are no well proven relationships for the important case of shallow beds.

For deep beds it is common to write:

$$U_B = (U - U_{mf}) + U_i \tag{4}$$

when an isolated single bubble moves with velocity:

$$U_i = \sqrt{gD_B/2} \tag{5}$$

and

$$D_B = K(U - U_{mf})L + D_0 \tag{6}$$

In this case X is related to L and the fraction of gas unburnt can be calculated as a function of U and L. For first estimates eqns. (3) and (4) become:

$$\frac{Q}{V_B} = 4 \cdot 5 \frac{U_{mf}}{D_B} \tag{7}$$

and

$$U_B = U - U_{mf} \tag{8}$$

Thus, it is easy to relate f_u to U_{mf} (or $\bar{d}_p$), U and L_{mf}. Two cases are considered below, first for a constant bubble size:

$$f_u = Z \exp\left[-\frac{4 \cdot 5 L_{mf}}{D_B Z}\left(\frac{U_{mf}}{U}\right)\right] \tag{9}$$

and second, using eqns. (6) and (2) to relate L and D_B, we obtain:

$$X = \frac{4 \cdot 5 U_{mf}}{K(U - U_{mf})^2} \ln\left[\frac{K(U - U_{mf})L_{mf} + D_0}{D_0}\right] \tag{10}$$

The simpler form (eqn. (9)) is often preferable, especially when the gas bypassing is high. Putting

$$\frac{L_{mf}}{D_B} = a \tag{11}$$

we obtain:

$$f = \frac{(U - U_{mf})}{U} \exp\left[-\frac{4 \cdot 5 U_{mf} a}{(U - U_{mf})}\right] \tag{12}$$

Since the mean bubble size is not readily measured for shallow beds, a first estimate for the mean bubble size in a vigorously bubbling bed is taken as being half of the bed depth; i.e. a is equal to 2.[3] This equation is plotted in Fig. 5 and would not be valid for $U - U_{mf}$ below 0·3 m/s, since the bubbling is not sufficiently vigorous and in-bed temperature gradients are significant. Further, at fluidizing velocities below 0·2 m/s, there is a risk of flashback into the distributor or windbox.[3]

J. Broughton

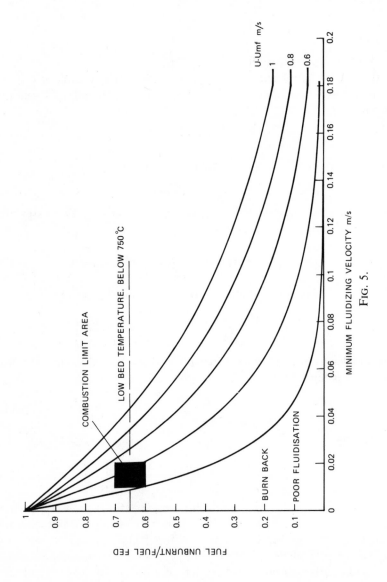

FIG. 5.

In general, combustion can not be stable, if more than 65 % of the fuel passes through the bed unburned.

4.2 Operating range—Shallow beds
Here the objective is to predict approximate operating limits, and the lowest minimum fluidizing velocity at which combustion can take place is about 0·015 m/s in a shallow bed.

Using the simple theory we can predict the limits for various particle types and the effect of higher gravity forces or pressure and compare with available data.[3-5] Unfortunately, few data have been published for comparison and these are of limited accuracy; the comparison is shown in Table 1, which shows the minimum particle size at which stable combustion

TABLE 1
Minimum particle size for stable combustion

Reference	Particles	'g' force	Experimental $\bar{d}_p$	Predicted $\bar{d}_p$
3	Silica sand	1	250	156–210
4	Silica sand	10	90	50–90
5	Zirconia	1	95	102–145

occurs. Thus, given the approximations used, the agreement is reasonable and illustrates trends.

4.3 Noise of combustion
A further application of this simplified theory is the prediction of when combustion noise ceases in a gas burning fluidized bed. Noisy gas/air combustion occurs when a bed is at low temperature (as at start-up) or when the particles are small. The work of Yanata *et al.*[7] shows that combustion noise occurs at 820 °C when particles of mean size 320 μm are used (minimum fluidization velocity = 0·057 m/s), for velocities above 0·2 m/s. This corresponds with less than 2 % of the fuel being unburnt in the dense phase; since this lies within the margin of error of the theory, the agreement can be considered to be reasonable. At higher bed temperatures the noise ceases at higher velocities. Also, more noise occurs with higher excess air.

4.4 Unmixed gas/air combustion
In this case the reaction rate is limited by the mixing of the gas and air in the bed. The mixing is very design dependent and a typical design is shown in

Fig. 2. For a hot fluidized bed the minimum fraction of the fuel burnt can be considered to be the fraction of the gas in the dense phase, i.e. (U_{mf}/U), since the dense phase gas mixes well in a bubbling bed of moderate depth. This does not apply to beds whose depth is less than the nozzle spacing, as then the gas cannot mix adequately in the bed.

The simplest approach is to model the bed as a mixing zone in series with a burn-out zone in which the gases are mixed. The mixing length can be taken as equal to the nozzle spacing as a first approximation; however, more detailed analyses are possible.[6]

4.5 Application of models

Generally, the designer uses the model in a semi-empirical form in conjunction with tests on a small-scale test rig. The constants in the equations are fitted from experimental data with feedback from larger units. The bubble formation and growth processes depend upon the distributor plate and windbox design. In addition, the gross bed behaviour resulting from the overall design used influences the bubble behaviour.

The combustion models are of the simple bubble phase to dense phase transfer limited type. These do not account for several important factors such as low temperature combustion where catalysis becomes important or combustion in the fuel rich regime where chemical equilibrium behaviour becomes significant.[7]

5 THE DESIGN OF DISTRIBUTORS FOR GAS COMBUSTORS

5.1 Introduction

A wide range of designs exist for the types of distributor plates used in fluidized beds.

The main three types are:

(i)   porous plates,
(ii) multiple orifice distributors, and
(iii) sparge pipe designs.

The most important functional requirements are to create uniform fluidization and ensure combustion. However, in most cases this is tied in with a requirement for low pressure losses, and a need for flow turn-down in operation. In addition to the functional requirements the distributor design must allow for thermal behaviour and possible particle flowback.

5.2 Pressure losses

When low pressure losses are essential, as in gas incinerators and power-plant, a shallow bed is often required and a porous plate distributor is then the best type. A porous plate, or even a packed bed of large, more dense particles, also offers better turn-down characteristics—this is because the orifice or sparging design imposes a relationship of the form:

$$\Delta P = K_T U^2 \tag{13}$$

So that halving the flow, quarters the pressure drop over the plate; this produces a deterioration in fluidization uniformity. For a porous plate, the Ergun form of equation can be used,[2]

$$\Delta P = K_L U + K_T U^2 \tag{14}$$

and the plate can be made to have a linear characteristic by using sufficiently fine particles or wire mesh sizes to ensure that viscous flow dominates ($K_L \gg K_T$).

It should be noted that the distributor design is influenced greatly by the windbox used which influences the uniformity of flow and oscillatory characteristics. This is very significant when the pressure drops must be low for economic reasons and the oscillations can be adequately predicted using the theory proposed by Davidson.[9] This shows that the frequency relates to the bed depth (L), windbox depth (S) and solids bulk density, ρ_B, by:

$$n = \frac{1}{2\pi} \left\{ \frac{\gamma P_0}{\rho_B SL} \right\}^{1/2} \tag{15}$$

5.3 Thermal design

Heat transfer from the combustion region of the bed back to the distributor can have two important consequences:

(i)   The distributor can be distorted due to thermal stresses.
(ii) Combustion can spread back into the distributor.

The bed is isothermal above an entry zone that depends upon many factors; distributor design, bed temperature, gas velocity, mixture strength and particle size are the most significant. Particle flowback can also influence this.

The most amenable approach to predicting heat transfer back from the bed to the distributor is to assume normal heat transfer back to the plate due to solids circulation. However, this does not allow for the critical effect of non-uniform air flow which can create substantial temperature gradients. In the case of a porous plate distributor combustion cannot spread back into the distributor if the surface temperature is below the ignition temperature. This can be estimated using a simple heat balance between gas temperature rise and heat transfer:

$$\rho_0 U_0 \bar{C}_p (T_S - T_O) = h_D (T_B - T_S) \qquad (16)$$

Therefore

$$T_S = \frac{St T_B}{(1 + St)} \qquad (17)$$

h_D is taken to be the same as the heat transfer coefficient to bare tubes and T_S is taken as being equal to the gas ignition temperature. This method is very conservative[3] because even when the surface temperature greatly exceeds the ignition temperature of the gas the residence time in the surface layers is not sufficient for combustion to occur. Initial temperature gradients are discussed further in refs. 3 and 7.

6 CLOSURE

Many facets of gas burning in fluidized beds have not been covered in this chapter; however, the basic features covered are intended to be of general usefulness to designers. Circulating beds are not covered;[10,11] however, these are not usually as severely limited as shallow beds as they have a dense phase primary combustion zone. Two other important cases that are beyond the scope of this chapter are combustion in a packed-fluidized bed and delayed mixing for carburizing purposes. The reader is recommended to consult refs. 12 and 13 for information on these topics.

REFERENCES

1. ELLIOTT, D. E., Exploiting fluidised bed combustion, *Second International Conference on Fluid Bed Combustion*, Hueston Woods, Ohio, 1970.
2. DAVIDSON, J. F. and HARRISON, D. (eds.), *Fluidisation*, Academic Press, London, 1971.
3. BROUGHTON, J., Gas combustion in shallow fluidised beds, *Applied Energy*, **1**, 61 (1975).
4. HOWARD, J. and METCALF, C. J., Fluidisation and gas combustion in a rotating fluidised bed, *Applied Energy*, **3**, 65 (1977).

5. PILLAI, K. K., Premixed gas combustion in shallow fluidised beds, *J. Inst. Fuel.*, 200 (1976).
6. STUBINGTON, J., The role of coal volatiles in fluidised bed combustion, *J. Inst. Energy*, 191 (December 1980).
7. YANATA, I., MAKHORIN, K. E. and GLUKHOMANYUK, A. M., Investigation and modelling of the combustion of natural gas in a fluidized bed of inert heat-carrier, *Int. Chem. Eng.*, **16**(1), 68 (1975).
8. ESSENHIGH, R. H. and COLE, W. E., Studies of the combustion of natural gas in a fluid bed, *Third International Conference on Fluidised Bed Combustion*, Hueston Woods, Ohio, *2/5/1*, October 1972.
9. DAVIDSON, J. F., Rapporteurs introduction, *I. Chem. Eng. Symp.*, *Tripartite Chem. Eng. Conf.*, Montreal, Canada, September 1968, p. 4.
10. REH, L., Trends in research and industrial application of fluidisation, *World Congress on Chemical Engineering*, Amsterdam, Holland, 30 June 1976.
11. DAVIS, J. S., YOUNG, W. W. and LYONS, C. J., Use of solid fuel possible for field steam generation, *Oil and Gas J.*, 129 (June 1981).
12. BASKAKOV, A. P., KIRNOS, V. I. and SVETLAKOV, V. I., Preparation of non-oxidising media by fluidised bed combustion, *Gasov. Prom.*, **13**(1), 25 (1968).
13. TAMALET, M., Applications of fluidised bed heat transfer in the metallurgical industries, *Symposium on Chemical Engineering in the Iron and Steel Industries*, Institute of Chemical Engineers, Swansea, 1968, p. 105.

Chapter 10

FLUIDIZED BED METALLURGICAL FURNACES

M. J. VIRR†

Stone-Platt Fluidfire Ltd, Brierley Hill, UK

1 METALLURGICAL APPLICATIONS

Using the premixed fuel injection method of firing natural gas or propane a range of metallurgical furnaces has been developed. These units usually use the ceramic tile as a distributor in the furnace and each type, for hardening, carburizing and tempering is described.

Such units have now gained widespread use in the metal processing industry; the main advantages being faster processing times and greater fuel economy in comparison with conventional techniques.[1]

1.1 The internally fired fluidized bed

This arrangement is shown in Fig. 1 in which the gas and air are mixed together in near stoichiometric proportions and passed through a porous ceramic plate over which the particles are fluidized in the gas stream. This bed is started by ignition of the combustible fluidizing gas–air mixture above the bed, the whole bed and freeboard region being contained in a metallic or refractory container. The established flamefront gradually moves down the depth of the bed until it stabilizes above the ceramic plate, combustion then taking place spontaneously within about 1 in (25 mm) of the plate surface. In order to control the atmosphere in the heating zone of the bath, the mixture is usually made slightly reducing by running sub-stoichiometric to give an atmosphere of 4–6 % carbon monoxide and 5–7 % carbon dioxide with no available free oxygen in the body of the bath.

† Present address: Johnston Boiler Co., Ferrysburg, Michigan 49409, USA.

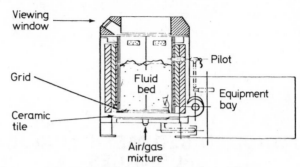

FIG. 1.   Schematic of fluidized bed hardening furnace.

The temperature of the bath is automatically controlled via a proportioning controller linked to a motorized valve that meters the appropriate amount of gas–air mixture to the bed. It is arranged so that the control range is above the minimum fluidization velocity of the particles and, in the case of batch furnaces, a double furnace skin with heat rejection by air cooling ensures that sufficient heat is lost during the no load or soaking operations to achieve the constant set temperature.

1.2 External gas fired fluidized beds

At temperatures below 750 °C it is not possible for combustion to stabilize (although it operates through the range on start-up) in internally fired baths. Therefore, for operations such as tempering, it is necessary to go to external firing, as shown in Fig. 2. This shows an excess air burner firing into a plenum chamber over which the fluidized bed is supported on a porous plate. The temperature of the bed is simply controlled by modulating the gas supply to a constant air throughput burner, and a temperature range of 250–700 °C may be obtained.

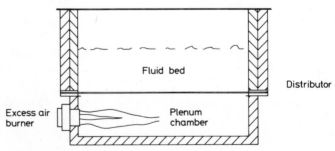

FIG. 2.   Fluidized bed tempering furnace.

1.3 Continuous rotary drum furnace

This line has been developed to harden small parts such as washers, ball bearings and bolts and nuts in an internally fired fluidized bed operating at 750–1000 °C, that generates its own protective atmosphere. The parts are conveyed through the bath within a rotary scroll, the outer drum portion of which is a pierced plate through which the fluidized particles may pass. The diameter of the holes is too small for the parts to drop through. The exit end of the rotary scroll includes a separation section where the heated parts are separated from the particles before being dropped down the chute into the quench tank (Fig. 3).

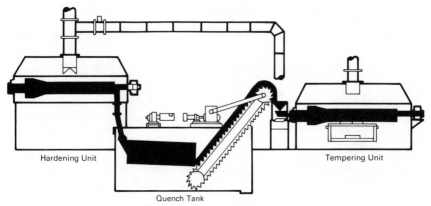

Hardening Unit   Tempering Unit

Quench Tank

FIG. 3.   Continuous rotary fluidized bed hardening and tempering unit.

The drop chute is continuously fed with a screen of cooled quenchant through which the parts drop into a rotary drum within the bulk quenching fluid, which prevents any tendency to back temper. The parts then pass on to the conveyor for transfer to the tempering furnace. If the quenchant used is oil, a washing machine is placed in the line at this point.

The hardened parts are then automatically conveyed into the drum of the tempering furnace which is heated primarily with the hot gas from the hardening furnace. This hot gas is drawn through a hot gas fan from the hardening furnace, attemperated as necessary before the fan and blown into the plenum chamber of the tempering furnace using the same principle as described in Section 2. At tempering temperatures above 350 °C a heat receiver is fitted after the fan into which a small excess air burner is fitted to give temperatures up to 700 °C within the tempering furnace.

The advantages of this arrangement are described more fully in

Section 4, but a fuel efficiency of 50 % (heat to steel) has been continuously achieved on lines of this type.

These furnaces have been installed for the heat treatment of washers, ball bearings and small automotive components.

2 PROPERTIES AND PERFORMANCE

2.1 Heating rates and hardening

The effects of heating on sections such as tools 2 to 3 in thick has been separately reported[2] but it is worth repeating some of the results, shown in Fig. 4, which show that the smaller sections can be heated at very rapid

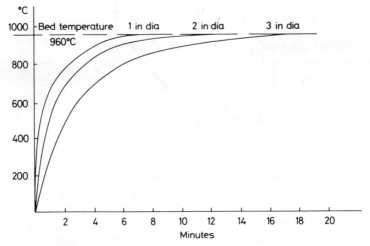

FIG. 4.   Heating rates of D3 steel samples in a fluidized bed at 950 °C.

rates. Typically, it has been found that 'through-hardness' is obtained by soaking the part for twice the time needed to get to temperature. Thus, 0·5, 1 and 2 in (12·5, 25 and 50 mm) tools or parts may be hardened by holding in the bath for 5, 10 and 20 min, respectively. All main tool steels have been hardened in different sizes and particularly successful results have been achieved with AISI designation, D, O, W, H and A types of tool steels. Any steel that can be hardened in the range 750–1250 °C can be satisfactorily treated in these baths.

The heat capacity of the internally fired fluid bed is indicated by the

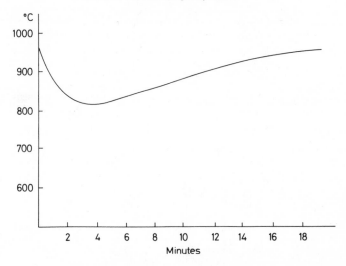

FIG. 5.   Furnace temperature recovery of a 6 ft³ Fluidfire furnace (type FH8) of a
300 lb (136 kg) load to 950 °C.

16 min recovery time (Fig. 5) of a 6 ft³ (0·17 m³) furnace, when loaded with
a 300 lb (136 kg) load of $\frac{5}{8}$ to 1 in (16–25 mm) diameter section.

The twin advantages of fast heating and short recovery time make the
batch furnace attractive, particularly if fitted with automatic sequential
handling mechanisms described later.

2.2 Atmosphere in gas fired hardening baths

The surface effects of holding tool steels in an internally fired bath at near
neutral or reducing conditions is shown in Fig. 6 (at 8 % carbon monoxide)
and indicates, as would be expected, that some decarburization occurs with
increasing immersion times, particularly at temperatures approaching
1000 °C. However, it is significant that because of the short immersion times
needed to obtain through hardness, sections up to 1 in (25 mm) show little
or no surface effect. Section sizes up to 2 or 3 in may be confidently treated,
provided that a normal grinding allowance is specified.

The resultant finish on wire and small sections are very similar to those
exhibited by induction hardening and because of the short immersion times
no apparent decarburization is experienced.

2.3 Carburizing in fluidized beds

It is possible in a special version of the hardening bed described, fitted with

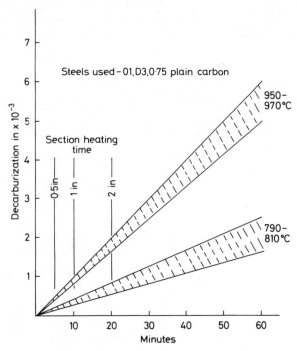

FIG. 6.   Representative decarburization band for steels held in internally fired
fluidized bed.

an inner zone, to supply this zone with a 'rich' gas/air ratio and obtain a
carburizing atmosphere. The first developments of this technique were
arranged so that the outer zone is run at near stoichiometric gas/air ratio
which therefore controls the heat input to the furnace (Fig. 7).

Results showed that using air/propane ratios of 4/1 to 8/1, carbon
potentials from 2 to 1 % are realized at 4/1 to 5/1 (Fig. 8). Using the high
carbon potential, rapid carburizing at 950 °C of 0·030 in (0·75 mm) in 1 h is
achieved. However, because of the high carbon potential and the difficulty
of controlling the atmosphere between 0·5 and 1·0 % carbon (due to the
steep slope of the curve at the required air/gas ratio) it has been found
necessary to use a boost-diffuse technique or re-hardening and quenching
to remove retained austenite.

This technique, although being quite satisfactory, is a little cumbersome
in using the few production furnaces that have been supplied. These are
described in detail in refs. 13 and 15.

In order to broaden the use of fluidized beds for surface treatments, an

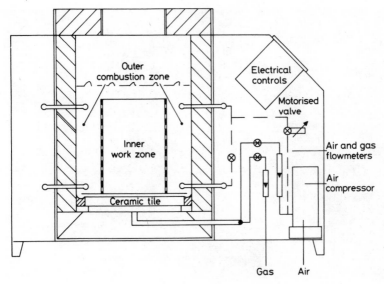

FIG. 7. Experimental fluidized bed carburizing furnace.

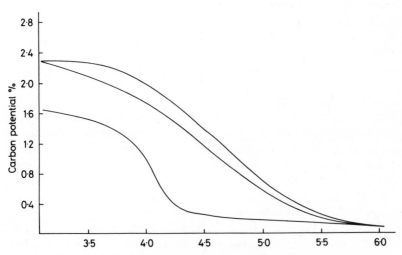

FIG. 8. Carbon potentials obtained with air/gas ratios at 975 °C using thin foils as the material being carburized.

FIG. 9.   All-atmosphere fluidized bed furnace for hardening and carburizing.

approach utilizing an all atmosphere furnace has been followed, using a
nitrogen or air based atmosphere to carry out carburizing, carbo-nitriding
and nitriding treatments which will have gas or electric heating (Fig. 9). A
commercial version of this furnace is now available.

3   APPLICATIONS OF FLUIDIZED BEDS

3.1 Batch type applications

The first application is straight replacement for neutral salt and lead baths
where the heating rates are similar but the environmental conditions are
improved. Table 1 shows the applications to process and Fig. 10 shows a
typical internal gas-fired bath for hardening. These furnaces are now a
standard product made in a range of sizes from 12 in diameter by 12 in deep
(300 mm × 300 mm) through to 5 ft long × 3 ft 3 in wide × 3 ft 3 in deep
(1·5 m × 1 m × 1 m) to process up to 1 ton loads.

However, all the larger industrial furnaces are made around a standard
burner module based on a 24 in × 16 in tile (0·6 m × 0·4 m) the air and gas
to which is mixed immediately below in a special mixing chamber.

Typically, a single module unit will process up to 220 lb/h (100 kg/h) for a

TABLE 1
Potential applications of fluidized bed technology for heat treating operations

	Process				
	Hardening range 750–1 200°C	Surface treatments 900–1 050°C	Isothermal and tempering treatments 250–700°C	Surface treatments 100–950°C	A continuous hardening and tempering
Type of fluidized bed preferred (in capitals)	GAS/Electric FIRED FH type	GAS FIRED FC type	GAS/Electric FIRED FT type	ALL ATMOSPHERE FURNACE	GAS FIRED ROTARY DRUM TYPE
Replacement for existing equipment (type)	Neutral salt baths. box furnances	Cyanide salt baths atmosphere	Forced air circulation furnace and/or ovens	Salt baths	Rotary retort shaker type mesh belt pan type

FIG. 10.   Batch type fluidized bed hardening furnace.

mean gas consumption of 2·5 therm/h (62 500 kcal/h) assuming heat treating at 960 °C.

These units are automatically controlled and are equipped with a full fail-safe panel having a semi-automatic start-up arrangement. As the units will start-up in 1 to 2 h (typically 1 h per 0·3 m bed depth) they may be run for single shifts and do not need to be left on a low fire all night as do salt or lead baths.

Both hardening and tempering baths are made in identical size, so a production line may be designed in-line and the mechanical handling equipment may traverse both furnaces as well as the quench tank and loading and unloading areas. Several batch processing lines have now been installed and are operating successfully in production environments.

The development of a similar unit for carburizing (already mentioned) is being developed for replacement of batch cyanide salt baths.

3.2 Continuous type applications

The application to continuous treatments with the fast process times already outlined make the fluidized bed particularly suitable for in-line annealing and patenting of wire and hardening of small parts.

Considerable work has already been done on patenting in fluidized beds[3] as an alternative to salt and lead and we have been very active in this field;

plants have been manufactured with a capacity of 3·5 tons/h of wire with a fluidized bed some 35 ft long × 4 ft wide.

The rotary drum furnace has been described and several lines are now in commercial operation.

The first line which went into operation in 1976 is being used for the hardening and tempering of washers at 850 and 450 °C, respectively. The material is a 0·63 % carbon steel which is quenched into oil from the hardening unit so a washer is used before the tempering unit. The atmosphere is monitored using flow meters, to give near neutral conditions using natural gas. A very good, slightly blued, finish is obtained.

This furnace has a total gas meter in the feed line and consumes 6·9 therms/800 lb material processed in 1 h. The total consumption on 1000 lb/h being 7·5 therms/h.

The second line installed was for the heat treatment of ball bearings, hardening at 860 °C and tempering at 180 °C. In this case the quenchant used is aquaquench and considerable care is taken to prevent back tempering on balls over 1 in diameter, by using a rotating drum in the quench tank. The ball material is 1 % carbon with 1–1·6 % chromium and 0·3–0·7 % manganese.

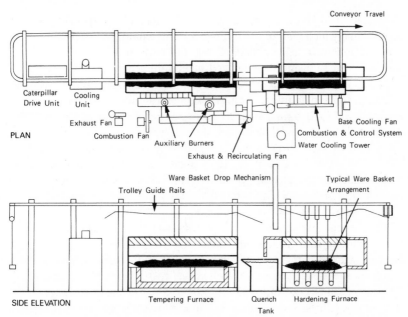

FIG. 11. Fluidized bed 1·5 ton/h split-roof overhead conveyor furnace line.

Finally, standard hardening and tempering furnaces may be used in conjunction with conveyor type automatic transfer gear that moves baskets or jigs in a predetermined cycle from the loading station to hardening bath, from the hardening bath to quench tank, from quench tank to tempering furnace and finally from tempering furnace to loading station (Fig. 11). This arrangement has the advantage of being able to accept a great variety of different components. The largest installation of this type is hardening and tempering agricultural components of all sizes at a throughput of up to 1·5 tons/h.

4   THE ECONOMIES

The main advantages gained from using fluidized beds in metal processing, in comparison to conventional equipment, are, first, economic and, second, environmental when directly compared with salt and lead baths.

A cost comparison of operating costs of a line using batch type equipment with automatic handling for fluidized beds and salt baths is shown in Table 2. This comparison has assumed an average of 600 lb/h (270 kg/h) over a typical year of 2000 h. The depreciation period chosen is seven years (which is common in Europe); this would tend to be kind to the salt bath line with its lower capital cost, but nevertheless a saving is shown over the year by using the fluidized bed equipment.

The main savings are in fuel consumption, salt or aluminium oxide consumption and maintenance and making no allowance for the extra costs involved with salt baths, such as cleaning and disposal of the salt. The comparison indicates that a considerable saving in processing cost is achievable by substituting a fluidized bed for a salt bath. The cost per lb of metal treated was 1·27 p/lb at 1980 prices and indicates a competitive processing cost which can be easily improved upon as the bed has considerable spare capacity. The tabled costs are process costs and do not include an overhead or profit element.

The cost comparison of using a continuous fluidized bed rotary furnace in comparison to a continuous shaker hearth furnace for hardening and tempering are even more interesting (Table 3). In this case the fluidized bed line is cheaper in both capital as well as fuel costs. The capital cost is about 18 % lower than a conventional line, part of which is due to the fact that the fluidized bed furnace does not need an endothermic generator because it generates its own protective atmosphere.

TABLE 2
A cost comparison of automated salt bath and fluidized bed hardening and tempering
(Tempering lines at 1980 prices)—Capacity 400–800 lb/h

No.	Salt bath		£	Fluidized bed		£
1	36 in dia × 24 in deep			48 in × 18 in × 18 in deep or		
	Preheater	1 500		36 in × 24 in × 18 in deep		
	Hardening	7 500		Hardening	10 700	
	Quench tank	3 000		Quench tank	3 000	
	Tempering	6 500		Tempering	9 800	
	Wash tank	4 500		Wash tank (if		
	Aut. line	14 000		oil quench)	4 500	
	Fume ext.	3 000		Aut. line	14 000	
		40 000		Flue	1 000	
	40 000			43 000	43 000	
	7 years =		5 714	7 years	=	6 142
2	Gas at 20p/therm					
	Heating up:					
	(17 therm/h) × 4 h	68		(13 therm/h) × 2 h	26	
	Idling:					
	(4 therm/h) × 12 h	48		Nil		
	Running:					
	(15 therm/h) × 8 h	120		(8·5 therm/h) × 8 h	68 therm	
	Total	236 therm		Total	94 therm	
	236/8 × 20p ×			94/8 × 20p ×		
	2 000 =		11 800	2 000	=	4 700
3	Salt pots			Liners		
	life 1 000 h			life 4 000 h		
	2 pots × 1 000 =		2 000	$\frac{1}{2}$ liner × 2 200	=	1 100
4	Salt consumption			Aluminium oxide and consumption		
	40 lb/8 h day			20 lb/8 h day		
	40 × 250 × 20p =		2 000	20 × 250 × 20p	=	1 000
5	Maintenance			Maintenance	=	300
	thermocouples,					
	etc. =		500			
6	Labour £2/h			Labour £2/h		
	at 750 h per			at 750 h per		
	2 000 h =		1 500	2 000 h	=	1 500
7	Jigs and			Jigs and		
	fixtures =		1 000	fixtures	=	500
8	Total cost =		24 514	Total cost	=	15 242
9	Saving			Saving		9 272
10	Cost per lb			Cost per lb		
	of metal			of metal		
	hardened and			hardened and		
	tempered			tempered		
	600 lb/h =		2·04p	600 lb/h	=	1·27p

Notes: These figures are based on single shift 2 000 h/pa. Individual factory overheads would be added.

TABLE 3

Cost comparison of fluidized bed rotary drum furnace with shaker hearth line for hardening and tempering (1980 prices)

No.	Shaker hearth line (Cap. 750 lb/h)		£	Fluidized bed rotary drum furnace (Cap. 800–1 000 lb/h)		£
1	Capital cost			Capital cost		
	Shaker hearth			Rotary drum		
	furnace, 36 in	61 500		furnace	45 000	
	Quench	14 000		Quench	14 000	
	Tempering F.	29 100		Tempering F.	30 000	
	Endothermic					
	generator	3 880				
		108 480			89 000	
	Amortized					
	over 7 years	=	15 500		=	12 715
2	Gas consumption			Gas consumption		
	Hardening F.	5·3 therm		Hardening F.	5·4 therm	
	Endo. atmos.	2·16 therm		Tempering F.	1·5 therm	
	Tempering	4·0 therm				
	Exo. atmos.	2·90 therm				
	Total	14·35 therm		Total	6·9 therm	
	Cost 6 000 × 14·35			Cost 6 000 × 6·9		
	× 20p	=	17 222	× 20p	=	8 280
3	Maintenance			Maintenance		
	Radiant tube replicate			3 sets tiles	360	
	8 tubes × £200			Labour	1 000	
		1 600		Thermocouples	500	
	Labour	1 000		Aluminium oxide	460	
	Thermocouples	500		Make up		
	Generators	400		29 kg/week		
	New trays	1 600		New liner	1 200	
			5 100			3 520
4	Labour			Labour		
	£2 h at 1 800 h			£2 h at 1 800 h		
	for 6 000 h		3 600	for 6 000 h		3 600
	Total cost		41 420			28 115
5	Saving			Saving		13 305
6	Cost per lb of			Cost per lb of		
	metal treated		41 420	metal treated		26 045
	750 lb × 6 000			800 × 6 000		
	× 0·75	=	1·23p/lb	× 0·75	=	0·78p/lb

The fuel cost is about one-half that of a conventional furnace line and the consistency of this fact has been proved over several years where the original furnace has given this fuel consumption. The overall saving is again in the order of one-third, with potentially less plant difficulties because of less equipment being involved.

One of the real advantages of these furnace types is the very rapid heating up rate from cold, the rotary drum type coming up to temperature in $1\cdot5$ h for the whole line and the small units being able to start in about 1 h. Since the method of combustion employed is a metallic heat resistant bath lining insulated by ceramic fibre, there is no refractory to heat up or repair.

The environmental advantages of using fluidized baths need hardly any further emphasis. A fluidized bed, whilst behaving like a liquid, has no melting or boiling point, and is non-wetting. This last characteristic is of particular significance to metal processors as it eliminates 'drag out' and further washing processes that are associated with salt baths.

5 QUENCHING INTO FLUIDIZED BEDS

There are special applications to heat treatment where austempering or patenting may be achieved by quenching from the austenizing furnace into a fluidized bed at an intermediate temperature.

Wire patenting is more traditionally carried out by plunging the wire, which is at $900–1000\,°C$, into a lead bath at $500–555\,°C$. The process cools the wire very quickly to the lower temperature. Holding it there for a reasonable time causes the formation of a fine pearlite structure with high ultimate tensile strength and good ductility. Plunging the same wire into a fluidized bed at this temperature unfortunately does not cool the wire quickly enough because of the lower heat transfer coefficient in a fluidized bed compared to lead (Fig. 12). However, if the wire is quenched into an intermediate fluid bed at $250–450\,°C$ it may be cooled at a similar rate to lead cooling. Then, by subsequently having the wire emerging into a bed at the higher temperature of $500\,°C$, the transformation to fine pearlite can be completed (Fig. 13). Similar processing to this can be used in the treating of steel shot.

Experience in attempting to use fluidized beds for martempering have usually been unsuccessful due to the lower heat transfer rates that can be achieved when plunging into a fluidized bed. Typically, $80–100\,Btu/ft^2\,h\,°F$ $(455–569\,W/m^2\,K)$ compared to a liquid-like salt or hot oil typically $500–$

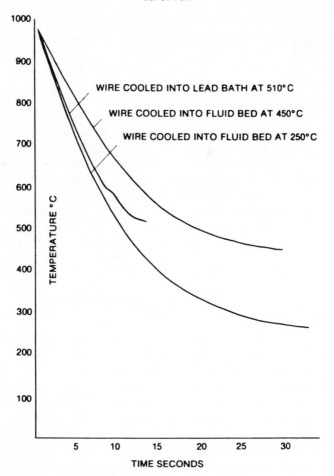

FIG. 12.   Quench curves for 5·5 mm diameter wire from 1000 °C.

1000 Btu/ft^2 h °F (2840–5690 W/m^2 K). Very thin section material less than 10 mm can be successfully treated but anything larger is a problem unless special techniques are adopted.

The running of fluidized beds is very simple and models ranging from the toolroom furnace upwards have automatic start-up sequences and even the large lines may be operational within 1·5 h. The fluidized solids are non-corrosive towards both the work pieces and the containment. The solid particles are also non-abrasive at the very low velocities, typically 2 ft/s (0·6 m/s), at which they are suspended.

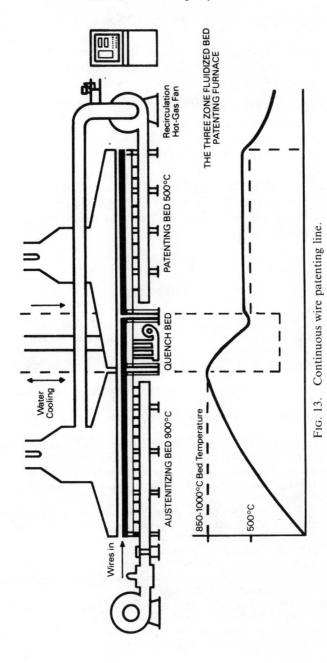

FIG. 13. Continuous wire patenting line.

6　CONCLUSIONS

In conclusion, the use of fluidized beds for component heat treatment offering favourable characteristics of compact plant, fuel economy and environmental control is most promising, with real economic and environmental advantages being realized in industrial conditions.

ACKNOWLEDGEMENTS

I am grateful to all the Directors and staff of Stone-Platt Fluidfire Ltd for their encouragement and assistance in preparing this chapter, in particular to Mr T. J. Keirle for his criticism and active help with preparation and reading of the final draft.

REFERENCES

1. MICKLEY, R. S. and FAIRBANKS, D. F., *A.I.Ch.E. Journal* (March 1974).
2. VIRR, M. J. and REYNOLDSON, R., Heat treatment in fluidized beds, *Industrial Process Heating* (December 1972).
3. BASKAKOV, A. P. *et al.*, The patenting of wire in a fluidized bed, *STAL* (in English), 574–9 (July 1964).
4. SUBOV, V. YA. *et al.*, Patenting in a fluidized bed on a semi-works plant, *STAL* (in English), 586–7 (July 1964).
5. YAMAKOSHI, N. *et al.*, Method for the direct patented rod in the fluidized bed, Japanese publication, 1972.
6. REYNOLDSON, R. and VIRR, M. J., The application of fluidized beds and vacuum heat treatment in the wire industry, *Wire Industry* (November 1972).
7. TAMALET, M., The application of fluidized bed heat transfer in metallurgical process, *Symposium of Chemical Engineering in the Iron and Steel Industry*, Swansea, March 1968.
8. ELLIOTT, D. E. and VIRR, M. J., Small scale applications of fluidized bed combustion and heat transfer, *Third International Conference on Fluidized Bed Combustion Session IV*, Hueston Woods, Ohio, 1972, Paper 1.
9. STOTT, R. J., The fluidized bed for patenting, *The US Wire Association Annual Convention*, October 20–24, 1968, Washington DC.
10. DAWES, C. and COOKSEY, R. J., *Surface Treatment of Engineering Components*, Iron and Steel Institute Special Report 95.
11. DAWES, C. and TRANTER, D. F., Application of gas carburising theory to practice, The Metals Society, *Metals Technology* (September 1974).
12. VIRR, M. J., Rapid heat treatment in fluidized beds, *XVth International Metallurgical Congress*, Caen, France, 29–31 May, 1974.

13. VIRR, M. J., Application of fluidized beds to heat treatment, *International Cold Forging Conference*, Brighton, England, October 1–3, 1975.
14. GOLDING, I. L. S., Heat treatment fuel cost can be halved, *Metallurgica and Metal Forming* (August 1974).
15. STOREY, G. G. and MOORE, J. T., Internally fired fluidized bed gas|carburising, *Metallurgia* (November 1979).
16. VIRR, M. J., Continuous heat treatment of wire using fluidized beds, *Wire Journal* (July 1978).

Chapter 11

FLUIDIZED BED HEAT RECOVERY SYSTEMS

M. J. VIRR†

Stone-Platt Fluidfire Ltd, Brierley Hill, UK

1 WASTE HEAT RECOVERY

Waste heat recovery is becoming an increasingly important part of energy system design as the availability and price of fuels become increasingly important. Heat may be recovered economically from a great number of waste gas sources in industry, such as oil and gas-fired gas turbines, diesel engines, metal processing furnaces and incinerators of waste gas, liquids or solids. Each presents the engineer with an opportunity to arrange the economical operation of energy systems. Naturally, the economic advantage from any form of heat recovery scheme depends not only on the cost of fuel, but also on how much of the heat recovered can be usefully used in some other part of the process, building heating or even power generation.

Ultimately the whole justification for waste heat recovery can be summed up in one word, 'payback'. The cost of fuel saved must be compared to the capital investment, installation, amortization, maintenance, operation costs, taxes, insurance and any other cost factors involved in owning and operating the equipment. The payback period must be determined for each case and a decision made by management as to what an acceptable period might be. Depending upon the economic climate operating in the country at the time, and the nature of the plant being installed, this is usually considered to be something between two and seven years. Examples of the calculation of payback will be referred to later on in this chapter.

Having once decided upon the need for a waste heat recovery system

† Present address: Johnston Boiler Co., Ferrysburg, Michigan 49409, USA.

there are many choices of heat exchanger that can be incorporated into the plant for recovery of the waste heat, such as plate, shell and tube heat exchangers, coil heat exchangers or gas-to-air units such as the plate or rotating wheel (Lungstrom) type of unit. All these heat exchangers have their advantages and disadvantages and books have been written alone on most of these types (ref. 1), but in this chapter it is the author's intention to just outline the development of the shallow fluidized bed heat exchanger and its application to waste heat recovery. The advantages of the shallow fluidized bed heat exchanger of relatively low pressure drop with greatly enhanced heat transfer characteristics were first recognized by Professor Douglas Elliott in the early 1970s and it is worthwhile going back into the history of the development of fluidized bed heat exchangers to see how this development has logically taken place.

2   HEAT TRANSFER THEORY

The mixing characteristics of a fluidized bed give rise to high heat and mass transfer between particles and gas, and high heat transfer coefficients to surfaces immersed in the bed.[2] Various theories of heat transfer have been proposed which differ slightly in detail but which assume that the bed emulsion adjacent to the surfaces continues to be replaced by 'fresh' hot material, due to the bubble-induced circulation of solids within the bed. Research has indicated that most of the heat transfer takes place by conduction through the thin gas film between the surface and the particles, rather than by direct conduction of the particle/surface contact points. However, this gas conduction is not the limiting factor and the residence times of particles near the surface are of prime importance since their heat capacity is small and they soon assume surface temperature. Although on average, particle/fluid heat transfer coefficients based on the total particle surface area are often not large (of the order of 6–23 $W/m^2 K$), a fluidized bed of particles is capable of exchanging heat very effectively with the fluidizing gas, because of the very large surface area exposed by the particles (3000–45 000 m^2/m^3). The presence of adjacent particles affects the thickness of the film; gas bypassing zones of the bed will adversely affect rates of heat transfer between the fluidizing gas and the particles. The degree of bypassing is dependent on the bed material, degree of fluidization, design of the apparatus and the consequent gross mixing patterns which become established so that rigorous analysis is not feasible.

Although theory provides an insight into the mechanisms of heat transfer, the complexity of particle motion precludes the direct calculation of heat transfer coefficients; therefore, recourse must be made to pilot plant investigations carried out under conditions not too far removed from those of the proposed full-scale plant, to obtain realistic values of heat transfer.

The low temperature applications of heat transfer are governed by the equation:

$$H = h_c A \, \Delta T$$

where H is the heat transfer flux, h_c is the heat transfer coefficient, A is the surface area and ΔT is the temperature difference between the surface and the bulk of fluidized bed. Although the heat transfer coefficient, h_c, is not amenable to calculation. factors which might be expected to influence it include fluidizing velocity, particle size, thermal properties of the gas, geometrical arrangement of the heat transfer surface, arrangement of the gas flow through the distributor and flow and thermal properties of the particles.

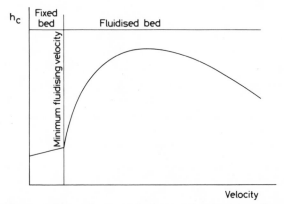

FIG. 1.   Influence of gas velocity on bed-to-surface heat transfer coefficient.

The influence of fluidizing velocity is shown diagrammatically in Fig. 1. Once the bed becomes fluidized the heat transfer coefficient rises rapidly then increases more slowly to a 'plateau' maximum. Eventually it begins to decrease and change to one of suspension flow. The initial rapid rise is due to increased solids circulation following the onset of fluidization, whilst the plateau region is the result of interaction between competing processes; increasing fluidizing velocity intensifies particle circulation which improves heat transfer, but the great bubble activity reduces the number of particles in contact with the surface which has the opposite effect.

M. J. Virr

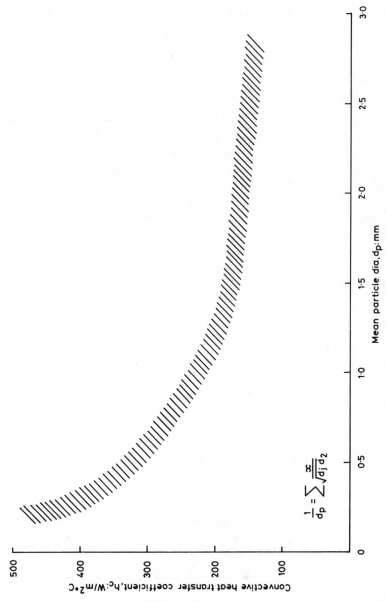

$$\frac{1}{d_p} = \sum \frac{x}{\sqrt{d_j d_2}}$$

FIG. 2. Empirical correlation between heat transfer coefficient and particle size.

In most cases the heat exchanger would be designed to operate in the plateau region, which fortunately is relatively flat over a wide range of fluidizing velocities.

3 PARTICLE SIZE

Particle size is the major influence. As might be expected from theory, reducing the particle size reduces the effective gas passage between immersed surfaces and increases the number of contacts at the surface. This results in higher values of h_c. Figure 2, which illustrates this effect, is based on the data of Baerg *et al.*[3] and Kim *et al.*,[4] together with measurements obtained from fluidized bed combustion systems. If the bed is composed of mono sized particles which do not degrade, there is no ambiguity in specifying the bed particle size. As in the case of the shallow bed heat exchanger, inert particles of silica sand, zirconium sand or alumina are used and the fluidizing velocities are relatively modest. Degradation does not occur in a serious fashion and therefore heat transfer coefficients can be assumed with some confidence. In presenting data, as in Fig. 2, it is usual to plot heat transfer coefficient against mean particle size calculated from the equation:

$$\frac{1}{d_p} = \frac{x}{\sqrt{d_{p_1} d_{p_2}}}$$

It is immediately obvious that these heat transfer coefficients are roughly ten times greater than could be achieved in conventional forced convection gas-to-surface heat exchangers. It was, therefore, surprising that in spite of this order of magnitude improvement in heat transfer coefficients, that up to the late 1960s there had been no examples of gas/liquid heat exchangers, such as economizers based on the process. The only applications had been limited to the extraction or input of heat to chemical engineering processes. Professor Elliott and Hulme pointed out[5] that the reason for this is obvious if an economic calculation is made of the relative saving in heat transfer surface area costs compared with the costs of power to force the gas through the fluidized bed. Initial calculations by the CEGB in the late 1950s using a bank of 25 mm diameter tubes on a 150 mm horizontal pitch at 75 mm vertical pitch (the packing limit it was thought prudent to assume in those early days) suggested that the pressure drop penalty was a number of times the saving in heat transfer tube cost.

4   OPTIMIZATION

A detailed study of the facts involved in heat exchanger design[6] showed
how the overall capital plus running costs could be optimized. Figure 3
plots the overall cost versus the diameter of a typical fluidized bed
economizer for various values of factor R where:

$$R = \frac{k^2 d\rho}{h\,\Delta T}$$

where k is equal to pitch/diameter ratio of the tube bundle, d is equal to the
diameter of the tubes, ρ is equal to the density of the fluidized bed, h is equal
to the overall heat transfer coefficient and ΔT is equal to the effective log
mean temperature difference.

In general the pumping power could be reduced to a negligible level by
building a very large fluidized bed with a single row of tubes to provide the
necessary heat transfer area. In this case the fluidized bed containment cost
would be very high, as shown by the rising curves on the right-hand-side of
Fig. 3. Containment costs can be reduced by accommodating the tubes in a
tall, small diameter vessel; in this case the pressure drop penalty becomes
excessive (left-hand-side of Fig. 3).

In all cases for a similar pitch/tube geometric arrangement the volume of
the bed to accommodate tube bundle was reduced in direct proportion to

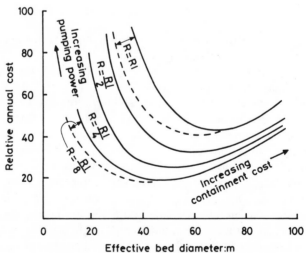

FIG. 3.   Interaction of design factors.

the tube diameter and inversely to the heat transfer coefficient and the temperature difference. The most significant factor is the pitch/diameter packing ratio, k, as this has the square law effect of reducing the bed volume requirement. As close packing of the tubes has such an overriding effect, research work was instituted to see how close together the tubes could be placed before heat transfer was impaired. Work by the NCB[7] showed that a 25 mm gap between tubes would only reduce heat transfer coefficients by about 20 %. Nevertheless, even using 25-mm tubes on a 50 mm triangular pitch giving a packing density of $36 \, m^2$ tube area per m^3 of bed volume compared with the $7 \, m^2/m^3$ assumed in the initial studies, did not produce an economic heat exchanger when all the engineering factors involved were taken into account.

5 EXTENDED SURFACE

The first indication that this intractable situation could be circumvented occurred during a short course on fluidized bed combustion and heat transfer organized by J. S. M. Botterill and D. E. Elliott when the former reviewed the work by Petrie *et al.*[2] which shows that by using finned tubes with a surface area of some 15 times greater than that of the base tube, a six or seven-fold gain in heat transfer flux could be obtained compared with that using plain tubes of the same diameter. These experiments, which were conducted in a 'deep' fluidized bed of approximately 1 m depth, show that the outside heat transfer coefficients between the bed and the finned surfaces was still some 50 % of the bare tube coefficients, even though very closely spaced fins were employed.

6 EARLY EXPERIMENTS ON SHALLOW FLUIDIZED BEDS

It occurred to Professor Elliott that this might give the clue for answering the problem of the pressure drop penalty. Professor Elliott's first experiments using a built-up finned tube bundle immersed in a shallow fluidized bed at a depth of 80 mm, gave somewhat surprising heat transfer coefficients which were significantly in excess of the existing data on bed to tube heat transfer, all of which had been obtained in deep fluidized beds (see Fig. 4). This bed/tube bundle configuration already had a packing factor of $80 \, m^2/m^3$, a gain in heat transfer per unit area, and most likely the bed could accommodate two or three times the amount of surface without detrimental effect. Even when using this configuration, a useful amount of heat exchanged between hot gases and water could be accomplished while still holding the pressure drop down to less than 150 mm of water, which is considered to be the upper limit for commercial applications. The value of Ah_0/MC_p (where A, h_0, M and C_p are, respectively, the area, heat transfer

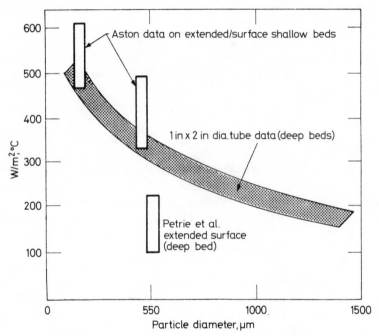

FIG. 4.   Comparison of shallow/deep bed heat transfer.

coefficient, mass flow of gas and gas specific heat) of just over 3 which was achieved would allow an effectiveness of 70 % to be obtained with a gas inlet temperature of 1000 °C (Fig. 5). At the equilibrium bed temperature of 300 °C, some 700 kJ/kg air would be transferred. This performance already appeared to be of the same order as normal gas-fired domestic central heating boilers and looked extremely promising.

Since these early experiments, quite a number of questions about the processes have been resolved and a great deal of information has been acquired on the heat transfer coefficients which could be achieved with commercial extended surface tubing of different materials and with specially designed tube bundles (Fig. 6). These show the superiority of

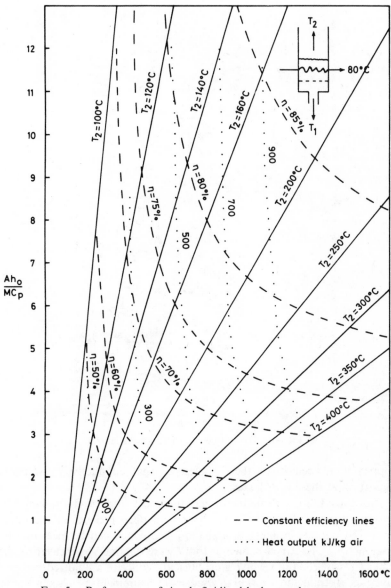

FIG. 5.   Performance of simple fluidized bed waste heat system.

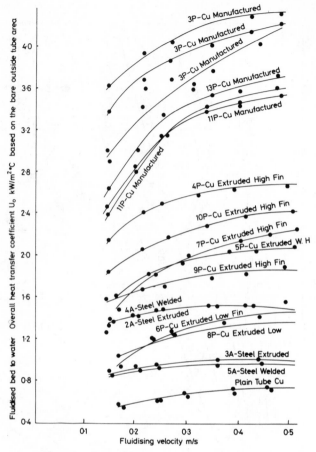

FIG. 6. Overall heat transfer coefficients.

copper tube because of their higher thermal conductivity, which is important at these high fluxes.

Professor Elliott pointed out that the most important fundamental reason why a shallow bed is superior to a deep bed lies in the much improved particle mobility in a shallow bed which can be measured by viscometry.[8] The curves shown in Fig. 7 were obtained by measuring the shear stress needed to revolve a cylindrical surface in a fluidized bed at a given shear strain and are akin to viscosity measurements in a liquid. It can be seen that the 'viscosity' of very shallow bed is significantly lower than that of a medium-shallow bed, suggesting that the mobility of the particles,

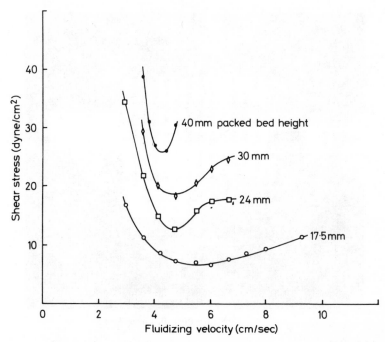

Fig. 7. Shear stress versus fluidization velocity for bauxite at varying bed depths.

and particularly that of the dense phase, reduces as the bed depth increases, thus leading to poorer heat transfer. This can also be demonstrated by photographing the passage of a bubble through a simulated finned surface in a deep bed and comparing it with particle motion in a shallow bed situation.

7 DEEP VERSUS SHALLOW BEDS

Figure 8 shows a series of fins and gaps built across a 25 mm wide fluidized bed some 200 mm above the distributor. In the absence of a bubble the dense incipiently fluidized phase remains relatively motionless within the fins (the dark vertical bands). The light, vertical bands are perspex strips spanning the bed to simulate the fins. When a bubble engulfs an area of the fins, the pressure drop in the top and bottom of the fins is reduced almost to zero and the solids which are in the fins drain out downwards because they

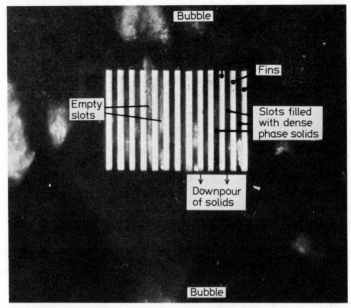

FIG. 8. Fins in a deep fluidized bed.

are unsupported. When the bubble has passed the fins the gap between them fills up again with a fresh charge of hot (or cold) particles which will then remain almost motionless until the next bubble appears. This situation conforms closely to the Mickley and Fairbanks model of fluidized bed heat transfer which considers the transient heat transfer from a surface after the passage of a bubble. The heat conduction from the wall into the first row of particles and hence into subsequent rows can be calculated by knowing thermal conductivity of package and particles. The process is terminated and re-started by the next bubble and, therefore, is predominantly affected by the nature of the bubbling process. In this context it is important to notice that the deeper and the larger the bed is, the bigger and less frequent will be the bubbles.

In a shallow bed (<100 mm), and even more so in a very shallow bed (<50 mm), bubbles are much more numerous and smaller. Moreover, when a series of vertical fins are placed immediately over the distributor, bubbles can only grow in two dimensions. This can readily be demonstrated by placing a densely packed row of vertical fins in part of the shallow bed and observing the difference in the bubbling pattern caused by the matrix (see Fig. 9). In the extreme, the heat exchanger process can be divided into a large number of completely independent 'fluidized' beds and in one particular application, a design for a domestic gas-fired fluidized bed water heater, consisted of a series of 50 mm wide by 50 mm deep by 50 mm thick modules, surrounded on five sides by copper cooling walls, where the only communication between separate beds is in a cloud of particles above the beds. In this design the distributor consists of 250 μm slots across the full width at the bottom of each bed. In this configuration the two-phase

Fig. 9.   Fins in a shallow fluidized bed placed just above the distributor.

concept is not applicable as the particles act independently in a highly expanded phase. The particle velocity is possibly an order of magnitude higher than those in the dense phase of a deep fluidized bed and this can easily be shown visually in a single cell perspex module. Bed expansions could be 400 % or more depending upon the fluidizing velocity, which further reduces the pressure drop needed across the model. In this case the Botterill model of heat transfer is much more appropriate. This model[9] considers the transient heat transfer to an individual particle which comes into contact, or very close proximity, with the surface. During the first few milliseconds the heat transfer rate is extremely high, the heat being conducted across the very thin but stationary air gap between the particles and surface. As the particle heats up the heat transfer coefficient drops to an order of a few hundred W/m^2 K after a period of about 100 μs. This theoretical model explains why it is that bed-to-surface heat transfer coefficients of the order of 500–700 W/m^2 K can be obtained in the highly expanded regime of the individual modules where, at any moment, only a small fraction of the heat transfer surface area is covered by an active particle. The extremely high mobility of the particles which gives rise to extremely short residence times, more than compensates for the reduced bed density.

8 COMMERCIAL DEVELOPMENT

Professor Elliott, having shown during the described research programme, that the pressure drop penalty could be overcome, applied himself to the intriguing problem of how to introduce the concept to the commercial market. There are some 7000 firms in the UK alone which waste an inordinate amount of heat from metal processing plant and the like, simply because in the past it has been uneconomic to recover the heat due to the high cost of heat transfer apparatus. After making a market survey of the various possibilities it was decided that the principles would lend themselves well to the recovery of heat from furnace exhaust gases. Fluidfire Development Ltd (now named Stone Fluidfire) was set up specifically to exploit fluidized combustion heat transfer technology and designed a range of package, waste heat fluidized bed boilers; the first commercial example is shown in Fig. 10. In the layout shown, hot gases from a brick firing kiln operating 24 h per day is fed, by a plenum chamber, through a pierced metal distributor plate into a 45 mm deep fluidized bed containing alumina particles in the size range $300-500\,\mu\text{m}$. Ordinary sand could have been used, but the designers were anxious to avoid any possible chance of silicosis which might result from the long-term exposure to a very fine abrasive material possibly being caused by the continual attrition due to particle interaction. The aluminium oxide material is particularly resistant to attrition and subsequent elutriation from the fluidized bed. A single row of commercially produced copper finned tubes is used to recover the heat from the bed to provide hot water for the central heating of a nearby office block. This copper tube has a fin height to thickness ratio of 14/1 and the tapered fins are integral with the base tube giving a 'Gardiner' fin efficiency of 86 % with a heat transfer coefficient applicable to 400-μm particles. The bed/tube configuration used gives a value of $Ah_0/MC_p = 4.5$ (see Fig. 5). Considering that the scale-up from the research results to the first commercial unit was by a factor of 50 or more, the heat transfer capability was estimated very accurately, the full-scale unit giving about 5 % more heat output than predicted. The unit operates automatically, switching itself on when the central heating system needs more hot water and shutting itself off when the water temperature exceeds a given level. The gases are sucked through the fluidized bed from the chimney by a standard blower and returned to the duct, thus imposing no back pressure on the furnace system which operates normally whether or not the heat exchanger is required for duty.

Operationally, the unit has had very few teething problems—in fact, the

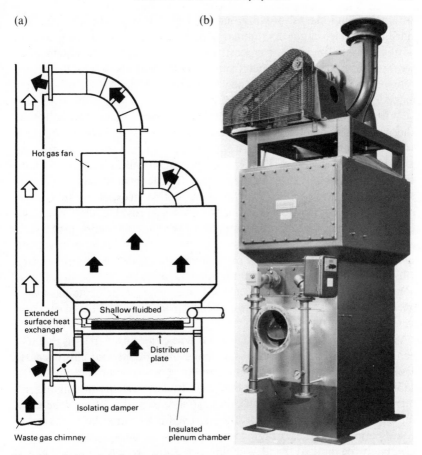

(a)   (b)

Hot gas fan

Extended surface heat exchanger

Shallow fluidbed

Distributor plate

Isolating damper

Waste gas chimney

Insulated plenum chamber

FIG. 10. Industrial fludized bed waste heat boiler. (a) Diagram of typical installation. (b) Packaged fluid bed waste heat exchanger.

only real problem was a slow build up of lightly bonded particulate deposits of sulphurous origin on the distributor plate, which caused the pressure drop across the plate to build up over a period of about a month to a level where the gas flow dropped appreciably. Light brushing of the distributor plate, which takes roughly half-an-hour, restores the unit to full capacity. The use of a more open distributor plate now allows the unit to operate for much longer periods between cleaning which is quite acceptable for a unit which has no fuel cost. The heat transfer surface and particles remain clear of deposit; indeed, after two years of operation, the copper fins show the original rolling marks, proving that the metal wastage is negligible. It is hard to convince engineers familiar with sand blasting techniques that the action of a fluidized bed at modest velocity is very gentle. Although a no-wear situation was predicted from laboratory trials it has been very reassuring to observe the 'as new' condition of the finned tubes after extended running.

With regard to the deposition problem it is pertinent to point out that thermal deposition of very fine material would probably have resulted in a very severe fouling problem had a conventional finned tube convective heat exchanger been installed. The very fine particles would soon have covered the surface to such an extent that heat transfer would have fallen off significantly in a few tens of hours running, so that an expensive-to-operate soot blowing system would have had to have been provided. One advantage of the fluidized bed heat exchanger is that it confines the fouling problem to a specific, easy to clean area, i.e. the distributor. This is subsequently being used to considerable advantage on the marine units and others in industry which have been installed in a situation where fouling has occurred. The solution in these cases is to use a 'windscreen wiper' type of stainless steel wire brush to clean the distributor at regular intervals and detail of this is described in ref. 1.

A further significant advantage of this type of heat exchanger is that it is economic and practicable to install units of high thermal effectiveness. Only a small fraction of the capital cost (10% or so) is in the heat transfer tubing so that the surface area could easily double without a significant increase in the overall cost. Thus, the cost/effectiveness relation is radically different from normal convective units. Even if a normal heat exchanger could approach these high effectivenesses economically, it would ultimately be limited by dew point corrosion troubles when the gas temperature falls to a critical level. This critical level is appreciably lower in a fluidized bed unit where metal temperature is higher than in a conventional heat exchanger due to the ten-fold increase in the thermal flux of the tubes. This permits the metal surfaces to be operated at a temperature above that of the dew point even when the gas temperature is quite low. However, care must be taken to keep the headers (tube leaders) out of the low temperature gas stream and also to insulate the exterior of the heat exchanger, because if the unit is mounted outside, condensation can occur, particularly during winter months and this can cause agglomeration of the bed material which is difficult to clear. Proper attention to the design, installation and start-up techniques completely eliminates such problems.

9 MARINE STEAM BOILER APPLICATION

Resulting from a number of successful industrial applications using fluidized bed heat exchangers, Shell International Marine Ltd approached Stone-Platt Fluidfire Ltd with the request to custom design a shallow fluidized bed heat exchanger to accept the waste heat from a BHP Sulzer diesel engine on board their product carrier, the M.V. Fjordshell. The Fjordshell was built for trading around the Scandinavian coast, loading in one of three refineries, discharging at numerous customer distribution installations. Consequently, voyages were short and port time relatively high. The designed trading pattern for the vessel was such that cargo requiring heating would be few and the then relatively cheap cost of fuel made installation of an exhaust gas waste heat boiler uneconomical when the ship was built in 1974. However, during building, extra cargo heating coils were installed following a review of the anticipated trading pattern, necessitating additional cargo heating and tank cleaning capacity. The operators of the vessel (Norsk Shell) were thus considering that an exhaust gas waste heat recovery boiler would be beneficial in reducing the additional boiler fuel consumption at sea. Studies were made of the economics of retrofitting such a unit.

It was evident from these feasibility studies that the fluidized bed principle offers some advantages when compared with conventional extended surface heat exchangers. These were:

(1)   The heat transfer rate between the gas, bed particles and tube material is some four to five times that of the conventional gas to extended surface metal heat exchanger, thus necessitating less heat transfer surface for the same steam production.

(2)   Because of the shallow bed design and high heat transfer rate, the heat transfer takes place in one row of tubes. This one row can be divided into parallel beds, as described later, and separated by gas spaces allowing easy access for inspection and maintenance.

(3)   The continual motion of the bed particles around the finned tubes in the bed prevents the build up of any fouling on the tubes. Fouling will occur on the underside of the gas distributor plates supporting the bed which, being flat, are easier to clean than finned tubes.

(4)   The use of fluidized beds for the combustion of coal and oil in marine use is being investigated by several organizations but no fluidized bed had previously been to sea. It was considered by Shell that building and operating a fluidized bed exhaust gas waste heat

boiler would demonstrate the advantages and disadvantages of this type of unit in comparison to the conventional waste heat boiler. It would also demonstrate the reaction of a simple fluidized bed to the marine environment before more complex coal- or oil-fired units were constructed.

Ideally it would have been preferable to build a prototype waste heat unit for use with a land based diesel engine. However, the urgent requirements of the Fjordshell for a retrofit exhaust gas boiler offered the opportunity for an excellent floating test bed facility.

10   MARINE BOILER DESIGN CONSIDERATIONS

Having considered, as described previously, that the main consideration is to reach a balance between a suitably sized particle to give a reasonably high heat transfer coefficient within the fluidized bed and a particle size density which will then accept the relatively serious impediment of heat transfer under these conditions, the problem was to package the unit into a suitable arrangement to make a marine boiler.[10]

The unit consists of three fluidized beds (approximately 3 m by 5 m) which are split up by a series of baffles into panels of 810 mm by 900 mm, thus ensuring that the bed material (aluminium oxide) does not slide uncontrollably from side to side across the boiler and end to end as the ship rolls and pitches.

The final boiler arrangement allows the rectangular water inlet headers to be bolted to the boiler structure enabling each of the nine tube panels to be removed from the boiler through the deck house after bulkhead, the latter being provided with removal access plates. Each tube nest has a return header which expands towards the forward casing of the boiler, thus enabling all external pipe headers to be at one end of the boiler. The boiler layout is shown in Fig. 10.

The boiler specification was as follows:

Engine load	85% of maximum continuous rating	
Gas flow rate, kg/h	74 460	
Gas inlet temperature, °C	320	
Gas outlet temperature, °C	238	
Steam pressure bar	8	12
Steam flow rate, kg/h	2 800	2 460
Thermal recovery, kW	2 073	1 829

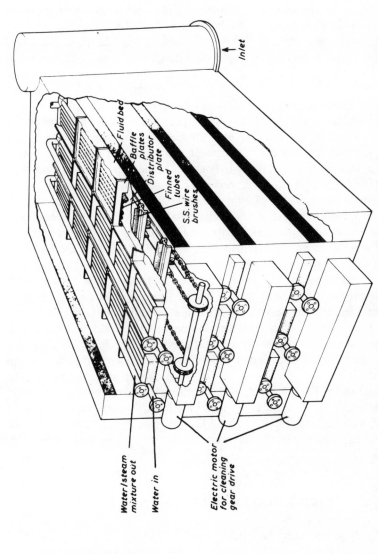

Inlet

Fluid bed
Baffle plates
Distributor plate
Finned tubes
S.S. wire brushes

Water/steam mixture out
Water in
Electric motor for cleaning gear drive

Fig. 11. Marine fluidized bed waste heat boiler. (From Cusdin and Virr[10] by courtesy of Marine Management(Holdings) Ltd.)

One of the major design considerations was the problem of cleaning the distributor plates which were receiving exhaust gas from a large marine diesel engine burning heavy fuel oil of viscosity Redwood 3500 s and a sulphur content of up to 3 %. Naturally, deposition of the sulphurous ash from this diesel engine was to be expected on the bottom face of the distributors and after a number of systems were considered, stainless steel brushes mounted on moving bogies were designed to operate this relatively high temperature environment. The bearings supporting these bogies were of a special plain bearing design using molybdenum disulphide grease. Most of the structural boiler parts are mild steel.

The problems relating to the performance of the fluidized bed when the ship was rolling and pitching were studied in some depth by making a complete full size fluidized bed panel and mounting it on the motion simulator at the Warren Spring Laboratory of the Department of Industry. The behaviour of the fluidized bed was recorded while it was subjected to the various forms of ship motion encountered in a simulated force 10 gale.

With a permanent list of 8 %, some elutriation occurred but a re-design of the distributor plate eliminated this problem and the model was subsequently tested satisfactorily with a list of 15 % when there was no elutriation or serious impediment of heat transfer. The optimum gas velocity range was 1·2 to 1·5 m/s without elutriation. The heat transfer coefficient to the extended surface tubing at these velocities is quite high. The tube spacing can be optimized to achieve maximum steam output for a given area of fluidized bed. It is the gas velocity which fixed the total area of the fluidized bed and therefore the total size of the unit. In fact, the high gas throughput of this particular boiler made a single fluidized bed an unrealistic size and it was thus decided to stack the beds as shown on the arrangement in Fig. 11, the final choice being three beds giving a relatively simple design within space constraints. Each bed comprises a single row of 36 tubes divided into three panels of 12 tubes each. Each panel has its own rectangular water inlet header and can be withdrawn from the boiler individually.

11.   MARINE BOILER OPERATIONAL EXPERIENCE

The fluidized bed waste heat boiler was started up for the first time on 10 July, 1977 and after the overcoming of some commissioning problems the boiler was put into regular service on 7 August, 1977 and tests carried out at sea up to the end of June, 1978 when the boiler had been operational for

2100 h. During this period it had been standard practice to use this waste heat boiler whenever the ship was at sea and the engine was operating at more than 60% of MCR.

During this period some problems were experienced in the following areas, all of which have subsequently been overcome:

(1) Some gas and consequential grit leakage was caused by expansion problems; this was overcome by lifting the floating header out of the fluidized bed to allow sufficient room for expansion.

(2) The cleaning action of the brush gear on the distributor plates had been found to remove burrs left by the punching method and this resulted in further grit loss through the plates into the catchment area. These plates have since been replaced and the quality control during plate manufacture improved so this no longer occurs.

(3) Some improvements had to be made initially to the drive mechanism of the distributor plate cleaning gear to ensure reliable and even cleaning of the plates. The effectiveness of this equipment is now most marked and only a few strokes are required to clean the plate and thus reduce the exhaust pressure before the boiler reduces from 350 mm down to 290 mm of water. The cleaning brushes are now used automatically every 2 h. During a major ship re-fit two years after installation, it was found that it was not necessary to replace the stainless steel brushes.

(4) It had been found that there was a general loss of bed material from the boiler over a period of time and it was apparent that this material had been elutriated from the boiler. Subsequent investigation indicated that the gas velocity through the boiler was some 25% higher than the original design velocity due to the exhaust gas temperature being some 55 K higher and underestimation by the manufacturers of the mass flow of gas through the engine. Subsequent work using a grit arrester arrangement immediately over the tubes in the fluidized bed has eliminated these difficulties and also shown the way to using higher gas velocities than currently used by restricting the height to which particles lift in the freeboard above the fluidized bed.

(5) The steam output from the boiler was generally higher than the specification at 2900 kg/h of steam at 85% load and 10·5 bar pressure.

The total waste heat boiler performance is shown in Fig. 12 and is found to be completely satisfactory.

The boiler is found to make a measured fuel saving of between 4·95 and

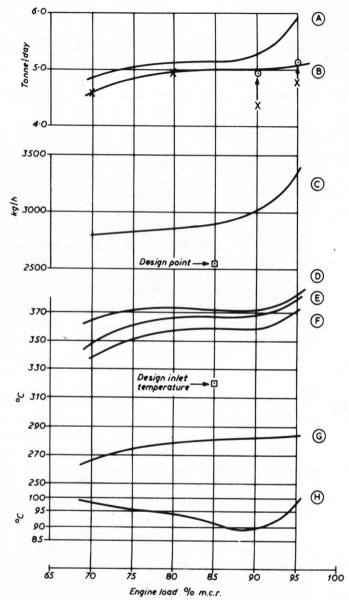

Fig. 12. The performance of a marine waste heat boiler. A, calculated fuel saving; B, measured fuel saving ($\times$, measured point; $\odot$, corrected for change in steam flow); C, calculated steam production from fluidized bed boiler (steam pressure, $10.8\,kg/cm^2$); D, boiler gas inlet temperature; E, M.E. cylinder exhaust temperature—boiler on; F, M.E. cylinder exhaust temperature—boiler off; G, boiler gas outlet temperature; H, differential gas temperature across boiler. (From Cusdin and Virr[10] by courtesy of Marine Management (Holdings) Ltd.)

5·25 tonne of oil per day which at 1978 values gave a saving during the year of £40 000.

The noise levels have been measured on the navigation bridge wings when the engine has been exhausting through the original silencer and repeated when exhausting through the fluidized bed unit. The result of using the fluidized bed boiler gives a reduction in sound level of 6 dBA which would make the unit particularly attractive to be used on land based installations if it was used in conjunction with power generating diesel engines.

12 POWER GENERATION USING FLUIDIZED BED WASTE HEAT BOILERS

A similar waste heat boiler has now been designed which would generate 6668 kg/h of steam from the exhaust gas of a 15 000 kW engine. Such a waste heat boiler has been designed which would have a superheat steam temperature of 246 °C and exhaust into a multi-bladed radial turbine with a vacuum condenser. The unit would generate 1 MW of electricity for on-board ship use.

It is normal practice on board ships to have three diesel electric generators and by powering one of the generators with the steam turbine accepting steam from the waste heat boiler, it would be possible to make one of the diesel electric units redundant, thus saving considerable capital cost.

13 PAYBACK CALCULATIONS

The capital cost recovery for a packaged waste heat unit should be based on the running costs of a conventionally fired boiler plant as an alternative heat source (Fig. 13). This particular example was carried out assuming conventional boiler operating efficiency of 70 %, the natural gas costs of 16p/therm and a gas oil cost of 32p/therm.

It should be noticed that if the unit is in operation for some 8000 h/year a very fast payback period is naturally achieved whereas on single shift heating operation of some 2000 h, a payback period approaching two years is more usual.

The capital cost should not only include the waste heat unit but also its

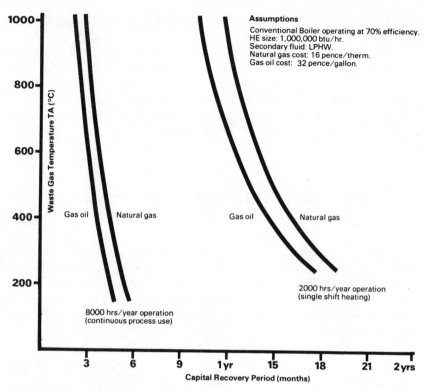

Fig. 13. Capital cost recovery for packaged waste heat recovery units.

installation and any consequential cost such as finance charges that might be applicable.

A more sophisticated look at payback can be seen in the example shown in Table 1, which refers to the marine waste heat boiler which replaces a diesel electric generator and has a capital cost bonus. This can also apply to the waste heat boiler where it completely replaces the conventional boiler, and this aspect should not be overlooked.

In all these simple calculations, a constant price of fuel has been assumed and we all know today that this is totally untrue. Therefore, if a history of fuel cost in the last year or two indicates the probable trends of increase of fuel cost in the next few years, the payback period should be adjusted every 3 or 6 months to take account of the increased fuel cost that is likely to be experienced, and this can have a fairly dramatic effect upon the payback period.

TABLE 1

| | Capital costs (£) | |
	500 kW unit (actually 593 kW output)	1 000 kW unit
Boiler	92 000	92 000
Steam turbine	45 000	230 000
Condenser (say)	10 000	20 000
Total	147 000	342 000
Saving	69 900	134 700
Nett capital cost	77 100	207 300
Fuel saving based on 30 weeks/year @ 55p/gal	80 000	—
Fuel saving/annum	—	120 000
Payback period, years	1	1·72

A proper study of the economies of waste heat boiler installations is essential in presenting a case to management for investment in such schemes which are inevitably expensive when purchasing modern industrial equipment of this type. However, a proper investigation of such costs brings home the very true advantages of waste heat recovery which can be enjoyed for many years in the future when availability of fuel is restricted and fuel costs are likely to increase at a much faster rate than the general rate of inflation. These facts alone make buying waste heat equipment earlier, rather than later, an attractive proposition.

REFERENCES

1. BOYEN, J. L., *Practical Heat Recovery*, John Wiley and Sons, New York, 1980.
2. PETRIE, J. C. *et al.*, In bed heat exchangers, *Chemical Engineering Progress*, 64(7) (1968).
3. BAERG, A., KLASSEN, J. and GISHLER, J., *Con. J. Res. F.*, 28, 287 (1950).
4. KIM, K. J., KIM, D. J., CHUN, K. S. and CHOO, S. S., *Institute of Chemical Engineers*, 8, 472 (1968).
5. ELLIOTT, D. E. and HULME, B. G., Fluidized bed heat exchangers, *Midlands Branch Symposium*, Institute of Chemical Engineers, March 1976.
6. BRUNDRETT, G. W. and GLENVILLE, R., Optimisation of fluidized bed heat exchangers, CEGB R&D Department, RD/M/W186, April 1966.
7. ELLIOTT, D. E., HEALEY, E. M. and ROBERTS, A. G., Fluidized bed heat exchangers, Institute of Fuel, Paris, June 1971.

8. McGUIGAN, S. J. and ELLIOTT, D. E., The viscosity of shallow fluidized beds, *Fourth International Congress of Chemical Engineers, Design and Automation*, September 1972.
9. BOTTERILL, J. S. M., CAIN, G. L., BRUNDRETT, G. W. and ELLIOTT, D. E., Heat transfer to gas fluidized beds, *Symposium on Development in Fluid Particle Technology*, Trans. American Institute of Chemical Engineers, December 1964.
10. CUSDIN, D. R. and VIRR, M. J., A marine fluidized bed waste heat boiler—Design and operating experience, *Proceedings Institute of Marine Engineers*, 10 October, 1978.

INDEX

A-group materials, 6–9, 13–15, 22–4
A-type boilers, 80–1, 129, 131
Activated carbon, 278
Additives, 112, 221
Adiabatic combustors, 173, 180, 192
Advantages and disadvantages of
    fluidization, 2–3, 78, 338, 358
Agglomeration, 71, 288, 294, 297, 356
Aggregative fluidization, 6–7
Air heater cycles, 172–4, 177–8
Airstore systems, 174
Alkali metals, 89, 171, 189–90
All-atmosphere furnaces, 328–9
Allied Boilers Ltd, 144
Alumina, 28, 135, 231, 345, 354, 358
American Electric Power Co., 181,
    183
Ammonia, 228, 230, 238, 279–80
Analyses (of coal and ash), 54–5,
    87–90, 295
Annealing processes, 330
Anthracite, 53, 88, 90, 122, 284, 286,
    295–6
Applied Combustion Systems, 164–5
Aquaquench, 331
Argentina, 277, 300–2
Argonne National Laboratory, 180–1,
    201, 210, 212, 218, 221–2
Arrhenius equation, 62, 208
As-fired calorific values, 53–4
As-mined coal, 52, 88, 191, 290

Ash
    contamination by, 155–6, 166
    content (of coal), 52–5, 74, 88
    fusion temperatures, 41, 43, 55–6,
        75, 81, 89, 171, 297, 300
    layers, 45, 67
    removal of, 42, 92, 117, 123–4, 127,
        135, 274, 280, 300
    sticking of, 81–2, 90, 280–2, 288–9,
        299
Atmospheric pressure beds (AFBC),
    4, 33, 42–75, 96–155, 174, 179,
    185–6
    coal gasification, for, 279
    emission of nitrogen oxides, and,
        227–40
    sulphur retention in, 201–3, 206–7,
        215, 217, 222
Atomization of fuels, 87
Attrition damage, 11, 24, 215–16, 354
Auscoke process, 306
Automated start-ups, 130, 151, 354
Automatically controlled boilers, 86,
    121, 135, 146, 157–8, 162
Automation, 78, 92, 330, 361

B-group materials, 6–7, 9, 15–16, 19,
    22–5, 28, 32, 280, 287
Babcock Power Ltd, 114–16, 123,
    151–3, 227